ONE-DIMENSIONAL TWO-PHASE FLOW

GRAHAM B. WALLIS

Sherman Fairchild Professor of Engineering Emeritus
Thayer School of Engineering
Dartmouth College

DOVER PUBLICATIONS, INC.
Garden City, New York

Bibliographical Note

This Dover edition, first published in 2020, is a revised republication of the work originally published by McGraw-Hill, Inc., New York, in 1969. This edition includes a new preface and two new appendixes.

Library of Congress Cataloging-in-Publication Data

Names: Wallis, Graham B., 1936– author.
Title: One-dimensional two-phase flow / Graham B. Wallis, Sherman Fairchild Professor of Engineering, Emeritus, Thayer School of Engineering, Dartmouth College.
Description: Garden City, New York : Dover Publications, Inc., 2020. | Series: Dover books on engineering | This Dover edition, first published in 2020, is a revised republication of the work originally published by McGraw-Hill, Inc., New York, in 1969. This edition includes a new preface and two new appendixes. | Summary: "Geared toward graduate students in engineering, this widely used monograph presents a two-part approach that focuses on both analytical techniques and practical applications. Newly updated by the author"–Provided by publisher.
Identifiers: LCCN 2020016610 | ISBN 9780486842820 (trade paperback)
Subjects: LCSH: Two-phase flow.
Classification: LCC QA922 .W35 2020 | DDC 531/.1134–dc23
LC record available at https://lccn.loc.gov/2020016610

Manufactured in the United States by LSC Communications
84282701
www.doverpublications.com

2 4 6 8 10 9 7 5 3 1

2020

Contents

Preface to the Revised Edition

I am grateful to Dover Publications for reprinting this book, which was originally published fifty years ago by the McGraw-Hill Book Company. This opportunity has allowed me to correct typos and some numbers in examples or problems if they were significantly inaccurate, add a few new references, and provide clarification in two new brief appendixes. The astute reader may discover some remaining anomalous features, such as where the signs of fluxes or forces are redefined to conform to the prevailing common usage.

Numerous technical papers describing research results and engineering applications have been published since 1969. They have significantly broadened the knowledge base. Three-dimensional analyses and numerical methods have greatly extended the range and ease of practical applications, such as nuclear reactor design and safety, without changing the basic analytical approaches.

Because of the complexity and uncertainty in fully describing the physics and chemistry of the vast variety of multiphase flows, the caveats mentioned in this book still apply. All of the analytical approaches, except for a few simple, ideal examples, are approximations that provide guidance and a skeletal structure. They often must be filled out with a suitable body of experimental and empirical evidence, and perhaps additional analysis, to provide appropriate confidence in predictions of actual performance. They may also provide clues to resolving surprises in observations, such as when bubbles in the foam at the top of a glass of Guinness appear to move downward. (The concentration waves do.) There are the usual pitfalls to be avoided in relying on the predictions of an analytical method or a computer code that was only qualified in some other realm of applications or using a correlation describing a different situation.

Graham B. Wallis
Norwich, Vermont
December 2019

Preface

The purpose of this book is to make a thorough presentation of the basic techniques for analyzing one-dimensional two-phase flows and to show how they can be applied to a wide variety of practical problems. The subject has immense importance in a large variety of traditional engineering disciplines. It is ripe for development to the point where it can rank on a par with compressible fluid flow and boundary layer theory as a mature branch of fluid mechanics. Eventually some of the simpler material should diffuse down into early undergraduate courses.

Since it would take many volumes to do justice to all aspects of two-phase flows, the scope of this book has been limited to flows which are essentially one dimensional, occurring in ducts and channels or on continuous surfaces. The use of velocity and concentration profiles permits variations across the direction of flow. Multidimensional phenomena are introduced only when they are needed to put the one-dimensional assumptions in perspective or to provide results which are needed before the problem can be adequately specified. For example, the two-dimensional flow field around and within a bubble rising in a liquid is not derived or discussed, but the resulting dependence of velocity on diameter is quoted. Of course, many of the one-dimensional equations and techniques are readily extended to the general case by substituting the appropriate vectors for scalars in the mathematics.

In the interest of brevity, attention has been confined almost entirely to the fluid mechanics of two-phase flow. Detailed discussions of heat and mass transfer phenomena and of the mechanics of nucleation during phase change have been omitted. The reader who is interested in certain problems of boiling, condensation, flashing, freezing, combustion, and flows with chemical reaction will therefore find that he may need to gather some supplementary information before a complete analysis can be performed.

Most of the analytical developments have been selected for their generality and usefulness for predicting parameters which are of engineering interest. Typical variables of concern are pressure and density variations along a duct, rates of discharge from containers, liquid film thickness, spatial variations in the concentrations of the phases, flow patterns, and the factors which limit the performance of equipment.

The book is designed for use both as a university text and as a reference for engineers and researchers. It is divided into two parts. Part 1, consisting of Chaps. 1 to 7, is concerned with analytical techniques which have a broad and quite general validity. These techniques are illustrated by examples from various fields of engineering. The emphasis is on generality, the mastering of key concepts, and the development of analytical skill, familiarity, and facility.

Part 2 is organized around particular phase combinations, flow regimes, and practical applications. For example, the specific topics of fluidization and sedimentation are considered as subsections under the heading of "Fluid-particle Systems" which is the title of Chap. 8. An entire chapter is devoted to each of the major gas-liquid flow patterns. All of the techniques which are developed in Part 1 are used in Part 2 when they are appropriate for the particular situation. Specific calculation procedures, correlations, flow regime boundaries, and the individual idiosyncrasies of certain systems are discussed. The emphasis is on the prediction of useful engineering parameters.

Single-phase flows are discussed at any length only in Chap. 6 in order to clarify the concepts of wave motion and interaction. Many textbooks in fluid mechanics are weak in this area. Few, if any, make any reference to continuity waves. Most of them fail to point out that compressibility and gravity waves are examples of a general class of dynamic waves which are produced by forces resulting from any form of concentration gradient.

In order to preserve a reasonable size, many theoretical developments are relegated to the status of problems in which the route is indicated or the answer is provided but the details of the derivation are omitted. The many fascinating aspects of multidimensional, multiphase flows are also excluded except when they have special relevance for completing the description of particular phenomena.

The reader is assumed to possess a basic knowledge of fluid mechanics and thermodynamics and to be familiar with the methods for deriving conservation equations from suitable control volumes. No attempt is made to reiterate the derivation of fundamental theorems and in most cases merely the results are written down. It may be useful for students who are unfamiliar with the particular form of some of these relationships to derive them from first principles and gain confidence in their validity.

From the academic point of view the book is probably too long for a single course. Parts 1 and 2 can be used for a two-term sequence for graduate students in mechanical, civil, or chemical engineering or with a particular interest in fluid mechanics in a flexible curriculum. A more elementary course, perhaps suitable for a senior elective following a general fluid mechanics course, can be constructed by combining Chaps. 1 to 4 with related material from Chaps. 8 to 12. The more advanced topics of wave motion, unsteady flow, velocity profiles, and interfacial phenomena can then be reserved for a later course. Honors students who are more interested in the overall theoretical picture might study Part 1 in detail, leaving Part 2 for independent reading.

For the professional engineer and researcher, Part 1 provides a theoretical perspective within which he can view the entire scope of the subject and perhaps broaden his interests. Part 2 provides the answers to more specific practical problems.

I have not tried to write a scientific or technological history. A detailed account of who did what, when, and how has been avoided. The book is intended to be read on its own and to be intelligible without frequent reference to the published literature. Bibliographical material is cited only in order to give credit where credit is due or to indicate where the reader may find more specific and detailed information.

The text has developed from parts of the lecture notes for summer courses which I offered jointly with John G. Collier at Dartmouth College (1965, 1966, 1968), The University of Glasgow (1967), and Stanford University (1967). This material has been considerably reorganized to be more suitable for use by students and by those who are unfamiliar with the field. Numerous worked examples and problems have been added to give physical meaning to the theoretical concepts and to promote an awareness of practical applications.

I am very much indebted to my students, J. Michael Turner,

Andrew Porteous, Philip E. Meyer, Dale E. Runge, Stephen S. MacVean, Stanley F. Birch, Gary Grulich, Thomas E. Brady, and Donald A. Steen who worked problems, searched the literature, wrote computer programs, and performed experiments which contributed immensely to the text. The work would also have been impossible without the administrative and clerical support of the Thayer School of Engineering at Dartmouth College, in particular the thoroughly reliable and painstaking secretarial work of Edith G. Henson.

The part of Chap. 3 describing entropy generation in separated flow is mostly taken from my article in the *International Journal of Heat and Mass Transfer* (vol. 11, pp. 459–472, 1968) and is reprinted by permission. I am also indebted to the Institution of Chemical Engineers (London, England) for allowing me to quote the material on sedimentation in Chap. 8 from my paper at the Symposium on the Interaction between Fluids and Particles, London, 1962.

The choice of nomenclature was mostly decided during conversations with Dr. Novak Zuber in 1964, during which we tried to develop a consistent set from the prolific symbols used by numerous authors in many fields.

GRAHAM B. WALLIS

List of Symbols

A	area; a constant or coefficient
a	amplification factor in Eq. (6.124); a constant or coefficient; acceleration
B	body force field; damping factor; a constant or coefficient; correction factor
b	body force per unit volume; breadth; index; constant
C	a constant; coefficient of apparent mass
C_D	drag coefficient
C_f	friction factor
C_0	defined in Eq. (4.21)
C_1, C_2	correction factors
c	velocity of dynamic wave relative to average or weighted average velocity; velocity of sound; specific heat; coefficient of consolidation
c_a	molal concentration of non-condensable gas
c_c	velocity of compressibility wave
c_{ch}, c_{cs}	wave velocities in homogeneous and stratified flow
c_p	specific heat at constant pressure
c_v	specific heat at constant volume
D	diameter of pipe
D_b	bubble diameter
D_d	drop diameter
D_0	orifice diameter

$\mathfrak{D}$	energy dissipation per unit volume
d	particle diameter
E	defined by Eq. (8.80); fraction or percent entrainment
e	base of natural logarithms; thermodynamic energy per unit mass
F	force per unit volume of entire flow field
$\mathbf{F}$	force; function defined in Eq. (11.25)
f	force per unit volume of component
G	mass flux
g	acceleration due to gravity
H	height; sign of shear stress
h	enthalpy; heat transfer coefficient; height
j	volumetric flux
j_{12} or j_{21}	drift flux
K	a constant; Bankoff parameter [Eq. (5.49)]; velocity ratio [Eq. (8.125)]
k	a constant; thermal conductivity
k_s	roughness
L	length; fractional thermal lag [Eq. (8.126)]
l	mixing length
ln	natural logarithm
M	Mach number; mass exchange per unit volume per unit time
m	mass flux in boiling or condensation; index; mass flow rate ratio; fraction of cylindrical bubble cross section occupied by liquid; a coefficient
N	dimensionless inverse viscosity; index
n	index in correlations; velocity ratio
p	pressure
Δp	pressure drop
P	perimeter
Q	volumetric flow rate
q	volumetric flow rate per unit width
q_e	heat transfer rate
R	radius, radius of curvature; gas constant
r	radius
$\mathbf{r}$	vector coordinate of a point in space
r_0	pipe or duct radius
r_2	parameter defined in Eq. (3.37)
S	source of mass
s	entropy
T	temperature
ΔT	temperature difference
t	time
U	wave velocity
u	wave velocity relative to V_0; local velocity
u^*	friction velocity
u_0	characteristic transverse velocity in the Reynolds flux model
v_1, v_2	component velocities
V_1, V_2	unperturbed component velocities
V	average velocity; single-phase velocity
v_{12}	relative velocity
v_{1j}	drift velocity of component 1 relative to the volumetric average
$\bar{v}$	average single-phase velocity
V_0	weighted average velocity [Eq. (6.75)]

V_w	continuity wave velocity
v_w	continuity wave velocity relative to average or weighted average velocity
V_s	shock-wave velocity
$\mathcal{v}$	specific volume
$\mathcal{V}$	volume
W	mass flow rate
w	work
w_s	shaft work
w_τ	shear work
X	generalized sedimentation coordinate [Eq. (8.68)]; Martinelli parameter [Eq. (3.31)]
x	quality [Eq. (1.7)]
Y	generalized sedimentation coordinate
y	spatial coordinate; distance from wall
z	coordinate in direction of motion
α	void fraction; volumetric fraction of component 2
β	contact angle (degrees); shock angle; source of matter per unit length; dimensionless pressure drop
γ	isentropic exponent
δ	film thickness; volumetric flow rate ratio
ϵ	volumetric fraction of component 1; Reynolds flux
ζ	fraction of dissipated energy transferred to component 2
η	limiting viscosity at high rates of shear; fraction of force due to phase change acting on component 2
θ	angle to vertical; wedge angle
λ	wavelength; parameter defined by Eq. (10.57)
μ	viscosity
ν	kinematic viscosity, μ/ρ
π	natural constant
ρ	density
σ	surface tension
τ	shear stress
ϕ	heat flux; function; velocity potential
ϕ_f, ϕ_g	Martinelli parameters [Eqs. (3.24) and (3.25)]
Φ	shape factor
ω	frequency
Ω	correction factor (Fig. 3.13); reaction frequency [Eq. (2.92)]
ψ	a function defined by Eq. (8.18)
Γ	mass flow per unit width

Subscripts

1	component 1; state 1; location 1
2	component 2; state 2; location 2
a	air
A	acceleration
b	bubble
c	continuous phase; curvature; core in annular flow
d	drop; discontinuous phase
e	exit; effective value of
f	liquid or fluid

F	friction
g	gas
G	gravitational
i	component i; interface; inlet; inside; inflexion; quantum state
j	relative to or pertaining to average volumetric flux j
m	mixed mean in homogeneous flow; value at tube center in single-phase flow
n, N	normal
0	initial or boundary value; orifice; outside; zero quality; zero flow rate; stagnation; relative to stationary fluid or particles
p	pipe; particle
s	solid; slug; shock; entropy
t	tangential; turbulent
w	wave; wall
y	yield
∞	single particle; drop or bubble in an infinite medium; limiting value
TP	two phase
ϵ	when the concentration of the continuous phase is ϵ
θ	at angle θ to vertical

Note: Subscripts in Chap. 6 denote partial differentiation while subscripts outside brackets enclosing partial derivatives denote variables which are kept constant

accel	acceleration
crit	critical
mf	minimum fluidization
frict	friction
max	maximum
min	minimum

Dimensionless Groups

Fr	Froude number
M	Mach number
Nu	Nusselt number
Pr	Prandtl number
Re	Reynolds number
We	Weber number
N_i	$\dfrac{D\rho_i^{1/2}[Dg(\rho_1 - \rho_2)]^{1/2}}{\mu_i}$
N_{Ar}	Archimedes number: $\dfrac{\rho_f\sigma^{3/2}}{\mu_f^2 g^{1/2}(\rho_f - \rho_g)^{1/2}}$
N_{Bo}	Bond number: $\dfrac{gR^2(\rho_f - \rho_g)}{\sigma}$
$N_{\text{Eö}}$	Eötvös number: $\dfrac{gD^2(\rho_f - \rho_g)}{\sigma}$
Y	Property group: $\dfrac{g\mu^4}{\sigma^3\rho}$
j_f^*	$\dfrac{j_f\rho_f^{1/2}}{[gD(\rho_f - \rho_g)]^{1/2}}$

j_g^*	$\dfrac{j_g \rho_g^{1/2}}{[gD(\rho_f - \rho_g)]^{1/2}}$
$j_f'^*$	$\dfrac{32 j_f \mu_f}{D^2 g(\rho_f - \rho_g)}$
Δp^*	$\dfrac{-(dp/dz) - \rho_g g \cos\theta}{g(\rho_f - \rho_g)}$
π_2	$\dfrac{j_g \mu_g}{\sigma}\left(\dfrac{\rho_g}{\rho_f}\right)^{1/2}$
R_i	$\left(\dfrac{\rho_i}{\rho}\right)^{1/2}$

Prefixes

d	differential
∂	partial differential
δ	small change
Δ	negative increment
∇	gradient operator
$\nabla .$	divergence
$d\mathbf{r}$	element of volume at $\mathbf{r}$
$d\mathbf{A}$	element of area

Superscripts

*	dimensionless form of; at point where the Mach number is unity
′	modified; perturbed; in new frame of reference
+	dimensionless form of velocity profiles

Abbreviations

Btu	British thermal unit
cfm	cubic feet per minute
cm	centimeter
fps	feet per second
ft	feet
hr	hour
lb	pound
mm	millimeter
psia	pounds per square inch absolute
°F	degrees Fahrenheit

Some Uncommon Abbreviations Used in References

AEC	U.S. Atomic Energy Commission
AEEW	Atomic Energy Establishment, Winfrith, UKAEA
AERE	Atomic Energy Research Establishment, UKAEA
ANL	Argonne National Laboratory
CISE	Centro Informazioni Studi Esperienze, Milan, Italy
EURAEC	Euratom-Atomic Energy Commission Joint Program
NYO	New York Operations Office of U.S. Atomic Energy Commission
UKAEA	United Kingdom Atomic Energy Authority

part one

Analytical Techniques

1
Introduction

1.1 WHAT IS TWO-PHASE FLOW?

A *phase* is simply one of the states of matter and can be either a gas, a liquid, or a solid. *Multiphase flow* is the simultaneous flow of several phases. *Two-phase flow* is the simplest case of multiphase flow.

The term *two-component* is sometimes used to describe flows in which the phases do not consist of the same chemical substance. For example, steam-water flows are two-phase, while air-water flows are two-component. Some two-component flows (mostly liquid-liquid) consist of a single phase but are often called two-phase flows in which the phases are identified as the continuous or discontinuous components.

Since the mathematics which describe two-phase or two-component flows are identical, it does not really matter which definitions are chosen. The two expressions will therefore be treated as synonyms in most developments in this book.

There are many common examples of two-phase flows. Some, such as fog, smog, smoke, rain, clouds, snow, icebergs, quicksands, dust storms, and mud, occur in nature. Others, such as boiling water, tea making, egg scrambling, salad tossing, jam spreading, cream whipping, sugar stirring, and sphaghetti twirling, are frequent occurrences in kitchens and dining rooms.

Several everyday processes involve a sequence of different two-phase flow configurations or flow patterns. In a coffee percolator, for example, the water is first boiled to form steam bubbles, alternate slugs of liquid and vapor then rise through the central tube, and the hot water percolates through the coffee grounds and eventually drips down into the pot. When beer is poured from a bottle, the rate of discharge is limited by the rise velocity of *slug-flow* bubbles in the neck; subsequently bubbles nucleating from defects in the walls of the glass rise to form a pleasing foam at the surface. Bread and cakes begin with a multiphase mixing process, are cooked with the release of bubbles, except when the appropriate ingredient is forgotten, and are eventually consumed orally in one of the most common multiphase phenomena of all.

The subtle blend of flavor, texture, and temperature that is achieved in a perfect Martini is the result of skillful control of a two-phase chemical-engineering process.

Biological systems contain very few pure liquids. Body fluids, such as blood, semen, and milk, are all multiphase, containing a variety of cells, particles, or droplets in suspension. Their behavior can be described by much the same equations as are used for analyzing paints, inks, pastes, and nuclear fuel slurries.

A more technical example can be taken from the familiar area of fire prevention and control. Almost without exception the various methods of fire extinguishing are all multiphase processes, involving sprays, jets, foams, or powders. Even the extinguishers which use pure gas cannot be analyzed without considering the flash evaporation which occurs as the material is expelled from the high-pressure storage cylinder. Moreover, the fires themselves usually result from a reaction between solid or liquid fuels and oxygen in the air, produce smoke and steam, which are invisible unless they are two-phase, and cause death by irritating the nose and throat until the victim drowns in his own multiphase secretions. Deliberate fires in boilers, automobile engines, and rockets are designed to burn two-phase dispersions.

Examples are equally profuse in the industrial field. Over half of all chemical engineering is concerned with multiphase flows. Many industrial processes such as power generation, refrigeration, and distillation depend on evaporation and condensation cycles. The performance of desalination plants is limited by the "state of the art" in two-phase technology. Steelmaking, paper manufacturing, and food processing all contain critical steps which depend on the proper functioning of multiphase devices. Many problems of air and water pollution are due to unwanted two-phase flows.

1.2 METHODS OF ANALYSIS

Two-phase flows obey all of the basic laws of fluid mechanics. The equations are merely more complicated or more numerous than those of single-phase flows.

The techniques for analyzing one-dimensional flows fall into several classes which can conveniently be arranged in ascending order of sophistication, depending on the amount of information which is needed to describe the flow, as shown in the following paragraphs.

CORRELATIONS

Correlation of experimental data in terms of chosen variables is a convenient way of obtaining design equations with a minimum of analytical work. The crudest correlations are mere mathematical exercises, readily performed with modern computers, while more advanced techniques use dimensional analysis or a grouping of several variables together on a logical basis.

A virtue of correlations is that they are easy to use. As long as they are applied to situations similar to those that were used to obtain the original data, they can be quite satisfactory, within statistical limits which are usually known. However, they can be quite misleading if used indiscriminately in a variety of applications. Furthermore, since little insight into the basic phenomena is achieved by data correlation, no indication is given of ways in which performance can be improved or accuracy of prediction increased.

In general, correlations will be avoided in this book unless they possess a viable claim to generality. Those which are quoted will be dimensionless and have some theoretical basis or have been tested against a variety of data.

SIMPLE ANALYTICAL MODELS

Very simple analytical models which take no account of the details of the flow can be quite successful, both for organizing experimental results and for predicting design parameters. For example, in the *homogeneous* model the components are treated as a pseudofluid with average properties, without bothering with a detailed description of the flow pattern. A suspension of droplets in a gas, a foam, or the stratified flow of a gas over a liquid are all treated exactly alike. In the *separated-flow* model the phases are assumed to flow side by side. Separate equations are written for each phase and the interaction between the phases is also considered. In the *drift-flux* model attention is focused on the relative motion.

Each of these simple models is accorded an entire chapter in Part 1 and is used extensively throughout Part 2 of this book.

INTEGRAL ANALYSIS

A one-dimensional integral analysis starts from the assumption of the form of certain functions which describe, for example, the velocity or concentration distributions in a duct. These functions are then made to satisfy appropriate boundary conditions and the basic fluid-mechanics equations in integral form. Similar techniques are quite commonly used for analyzing single-phase boundary layers.

DIFFERENTIAL ANALYSIS

In a differential analysis the velocity and concentration fields are deduced from suitable differential equations. Usually, following the one-dimensional flow idealization, the equations are written for time-averaged quantities, as in single-phase theories of turbulence. More sophisticated versions of the theory may even consider temporal variations.

Efforts will be made throughout the book to show how the various levels of analysis are related. Usually the more complex theories lead to the inclusion of additional effects and the prediction of numerical values of *correction factors*, which can be applied to the simpler theories in order to increase their accuracy. The complex theory may also lead to an analytical rather than empirical relationship between the important variables. Thus these sequential levels of analysis resemble a pyramid in which the broader and more general theories serve to support the more approximate and simpler techniques.

UNIVERSAL PHENOMENA

In addition to this hierarchy of analytical methods, there is a class of very powerful techniques based on universal phenomena that are independent of the flow regime, the analytical model, or the particular system. Typical of these methods are the various theories of wave motion and extremum techniques for obtaining the locus of the limiting behavior of a system. These ideas thread their way through the various chapters and help to bind them together into a uniform conceptual framework. Understanding of these concepts should develop as each new application is described.

1.3 FLOW REGIMES

The price that is paid for a greater accuracy in prediction of results is an increase in complexity. In two-phase flow the amount of knowledge which is needed in order to perform a detailed analysis is often surprisingly great. For example, in studying the motion of a single gas bubble rising in a stagnant liquid, one is concerned with all of the following effects:

Inertia of the gas and the liquid
Viscosity of the gas and the liquid
Density difference and buoyancy
Surface tension and surface contamination

The last item above is itself extremely complicated since "contamination" can take the form of dirt, dissolved matter, or surface-active agents. Heat transfer and mass transfer to the bubble also alter its motion.

Perhaps the first step in rendering this hydralike problem tractable is to break it up into various regimes which are each governed by certain dominant geometrical or dynamic parameters.

Part of the definition of the flow regime is a description of the morphological arrangement of the components, or *flow pattern*. The flow pattern is often obvious from visual or photographic observations but is not adequate to define the regime completely because of additional distinguishing criteria, such as the difference between laminar and turbulent flow or the relative importance of various forces. In order to keep the terminology manageable, the numerous imaginative expressions which have been used throughout the literature to describe flow patterns will not be quoted. It is far simpler to restrict classifications to the morphological flow patterns (for example, bubbly, slug, annular, and drop flow in gas-liquid systems) and create further subdivisions into distinct regimes within each of these classifications. Hybrid flow patterns, usually representing a region of transition from one pattern to another, are denoted by hyphenated expressions (thus, slug-annular and annular-drop flows). Some synonyms (e.g., "fog" or "mist" instead of "drop") may be used when perfunctory repetition of a single word becomes monotonous.

As an example of the complexity of two-phase flows, Fig. 1.1 shows a sequence of flow patterns occurring in an evaporator as more and more liquid is converted to vapor. Obviously different parts of the evaporator require different methods of analysis, and the problem of how one regime develops from another has to be considered also.

Numerous authors have presented flow-pattern and flow-regime maps in which various areas are indicated on a graph for which there are two independent coordinates. For a given apparatus and specified components this is readily done in terms of the flow rates, as shown in Figs. 1.2 and 1.3. However, since the flow regime is governed by about a

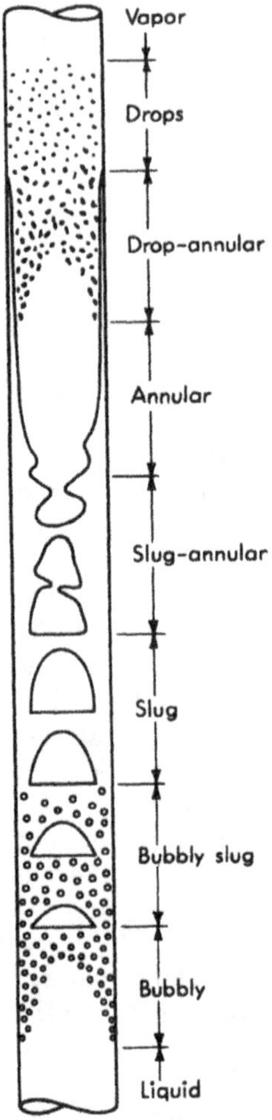

Fig. 1.1 Approximate sequence of flow patterns in a vertical tube evaporator.

dozen variables, a two-dimensional plot is quite inadequate for general representation. In this text the criteria for regime boundaries, when they are known, will be discussed in detail as each regime is analyzed. When these criteria are represented on a two-dimensional plot for particular applications, it will be found that the areas covered by a particular regime change drastically both in size and shape, as variables such as pressure and diameter are changed. For certain combinations of parameters, whole regimes disappear from the map altogether.

1.4 NOTATION

Before proceeding with the analysis of two-phase flow it will be necessary to define some of the relevant terminology. Although a detailed list of nomenclature is given at the front of this volume, it may require some explanation. In addition, a certain familiarity with the simple relationships among some of the parameters will enable the analysis to be understood more rapidly.

SIMPLE DEFINITIONS

The two components are distinguished by subscripts 1 and 2 in general, or by subscripts f and g for a liquid-gas system or f and s for a fluid-solid system. Component 2 is usually chosen to be the dispersed phase or the lighter phase in a stratified flow.

The total mass rate of flow (in pounds per second) is represented by the symbol W. The total flow is the sum of the component flows.

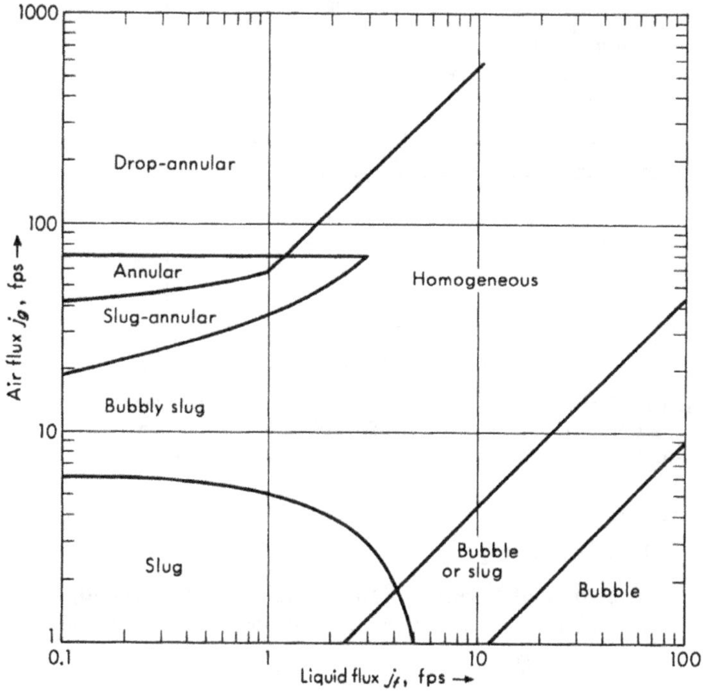

Fig. 1.2 Flow-pattern boundaries for vertical upflow of air and water at 15 psia in a 1-in.-diam tube deduced from equations in the text.

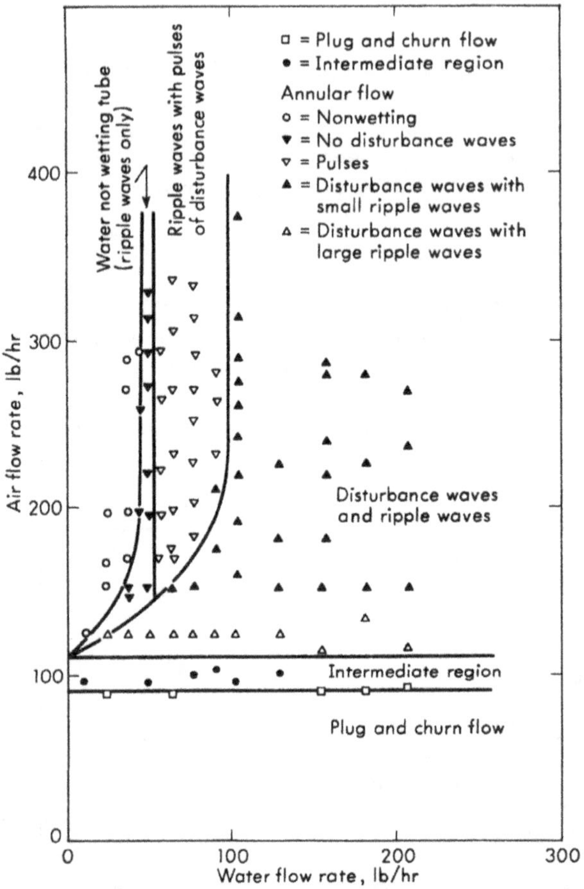

Fig. 1.3 Various "regimes" or subdivisions of the annular flow pattern for cocurrent upward flow of air and water in a 1¼-in.-diam pipe at 15 psia. (*Hall-Taylor and Hewitt.*[1])

Therefore

$$W = W_1 + W_2 \tag{1.1}$$

The volumetric rate of flow (in cubic feet per second) is represented by the symbol Q. The following relationships are obvious:

$$Q = Q_1 + Q_2 \tag{1.2}$$

$$Q_1 = \frac{W_1}{\rho_1} \tag{1.3}$$

$$Q_2 = \frac{W_2}{\rho_2} \tag{1.4}$$

Every part of the flow field is occupied by one or other component. If α represents the fraction of an element of volume which is occupied at any instant by component 2, then, evidently, if the element is chosen small enough, α can only be 0 or 1. However, for most purposes a volume much larger than the discrete particles (drops or bubbles) is chosen, and α then represents an average volumetric concentration. Usually α is measured as an average over the whole flow cross section and a sufficient length of duct to eliminate local fluctuations. Thus, if a pipe of length L and cross-sectional area A is suddenly isolated by closing valves at both ends, the contents can be analyzed and the total volume $\mathcal{V}_2$ of component 2 in the pipe can be determined. The average value of α is then

$$\langle\alpha\rangle = \frac{\mathcal{V}_2}{AL} \tag{1.5}$$

Often it is not possible to measure $\langle\alpha\rangle$ over a long length of pipe because the flow is not uniform. In this case a large number of instantaneous readings over a length δL give the time average of α at a given location. The average value of α both in space and time is then

$$\langle\alpha\rangle = \frac{\int\int \alpha(L,t)\, dL\, dt}{\int dL \int dt} \tag{1.6}$$

Usually the symbol α is used loosely to represent an *average volumetric concentration* without bothering to define exactly how the average is to be taken. Thus, extra care is needed when periodic phenomena and nonuniform concentrations are important. In gas-liquid flows, α usually represents the *void fraction*, or volumetric concentration of the gas.

It is often convenient, particularly in boiling or condensing applications, to have a measure of the fraction of the total mass flow across a given area which is composed of each component. The *quality* is therefore defined as

$$x = \frac{W_2}{W} \tag{1.7}$$

Evidently x is subject to averaging laws when flow is unsteady or nonuniform. The average is to be taken over a specified surface and for a period of time. Therefore

$$\langle x\rangle = \frac{\int G_2\, dA\, dt}{\int G\, dA\, dt} \tag{1.8}$$

The symbol j is used to represent *volumetric flux* (in feet per second) or volumetric flow rate per unit area. The flux is really a vector quantity but at the level of sophistication which is appropriate at present, j will be used exclusively to represent the scalar component in the direction of motion along a pipe or duct. The flux is related to the local component concentration and velocity as follows:

$$j_1 = (1 - \alpha)v_1 \tag{1.9}$$
$$j_2 = \alpha v_2 \tag{1.10}$$

The total local flux is

$$j = j_1 + j_2 \tag{1.11}$$

The following results are self-evident

$$Q_1 = \int j_1 \, dA \tag{1.12}$$
$$Q_2 = \int j_2 \, dA \tag{1.13}$$

The average volumetric flux across an area A is then

$$\langle j_2 \rangle = \frac{Q_2}{A} \tag{1.14}$$

Unless variations across the flow are being considered, the brackets are usually omitted from Eq. (1.14).

The mass flux is represented by the symbol G (in pounds per hour per square foot).

Clearly, for a small element in which the density of each component can be regarded as uniform:

$$G_1 = \rho_1 j_1 \tag{1.15}$$
$$G_2 = \rho_2 j_2 \tag{1.16}$$
$$G = G_1 + G_2 \tag{1.17}$$

The average mass flux of component 2 across an area A is

$$\langle G_2 \rangle = \frac{W_2}{A} \tag{1.18}$$

Although the most general description of two-component flow involves a consideration of the three-dimensional and temporal variation of all the above quantities, we shall usually be content with the one-dimensional flow assumptions and work entirely in terms of averages across the duct. Under certain circumstances, however, when large variations across the duct occur, the theory will prove inadequate and more detailed analysis will be necessary.

Some useful relationships for one-dimensional flow are summarized below.

$$j_1 = \frac{Q_1}{A} \tag{1.19}$$

$$j_2 = \frac{Q_2}{A} \tag{1.20}$$

$$j = \frac{Q_1 + Q_2}{A} \tag{1.21}$$

$$v_1 = \frac{j_1}{1 - \alpha} \tag{1.22}$$

$$v_2 = \frac{j_2}{\alpha} \tag{1.23}$$

$$G_1 = \frac{W_1}{A} \tag{1.24}$$

$$G_2 = \frac{W_2}{A} \tag{1.25}$$

$$W_1 = Q_1\rho_1 \tag{1.26}$$

$$W_2 = Q_2\rho_2 \tag{1.27}$$

$$\frac{j_1}{j_2} = \frac{Q_1}{Q_2} = \frac{v_1}{v_2}\frac{1 - \alpha}{\alpha} \tag{1.28}$$

$$\frac{G_1}{G_2} = \frac{W_1}{W_2} = \frac{1 - x}{x} \tag{1.29}$$

From Eqs. (1.26) through (1.29),

$$\frac{1 - x}{x} = \frac{v_1}{v_2}\frac{\rho_1}{\rho_2}\frac{1 - \alpha}{\alpha} \tag{1.30}$$

The *relative velocity* is defined as

$$v_{21} = (v_2 - v_1) = -v_{12} \tag{1.31}$$

Drift velocities are defined as the difference between the component velocities and the average as follows:

$$v_{1j} = v_1 - j \tag{1.32}$$

$$v_{2j} = v_2 - j \tag{1.33}$$

The *drift flux* represents the volumetric flux of a component relative to a surface moving at the average velocity, i.e.,

$$j_{21} = \alpha(v_2 - j) \tag{1.34}$$

$$j_{12} = (1 - \alpha)(v_1 - j) \tag{1.35}$$

Substituting Eq. (1.11) into Eq. (1.34) and using Eq. (1.10), we obtain

$$j_{21} = j_2 - \alpha(j_1 + j_2) = j_2(1 - \alpha) - \alpha j_1 \tag{1.36}$$

Similarly,

$$j_{12} = j_1\alpha - (1 - \alpha)j_2 \tag{1.37}$$

Therefore

$$j_{21} = -j_{12} \tag{1.38}$$

This symmetry is an important and useful property of the drift flux.

Substituting for j_1 and j_2 in Eq. (1.37) by using Eqs. (1.22) and (1.23), we get

$$j_{12} = \alpha(1 - \alpha)(v_1 - v_2) = \alpha(1 - \alpha)v_{12} \tag{1.39}$$

Therefore the drift flux is proportional to the relative velocity.

Any system of units may be used in the above equations as long as the requirements of consistency and compatibility are satisfied.

Velocities which are characteristic of the overall flow, such as wave velocities, are represented by capital letters with appropriate subscripts. For example, the continuity wave velocity is given the symbol V_w and the shock-wave velocity the symbol V_s.

PROPERTIES

To distinguish between the symbol for volume and the symbol for velocity, the former is written in script form. Thus, the volume of a bubble is $\mathcal{V}_b$ and the specific volume of gas is $\mathscr{v}_g$. Care should be taken in distinguishing these symbols; for example v_{fg} is the liquid velocity relative to the gas, whereas $\mathscr{v}_{fg}$ represents the change in specific volume on vaporization. The subscript convention is reversed in the case of thermodynamic properties; thus $\mathscr{v}_{fg} = \mathscr{v}_g - \mathscr{v}_f$ and $h_{fg} = h_g - h_f$.†

There should be no difficulty in identifying the common symbols for properties such as ρ for density and μ for viscosity. Surface tension is not so widely used and does not have a standard symbol; it will be represented by σ with the units of force per unit length. Both enthalpy and heat-transfer coefficient have the symbol h but the context should make the distinction clear.

PRESSURE DROP

The symbol for pressure drop in a pipe is Δp. On the other hand dp/dz represents the rate at which the pressure increases with distance in the z direction. Therefore if z denotes the coordinate down the axis of a pipe measured in the direction of flow, the pressure drop over a length L will be

$$\Delta p = -\int_0^L \frac{dp}{dz}\,dz \tag{1.40}$$

COORDINATES

Since the symbol x has already been chosen to represent the quality, it will be avoided for use in describing the coordinate system. Usually z will be the coordinate measured in the direction of flow and y the coordinate measured from a boundary, such as a wall. The radial distance from a pipe axis will be denoted by r.

† In hand work it is often convenient to use the density rather than specific volume as the basic variable in order to avoid possible confusion between the $\mathscr{v}$'s and the v's.

UNITS

All equations are written in a consistent dimensional form and are suitable for use with any convenient system of units. Superfluous factors which represent the ratio between various conventional units have been omitted entirely. Any student who has learned to cope with the plethora of units (which have evolved by historical accident) in elementary fluid mechanics and thermodynamics courses should have developed sufficient maturity to have no difficulty using the equations in this book.

PROBLEMS

1.1. A bubbly mixture flows in a 1-in.-diam pipe. The gas flow rate is 30 cfm and the bubble velocity is determined photographically to be 100 fps. What is the void fraction? What is the liquid velocity if the liquid flow rate is 5 cfm?

1.2. 300 lb/hr of air at 70°F and 20 psia flow together with 300 lb/hr of water in a 1.25-in.-diam pipe. What is the overall volumetric flux j? If the drift flux j_{gf} is 10 fps, what are the average velocities of the phases?

1.3. A steam-water mixture with 1% quality flows at atmospheric pressure in the riser of a coffee percolator. The void fraction is measured to be 80%. What is the ratio of the average steam velocity to the average water velocity?

1.4. In a certain fluid-solid system (a quicksand) the drift flux is related to the volumetric concentration ϵ of the fluid by the equation

$$j_{fs} = 5\epsilon^3(1 - \epsilon) \qquad \text{fps}$$

If a flux $j_f = 1$ fps of fluid flows upward through a stationary bed of particles, what is the value of ϵ? If the particles are spheres, which have a random packing value of $\epsilon_0 = 0.4$, is the system "fluidized" or do the particles rest on one another?

1.5. Express G in terms of the quality, the individual phase velocities, and the densities.

1.6. Express j in terms of the individual mass flow rates, the pipe diameter, and the phase densities.

1.7. Show that the drift flux is independent of the motion of an observer.

1.8. On a graph of j_2 versus j_1, for phases with given properties, show lines of constant j, constant G, and constant x. Can lines of constant α be drawn? Why not?

1.9. On the graph of j_2 versus j_1, show lines of constant α, if

(a) $\dfrac{v_1}{v_2} = \text{const}$

(b) $v_{12} = \text{const}$

(c) $j_{12} = \text{const}$

(d) $j_{12} = k\alpha(1 - \alpha)^n$

1.10. Express the momentum flux in one-dimensional flow in terms of G, x, α, and the densities of the phases. For what value of α will the momentum flux be a minimum if G and x are constant? For what values of α will G be a maximum if x and the momentum flux are fixed?

1.11. Solve Prob. 1.10 using the kinetic-energy flux instead of the momentum flux.

1.12. Show that for incompressible flow in a constant-area duct, j is constant with position, although the individual fluxes may vary.

1.13. Show that the drift flux is zero when $\alpha = 0$ or $\alpha = 1$.

1.14. Prove that $j_2 v_{12} = v_2 v_{1j}$.

1.15. A glass is filled to the brim with draft beer. After it has stood for a while until bubble activity ceases, the glass is observed to be 70% full. What was the original void fraction? Estimate the quality of the mixture issuing from the tap.

1.16. Estimate the maximum value of α in a stable foam which was made from bubbles 1 mm in diameter and has drained for a very long time until the liquid filaments are a few molecules thick.

1.17. What value of α corresponds to a close-packed array of spheres?

1.18. Water at 100 psia enters a straight evaporator tube. If the velocity ratio v_g/v_f is constant at the value 2.5, and the mass flux is 2×10^5 lb/(hr)(ft^2), what are the values of void fraction and momentum flux when $x = 0$, 0.1, and 0.5?

1.19. In a particular vertical flow regime the relative velocity is constant and equal to v_0. Draw the lines of constant α on a graph of j_1 versus j_2. Show that the lines envelop a curve in the quadrant which represents countercurrent flow. This envelope is the "flooding line" outside which operation is impossible. Show that the equation of the flooding line is

$$|j_1|^{1/2} + |j_2|^{1/2} = v_0^{1/2}.$$

REFERENCE

1. Hall-Taylor, N., and G. F. Hewitt: AERE-R3952, UKAEA, 1962.

2
Homogeneous Flow

2.1 INTRODUCTION

Homogeneous flow theory provides the simplest technique for analyzing two-phase (or multiphase) flows. Suitable average properties are determined and the mixture is treated as a pseudofluid that obeys the usual equations of single-component flow. All of the standard methods of fluid mechanics can then be applied.

The average properties which are required are velocity, thermodynamic properties (e.g., temperature and density), and transport properties (e.g., viscosity). These pseudo properties are weighted averages and are not necessarily the same as the properties of either phase. The method of determining suitable properties is often to start with more complex equations and to rearrange them until they resemble equivalent equations of single-phase flow. For example, the virtual viscosity of an emulsion can be obtained by analyzing the three-dimensional flow field in the two components. In another case, the apparent properties for a gas-particle mixture are found from the separate equations for each component by assuming a class of similar solutions (Chap. 8, p. 212).

Differences in velocity, temperature, and chemical potential between the phases will promote mutual momentum and heat and mass transfer. Often these processes proceed very rapidly, particularly when one phase is finely dispersed in the other, and it can be assumed that equilibrium is reached. In this case the average values of velocity, temperature, and chemical potential are the same as the values for each component and we have *homogeneous equilibrium flow*. The resulting equations are simple and easy to use, but it is often advisable to check the validity of the equilibrium assumptions by using the more accurate theories which will be presented in later chapters or by performing a detailed analysis of other transport processes which are beyond the scope of this book. For example, rapid acceleration and pressure changes make the equilibrium theory inaccurate for describing the discharge of flashing steam-water mixtures through short nozzles or orifices. It is necessary to consider the rates of bubble nucleation and growth in the superheated liquid. Nonequilibrium effects also occur when supercooled vapor condenses in high-velocity steam or when particles of solid propellant are burned in the nozzle of a rocket engine.

In some cases the use of homogeneous theory is obviously inappropriate. For example, countercurrent vertical flows, which are driven by gravity acting on the different densities of the phases, cannot be described by a suitable "average" velocity.

This chapter will develop homogeneous flow theory using a sufficient number of examples to illustrate the techniques. More specific and detailed applications will be found in later chapters.

2.2 ONE-DIMENSIONAL STEADY HOMOGENEOUS EQUILIBRIUM FLOW

The basic equations for steady one-dimensional homogeneous equilibrium flow in a duct are:

Continuity
$$W = \rho_m v A = \text{const} \tag{2.1}$$

Momentum
$$W\frac{dv}{dz} = -A\frac{dp}{dz} - P\tau_w - A\rho_m g \cos\theta \tag{2.2}$$

Energy
$$\frac{dq_e}{dz} - \frac{dw}{dz} = W\frac{d}{dz}\left(h + \frac{v^2}{2} + gz_g\right) \tag{2.3}$$

In the above equations A and P represent the duct area and perimeter, τ_w is the average wall shear stress, dq_e/dz is the heat transfer per unit length of duct, z_g is the vertical coordinate, and θ is the inclination of the duct to the vertical. Work terms are assumed to be zero in the energy equation in most cases. It is often possible to use the momentum and energy equations in integral form when one is only interested in changes between particular points in the duct.

Equation (2.2) is often rewritten as an explicit equation for the pressure gradient. Thus

$$\frac{dp}{dz} = -\frac{P}{A}\tau_w - \frac{W}{A}\frac{dv}{dz} - \rho_m g \cos\theta \tag{2.4}$$

The three terms on the right side can then be regarded as *frictional*, *accelerational*, and *gravitational* components of the pressure gradient. Since engineers (being pessimists) are mostly interested in pressure drops, the following definitions are usually adopted.

$$-\left(\frac{dp}{dz}\right)_F = \frac{P}{A}\tau_w \tag{2.5}$$

$$-\left(\frac{dp}{dz}\right)_A = \frac{W}{A}\frac{dv}{dz} \tag{2.6}$$

$$-\left(\frac{dp}{dz}\right)_G = \rho_m g \cos\theta \tag{2.7}$$

The total pressure gradient is then the sum of the components, as follows:

$$\frac{dp}{dz} = \left(\frac{dp}{dz}\right)_F + \left(\frac{dp}{dz}\right)_A + \left(\frac{dp}{dz}\right)_G \tag{2.8}$$

In addition to the above equations we usually have some knowledge about the equation of state for the components. For a steam-water mixture, for example, the steam tables or Mollier chart can be employed. For a mixture of a gas and a solid, an equivalent equation of state can be derived by assuming equilibrium at all times between the components or by making other appropriate assumptions.

The mean density can be expressed in various ways. In terms of the volume fraction α, it is

$$\rho_m = \alpha\rho_2 + (1 - \alpha)\rho_1 \tag{2.9}$$

whereas in terms of the quality or mass fraction specific volumes are additive. Therefore

$$\frac{1}{\rho_m} = \frac{x}{\rho_2} + \frac{1 - x}{\rho_1} \tag{2.10}$$

The mass of each component per unit volume can be expressed in terms of α or x to give the following equations:

$$x\rho_m = \alpha\rho_2 \tag{2.11}$$

$$(1 - x)\rho_m = (1 - \alpha)\rho_1 \tag{2.12}$$

For homogeneous steady flow with velocity equilibrium, the void fraction and quality are

$$\alpha = \frac{Q_2}{Q_1 + Q_2} = \frac{j_2}{j} \tag{2.13}$$

$$x = \frac{W_2}{W_1 + W_2} = \frac{G_2}{G} \tag{2.14}$$

An example of simple homogeneous flow can be taken from the case of isentropic expansion of a steam-water mixture through a nozzle with no friction or heat transfer at the walls. For expansion between two pressures at constant entropy the final state can be determined on a Mollier chart. Therefore the enthalpy change and final density can be determined. The exit velocity then follows from the integral of Eq. (2.3) and the area from Eq. (2.1). The technique is exactly the same as the equivalent technique that is used for predicting the expansion of dry steam. Errors, however, are likely to be introduced because of nonequilibrium effects. The water and steam are not in thermal equilibrium nor do they have equal velocities in a rapid expansion. Friction and heat transfer between the phases also introduce irreversibility which violates the constant-entropy assumption.

Example 2.1 Dry saturated steam from a large container at 100 psia is expanded through a frictionless adiabatic nozzle to a final pressure of 15 psia. What are the exit velocity and flow per unit area?

Solution From steam tables, for isentropic expansion as above, we find that the exit wetness is 10.7 percent. Therefore the exit specific volume and enthalpy are:

$$v_e = (0.893)(26.32) + (0.107)(0.0167) = 23.5 \text{ ft}^3/\text{lb}$$
$$h_e = 1150.7 - (0.107)(969.6) = 1047 \text{ Btu/lb}$$

The inlet enthalpy is $h_i = 1187.3$ Btu/lb.

Gravitational terms are negligible in Eq. (2.3), which reduces for adiabatic flow to the equation

$$\frac{v_e^2}{2} = h_1 - h_e = 140.3 \text{ Btu/lb}$$

Therefore

$$v_e = (2 \times 140.3 \times 32.2 \times 778)^{1/2} = 2660 \text{ fps}$$

The mass flow per unit area is derived from Eq. (2.1).

$$\frac{W}{A} = \frac{V_e}{v_e} = \frac{2660}{23.5} = 113 \text{ lb/(ft}^2\text{)(sec)}$$

Example 2.2 Formulate equations for the one-dimensional frictionless expansion of a dilute gas containing a dispersion of small solid particles. Assume homogeneous flow with uniform velocity, and consider the two limiting conditions which bound the possible thermal behavior, namely,

1. No heat transfer between the gas and the particles
2. Thermal equilibrium at all times between the components

Using these results, calculate the exit velocity, upstream stagnation pressure, and exit gas temperature for expansion of a mixture of 2 lb of sand [specific heat 0.21 Btu/(lb)(°F)] per pound of air to a Mach number of 2 at 15 psia. The upstream stagnation temperature is 1500°F.

Solution The simplest method of solution is to derive the equation of state and adiabatic expansion law for a pseudo gas consisting of the sand and the air combined.

Let T be the air temperature and let all the air properties be denoted by subscript a.

For 1 pound of the air alone the usual equation of state is

$$p_a v_a = R_a T \tag{2.15}$$

If the particle density is very large compared with the air density and if the two-phase mixture contains m pounds of particles per pound of air, the apparent density (or specific volume) will be related to the air density (or specific volume) by the approximate equations (see Prob. 2.1)

$$\frac{\rho}{\rho_a} = \frac{v_a}{v} = 1 + m \tag{2.16}$$

Using Eq. (2.16), Eq. (2.15) can be rewritten to give an equation of state for the pseudo gas consisting of both air and particles, as follows:

$$pv = \frac{R_a}{1 + m} T \tag{2.17}$$

Therefore the effect of adding the particles is simply to modify the appropriate value of R in the perfect-gas equation.

The expansion law will depend upon the mutual heat transfer. If there is no heat transfer and no friction between the air and the particles (since they are assumed to have the same velocity), the air will expand isentropically according to the law

$$p v_a^{\gamma_a} = \text{const} \tag{2.18}$$

Since $1 + m$ is a constant in Eq. (2.16), Eq. (2.18) is equivalent to the relation

$$p v^{\gamma_a} = \text{const} \tag{2.19}$$

Therefore for the zero heat-transfer case the mixture behaves exactly as a pseudo gas with the same value of the isentropic exponent γ_a as for air alone, but with a modified value of $R = R_a/(1 + m)$.

On the other hand, if the gas and particles are always in thermal equilibrium, the total entropy of them both taken together stays constant. If the particles have a specific heat of c Btu/(lb)(°F), their entropy increase accompanying a transfer of heat dQ to the air is

$$ds_p = -\frac{dQ}{T} = cm\frac{dT}{T} \qquad \text{per lb of air} \tag{2.20}$$

The entropy increase for the air is

$$ds_a = \frac{dQ}{T} = c_{p_a}\frac{dT}{T} - R_a\frac{dp}{p} \tag{2.21}$$

Combining Eqs. (2.20) and (2.21) we get

$$ds_a + ds_p = \frac{dT}{T}(c_{pa} + mc) - R_a\frac{dp}{p} = 0 \tag{2.22}$$

Therefore the pseudo gas obeys the expansion law

$$T^{-1}p^{(\gamma-1)/\gamma} = \text{const} \tag{2.23}$$

where

$$\frac{\gamma - 1}{\gamma} = \frac{R_a}{c_{pa} + mc} \tag{2.24}$$

Solving Eq. (2.24) for γ and using the relation $R_a = c_{pa} - c_{va}$, we have

$$\gamma = \frac{c_{pa} + mc}{c_{va} + mc} \tag{2.25}$$

This result could also have been deduced by realizing that the effective specific heats at constant pressure and constant volume are

$$\frac{c_{pa} + mc}{1 + m} \quad \text{and} \quad \frac{c_{va} + mc}{1 + m}$$

The effect of mutual heat transfer is therefore to modify the isentropic exponent to the value given by Eq. (2.25). The pseudo gas therefore behaves as if it possessed a value of $R = R_a/(1 + m)$ and a value of γ given by Eq. (2.25), but in other ways obeys all of the well-known one-dimensional flow relationships of gas dynamics that can be taken from a standard text such as Ref. 1. Alternatively these results can be deduced from Eqs. (2.1), (2.2), and (2.3), neglecting gravity and friction and assuming no external heat transfer.

For the numerical example we have to deal with a pseudo gas with a value of

$$R = \frac{53.3}{3} = 17.8 \text{ (ft)(lb}_f\text{)/(lb}_m\text{)(°F) and a value of } \gamma = 1.4$$

for assumption 1 and

$$\gamma = \frac{0.241 + (2)(0.21)}{0.173 + (2)(0.21)} = 1.11 \qquad \text{for assumption 2}$$

The stagnation pressure follows from the equation

$$p_0 = p_e\left(1 + \frac{\gamma - 1}{2} M_e^2\right)^{\gamma/(\gamma-1)} \tag{2.26}$$

and the exit air temperature from the equation

$$T_e = T_0\left(1 + \frac{\gamma - 1}{2} M_e^2\right)^{-1} \tag{2.27}$$

The exit velocity is

$$v_e = M_e \sqrt{\gamma R T_e} \tag{2.28}$$

Putting in the appropriate numbers we get:

1. For no mutual heat transfer

$P_0 = 117$ psia $\quad T_e = 1090\text{°R} = 630\text{°F}$
$V_e = 1870$ fps

2. For equilibrium between the components

$P_0 = 109.5$ psia $\quad T_e = 1605\text{°R} = 1145\text{°F}$
$V_e = 2020$ fps

FURTHER DEVELOPMENT OF THE MOMENTUM EQUATION

The momentum equation can be developed further by expressing the wall shear forces in terms of a friction factor and a hydraulic mean diameter D. The average wall shear stress is

$$\tau_w = C_f \tfrac{1}{2} \rho_m v^2 \tag{2.29}$$

The frictional pressure gradient is then

$$-\left(\frac{dp}{dz}\right)_F = 2C_f \rho_m \frac{v^2}{D} \tag{2.30}$$

A convenient modification to Eq. (2.30) can be made by substitution in terms of the volumetric and mass flow rates. Thus

$$v = j = \frac{Q_1 + Q_2}{A} \tag{2.31}$$

$$\rho_m v = G = \frac{W_1 + W_2}{A} \tag{2.32}$$

Therefore

$$-\left(\frac{dp}{dz}\right)_F = \frac{2C_f G j}{D} \tag{2.33}$$

Alternatively we may choose to work in terms of the specific volumes of the components and the quality. From Eqs. (2.10) and (2.32) we have

$$v = \frac{G}{\rho_m} = G[x\mathcal{v}_2 + (1 - x)\mathcal{v}_1] = G(\mathcal{v}_1 + x\mathcal{v}_{12}) \tag{2.34}$$

Using Eqs. (2.32) and (2.34) in Eq. (2.30) then gives

$$-\left(\frac{dp}{dz}\right)_F = \frac{2C_f G^2}{D}(\mathcal{v}_1 + x\mathcal{v}_{12}) \tag{2.35}$$

Since the mass flow rate is constant and each phase has the same velocity, the accelerational pressure gradient in Eq. (2.6) becomes

$$-\left(\frac{dp}{dz}\right)_A = G\frac{dv}{dz} \tag{2.36}$$

Substituting for v from Eq. (2.1) in (2.36) gives

$$-\left(\frac{dp}{dz}\right)_A = G\frac{d}{dz}\left(\frac{W}{A\rho_m}\right) \tag{2.37}$$

Expanding the differential we find

$$-\left(\frac{dp}{dz}\right)_A = G^2\frac{d}{dz}\left(\frac{1}{\rho_m}\right) - \frac{G^2}{\rho_m}\frac{1}{A}\frac{dA}{dz} \tag{2.38}$$

Further, from differentiating Eq. (2.10),

$$\frac{d}{dz}\left(\frac{1}{\rho_m}\right) = \frac{dx}{dz}\left(\frac{1}{\rho_2} - \frac{1}{\rho_1}\right) + x\frac{d}{dz}\left(\frac{1}{\rho_2}\right) + (1-x)\frac{d}{dz}\left(\frac{1}{\rho_1}\right) \tag{2.39}$$

or, in terms of the specific volumes of the phases,

$$\frac{d}{dz}\left(\frac{1}{\rho_m}\right) = v_{12}\frac{dx}{dz} + x\frac{dv_2}{dz} + (1-x)\frac{dv_1}{dz} \tag{2.40}$$

In the two-phase region (e.g., vapor-liquid) for a pure substance, v_f and v_g are only functions of pressure. For a two-component mixture, likewise, v_1 and v_2 can be expressed in terms of the pressure as long as the thermodynamic path can be determined. Equation (2.40) can then be rewritten as

$$\frac{d}{dz}\left(\frac{1}{\rho_m}\right) = v_{12}\frac{dx}{dz} + \frac{dp}{dz}\left[x\frac{dv_2}{dp} + (1-x)\frac{dv_1}{dp}\right] \tag{2.41}$$

The acceleration pressure drop, in terms of quality, flow rate, and property variations, is now found from Eq. (2.38), with the use of Eq. (2.41), to be

$$-\left(\frac{dp}{dz}\right)_A = G^2\left\{v_{12}\frac{dx}{dz} + \frac{dp}{dz}\left[x\frac{dv_2}{dp} + (1-x)\frac{dv_1}{dp}\right] - (v_1 + xv_{12})\frac{1}{A}\frac{dA}{dz}\right\} \tag{2.42}$$

The gravitational pressure drop is found in terms of quality by substituting for ρ_m from Eq. (2.10) in (2.7). The result is

$$-\left(\frac{dp}{dz}\right)_G = g\cos\theta\,\frac{1}{v_1 + xv_{12}} \tag{2.43}$$

Combining Eqs. (2.35), (2.42), and (2.43) in the form of Eq. (2.8) and rearranging, we eventually obtain an expression from which the pressure gradient can be calculated as follows:

$$-\frac{dp}{dz} = \frac{\dfrac{2C_f}{D}G^2(v_1 + xv_{12}) + G^2v_{12}\dfrac{dx}{dz} - G^2(v_1 + xv_{12})\dfrac{1}{A}\dfrac{dA}{dz} + \dfrac{g\cos\theta}{v_1 + xv_{12}}}{1 + G^2\left[x\dfrac{dv_2}{dp} + (1-x)\dfrac{dv_1}{dp}\right]} \tag{2.44}$$

This equation can be expressed in many other ways by making substitutions in terms of other chosen variables. However, the form of the equation will be retained and the physical significance of each term will be unchanged. In fact, any one-dimensional steady-flow analysis will eventually be found to lead to an equation of the form

$$-\frac{dp}{dz} = \frac{C_F + C_x \frac{dx}{dz} + C_A \frac{1}{A}\frac{dA}{dz} + C_g g \cos\theta}{1 - M^2} \tag{2.45}$$

In this equation C_F, C_x, C_A, and C_g are *influence coefficients*, which express the effect of friction, phase change, area change, and gravity on the pressure gradient. The term M^2 in the denominator has the same significance as the square of the Mach number in single component flow. Comparing Eq. (2.44) with (2.45) and using Eq. (2.32) we can therefore deduce that the velocity of a compressibility wave in a homogeneous two-phase mixture is

$$c = \left\{-\rho_m{}^2\left[x\frac{dv_2}{dp} + (1-x)\frac{dv_1}{dp}\right]\right\}^{-1/2} \tag{2.46}$$

This result will be deduced more directly in Chap. 6. Using Eqs. (2.9), (2.11), and (2.12) an alternative expression in terms of α is found to be

$$c = \left\{[\alpha\rho_2 + (1-\alpha)\rho_1]\left[\alpha\rho_2\left(-\frac{dv_2}{dp}\right) + (1-\alpha)\rho_1\left(-\frac{dv_1}{dp}\right)\right]\right\}^{-1/2} \tag{2.47}$$

The pseudo velocities of sound in the components taken alone and following the thermodynamic path consistent with the two-phase flow may be defined as

$$c_1{}^2 = \frac{dp}{d\rho_1} = \rho_1{}^{-2}\left(-\frac{dv_1}{dp}\right)^{-1} \tag{2.48}$$

$$c_2{}^2 = \frac{dp}{d\rho_2} = \rho_2{}^{-2}\left(\frac{-dv_2}{dp}\right)^{-1} \tag{2.49}$$

Using these definitions, Eq. (2.47) can be rearranged to give

$$\frac{1}{c^2} = [\alpha\rho_2 + (1-\alpha)\rho_1]\left[\frac{\alpha}{\rho_2 c_2{}^2} + \frac{1-\alpha}{\rho_1 c_1{}^2}\right] \tag{2.50}$$

In cases where $\rho_1 c_1{}^2 \gg \rho_2 c_2{}^2$ and $\rho_1 \gg \rho_2$ (e.g., an air-water mixture at atmospheric pressure) this equation reduces to the approximate expression

$$c^2 = \frac{\rho_2}{\rho_1}\frac{c_2{}^2}{\alpha(1-\alpha)} \tag{2.51}$$

Evidently the velocity of sound in a homogeneous mixture can be far less than the velocity of sound in the gas alone. A minimum occurs at $\alpha = \frac{1}{2}$ and for air and water at atmospheric pressure, this gives a sonic velocity of about 70 fps.

Usually the rate of change of quality is calculated from the energy equation by equating heat transfer to latent heat changes. However, if significant "flashing"† occurs because of pressure changes, the quality is not only a function of enthalpy, and a more correct way to proceed is as follows. Let the quality be a function of both enthalpy and pressure. Then for a given thermodynamic system we have

$$x = x(h,p) \tag{2.52}$$

Whence, by differentiation,

$$\frac{dx}{dz} = \left(\frac{\partial x}{\partial h}\right)_p \frac{dh}{dz} + \left(\frac{\partial x}{\partial p}\right)_h \frac{dp}{dz} \tag{2.53}$$

Moreover

$$\left(\frac{\partial x}{\partial h}\right)_p = \frac{1}{h_{12}} \tag{2.54}$$

Equation (2.44) now becomes

$$-\frac{dp}{dz} = \frac{\dfrac{2C_f}{D} G^2(v_1 + xv_{12}) + G^2 \dfrac{v_{12}}{h_{12}} \dfrac{dh}{dz} - G^2(v_1 + xv_{12}) \dfrac{1}{A} \dfrac{dA}{dz} + \dfrac{g \cos \theta}{v_1 + xv_{12}}}{1 + G^2 \left[x \dfrac{\partial v_2}{\partial p} + (1 - x) \dfrac{\partial v_1}{\partial p} + v_{12} \left(\dfrac{\partial x}{\partial p}\right)_h \right]} \tag{2.55}$$

Many alternative forms of Eq. (2.55) could have been obtained by using different thermodynamic properties in Eq. (2.52).

In practice the condition for a Mach number of unity is more likely to be given by Eq. (2.46) than by the equivalent result which would be obtained by using the whole denominator of Eq. (2.55). This is because of nonequilibrium effects which preclude rapid phase change caused by sudden pressure changes.

If kinetic-energy effects are appreciable, the enthalpy gradient cannot be calculated directly from the energy equation, and a more complex form of Eq. (2.55) results (see Prob. 2.25).

Nonequilibrium phase-change effects can be analyzed by assuming an effective thermodynamic path to replace Eq. (2.52). This might consist of only allowing a given fraction of the equilibrium quantity of vapor to be formed, or of a more thorough analysis in which the actual quality is related to nonequilibrium heat- and mass-transfer processes.[14]

2.3 THE HOMOGENEOUS FRICTION FACTOR

LAMINAR FLOW

Many methods have been proposed for evaluating the two-phase homogeneous friction factor, C_f, which is the only empirical parameter in Eqs.

† The term *flashing* is usually used to describe vapor formation caused by pressure changes, whereas *boiling* refers to vapor formation as a result of heat addition.

(2.44) and (2.55). In laminar flow the simplest technique is to find a suitable "virtual viscosity" for the mixture. For example, a theoretical solution for a suspension of fluid spheres at low concentrations is

$$\mu = \mu_1\left(1 + 2.5\alpha\,\frac{\mu_2 + \frac{2}{5}\mu_1}{\mu_2 + \mu_1}\right) \tag{2.56}$$

where the subscript 1 refers to the continuous phase. If the suspension consists of solid particles, μ_2 is very large and Eq. (2.56) becomes Einstein's equation[3]

$$\mu = \mu_1(1 + 2.5\alpha) \tag{2.57}$$

If, on the other hand, the emulsion consists of bubbles containing gas of a low viscosity, the result is

$$\mu = \mu_1(1 + \alpha) \tag{2.58}$$

Unfortunately, Eqs. (2.56), (2.57), and (2.58) are only valid at concentrations below about 5 percent for which the change in viscosity is small. Numerous rheological models for taking account of larger values of α and particles of various shapes and sizes have been proposed and will be discussed in later chapters. Many two-phase mixtures are nonnewtonian.

Often the details of the two-phase flow pattern are not known and an idealized rheological model cannot be defined. Faced with the necessity of choosing some expression for the viscosity, many workers have chosen averages which fit the limiting cases in which either phase is not present. Some common expressions for gas-liquid flow are

$$\frac{1}{\mu} = \frac{x}{\mu_g} + \frac{1-x}{\mu_f} \qquad \text{McAdams}^4 \tag{2.59}$$

$$\mu = x\mu_g + (1-x)\mu_f \qquad \text{Cicchitti}^5 \tag{2.60}$$

$$\mu = \frac{j_f}{j}\,\mu_f + \frac{j_g}{j}\,\mu_g \qquad \text{Dukler}^6 \tag{2.61}$$

It is often convenient to relate the viscosity, friction factor, and friction pressure drop in two-phase flow to the equivalent values for single-phase flow of one of the phases alone. For example, from Eq. (2.59) in laminar flow

$$\frac{\mu}{\mu_f} = \left[x\,\frac{\mu_f}{\mu_g} + (1-x)\right]^{-1} \tag{2.62}$$

For flows with change of phase, if the subscript f_0 is used to denote the case in which liquid flows in the same pipe with the same mass velocity as the combined flows, we have, using Eq. (2.62) for laminar flow

$$-\left(\frac{dp}{dz}\right)_F = -\left(\frac{dp}{dz}\right)_{Ff_0}\left[1 + x\left(\frac{\rho_f}{\rho_g} - 1\right)\right]\left[1 + x\left(\frac{\mu_f}{\mu_g} - 1\right)\right]^{-1} \tag{2.63}$$

The ratios between the frictional pressure gradient for the two-phase flow and the frictional pressure gradients for related single-phase flows are usually known as *two-phase multipliers* and are denoted by the symbol ϕ^2 with appropriate subscripts, for example,

$$\phi_{f0}^2 = \frac{-(dp/dz)_F}{-(dp/dz)_{Ff0}} \tag{2.64}$$

If ϕ_{f0}^2 can be determined, then the first term in the numerator of Eq. (2.44) or (2.55) can be replaced by

$$\frac{2C_f}{D} G^2(v_f + xv_{fg}) = \phi_{f0}^2 \frac{2C_{f0}}{D} G^2 v_f \tag{2.65}$$

TURBULENT FLOW

Single-phase friction factors in turbulent flow are usually correlated in terms of the Reynolds number and the pipe roughness. Except in extreme cases the actual value does not differ by more than a factor of 2 from the rule-of-thumb estimate of $C_f \approx 0.005$. In commercial situations where pipes are subject to corrosion, distortion, and scaling, the accuracy with which pressure drops can be computed in single-phase flow is often no better than 25 percent and it would be presumptuous to expect a better correlation in the case of two-phase flow.

The three common alternatives for estimating two-phase turbulent friction factors are

1. Use a constant value for all conditions. A good choice is

$$C_f = 0.005 \tag{2.66}$$

Figure 2.1 shows a comparison between this prediction and some data taken in the high-velocity annular-mist flow regime. Under these conditions a considerable fraction of the liquid flow is entrained in the form of droplets, and homogeneous theory provides a reasonable approximation.

2. Use a friction factor calculated from some equivalent single-phase flow. For example, for low-quality vapor-liquid mixtures it may be assumed that the friction factor is the same as it would be if the total mass flow (liquid plus vapor) flowed entirely as liquid. The appropriate Reynolds number is

$$\mathrm{Re}_f = \frac{GD}{\mu_f} \tag{2.67}$$

and it is readily shown that

$$\phi_{f0}^2 = 1 + x\left(\frac{\rho_f}{\rho_g} - 1\right) \tag{2.68}$$

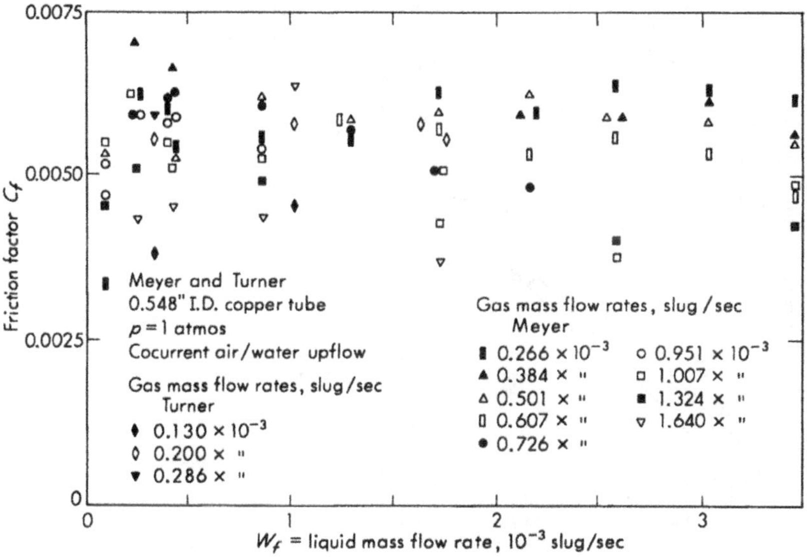

Fig. 2.1 Homogeneous friction factor for annular-mist flow. (*Meyer and Wallis.*[7])

This method was used by McAdams *et al.*[4] and Owens.[8] Typical results are shown in Fig. 2.2 which also represents the kind of accuracy which can be achieved using this method. At high qualities approaching the pure vapor condition, it is more accurate to replace Eq. (2.68) by an analogous expression based on the all-vapor flow condition.

3. Substitute one of the expressions for equivalent viscosity in the Reynolds number and use the single-phase friction-factor charts. For example, Blasius' equation for smooth pipe flow is

$$C_f = 0.079 \text{ Re}^{-0.25} \tag{2.69}$$

Using Eq. (2.59) and the definition of ϕ_{fo}^2 in Eq. (2.64), it is readily found that

$$\phi_{fo}^2 = \left[1 + x\left(\frac{\rho_f}{\rho_g} - 1\right)\right]\left[1 + x\left(\frac{\mu_f}{\mu_g} - 1\right)\right]^{-1/4} \tag{2.70}$$

The predictions of Eq. (2.70) for water are tabulated in Table 2.1*a*. Similar tables can be drawn up for cryogenic or refrigerant fluids to enable rapid design predictions to be made.

An alternative procedure for calculating frictional pressure drop is

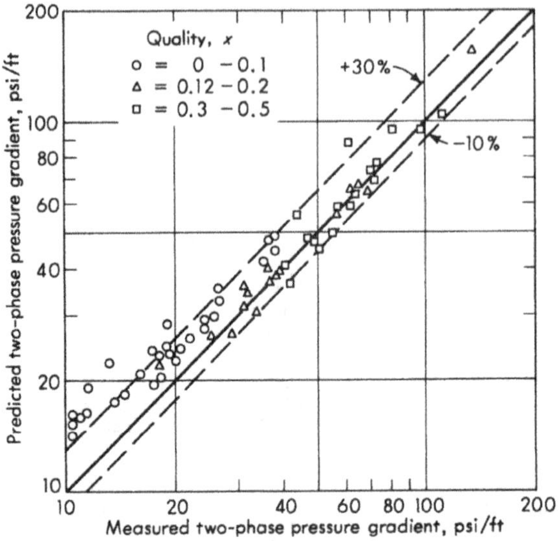

Fig. 2.2 Predicted versus measured two-phase pressure gradient for 0.118 in.-diam, 30-in.-long tube. Water at pressures from 71 to 374 psia, and qualities from 4 to 50 percent. (*Owens.*[8])

Table 2.1a Values of ϕ_{f0}^2 for steam-water mixtures predicted by the homogeneous model [Eq. (2.70)]

Steam quality, wt %	*Pressure, psia*								
	14.7	100	500	1000	1500	2000	2500	3000	3206
1	16.21	3.40	1.44	1.19	1.10	1.05	1.04	1.01	1.0
5	67.6	12.18	3.12	1.89	1.49	1.28	1.16	1.06	1.0
10	121.2	21.8	5.06	2.73	1.95	1.56	1.30	1.13	1.0
20	212.2	38.7	7.8	4.27	2.81	2.08	1.60	1.25	1.0
30	292.8.	53.5	11.74	5.71	3.60	2.57	1.87	1.36	1.0
40	366	67.3	14.7	7.03	4.36	3.04	2.14	1.48	1.0
50	435	80.2	17.45	8.30	5.08	3.48	2.41	1.60	1.0
60	500	92.4	20.14	9.50	5.76	3.91	2.67	1.71	1.0
70	563	104.2	22.7	10.70	6.44	4.33	2.89	1.82	1.0
80	623	115.7	25.1	11.81	7.08	4.74	3.14	1.93	1.0
90	682	127	27.5	12.9	7.75	5.21	3.37	2.04	1.0
100	738	137.4	29.8	13.98	8.32	5.52	3.60	2.14	1.0

Table 2.1b Values of ϕ_{fo}^2 for steam-water mixtures based on Martinelli's empirical correlation

Steam quality, wt %	*Pressure, psia*								
	14.7	100	500	1000	1500	2000	2500	3000	3206
1	5.6	3.5	1.8	1.6	1.35	1.2	1.1	1.05	1.00
5	30	15	5.3	3.6	2.4	1.75	1.43	1.17	1.00
10	69	28	8.9	5.4	3.4	2.45	1.75	1.30	1.00
20	150	56	16.2	8.6	5.1	3.25	2.19	1.51	1.00
30	245	83	23.0	11.6	6.8	4.04	2.62	1.68	1.00
40	350	115	29.2	14.4	8.4	4.82	3.02	1.83	1.00
50	450	145	34.9	17.0	9.9	5.59	3.38	1.97	1.00
60	545	174	40.0	19.4	11.1	6.34	3.70	2.10	1.00
70	625	199	44.6	21.4	12.1	7.05	3.96	2.23	1.00
80	685	216	48.6	22.9	12.8	7.70	4.15	2.35	1.00
90	720	210	48.0	22.3	13.0	7.95	4.20	2.38	1.00
100	525	130	30.0	15.0	8.6	5.90	3.70	2.15	1.00

to avoid use of the friction factor and to use correlations for the two-phase multipliers instead. The details of this technique will be left to the next chapter. For the particular case of boiling water the results of such a correlation scheme, evolved by Martinelli and Nelson,[9] are shown in Table 2.1*b* for comparison with the predictions of Eq. (2.70). Quite large

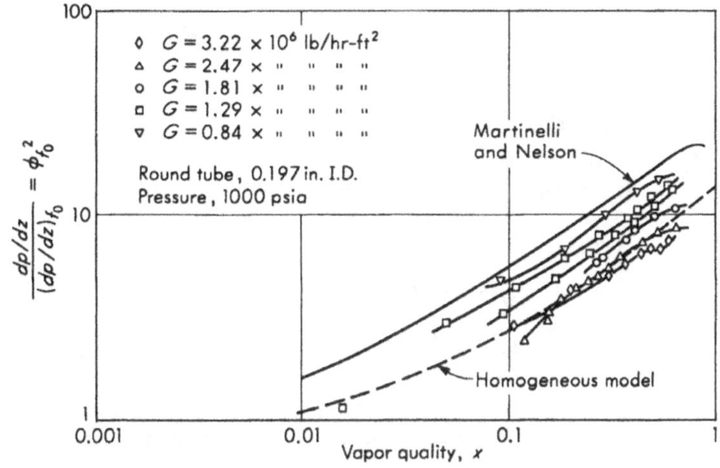

Fig. 2.3 Dependence of ϕ_{fo} on mass flux for water at 1000 psia. (*Muscettola.*[10])

differences between the tables are evident. Which of them is the more correct depends on the flow regime. The Martinelli-Nelson predictions tend to be better for separated flows, whereas homogeneous theory is better for dispersed flows. For example, Fig. 2.3 shows a drift from the Martinelli-Nelson predictions toward homogeneous theory as mass velocity, and hence entrainment, is increased in high-speed steam-water flows.

Example 2.3 Formulate equations for predicting the pressure drop during the homogeneous flow of a boiling liquid in a round tube of constant area with a uniform heat flux. Neglect kinetic- and potential-energy terms and the effects of flashing and compressibility (these assumptions are only valid at high pressures and low velocities when the overall pressure drop is small compared with the absolute pressure). Assume a constant friction factor.

Solution First we relate the rate of heat addition per unit length to the heat flux, as follows

$$\frac{dq_e}{dz} = \pi D \phi \tag{2.71}$$

where ϕ is the heat flux (in British thermal units per hour per square foot).

Using the definition of G, and remembering that W is constant, Eq. (2.3) becomes, with the given assumptions,

$$\frac{dh}{dz} = \frac{4\phi}{GD} \tag{2.72}$$

Substituting Eq. (2.72) into Eq. (2.55) we get, neglecting flashing and compressibility,

$$-\frac{dp}{dz} = \frac{2C_f}{D} G^2(v_f + xv_{fg}) + G^2 \frac{v_{fg}}{h_{fg}} \frac{4\phi}{GD} + \frac{g \cos \theta}{v_f + xv_{fg}} \tag{2.73}$$

The enthalpy of the mixture is given by the equation

$$h = h_f + xh_{fg} \tag{2.74}$$

If pressure changes are small, h_f and h_{fg} may be treated as constants. Equation (2.72) is now used in conjunction with Eq. (2.74) to give

$$\frac{dx}{dz} = \frac{4\phi}{GDh_{fg}} \tag{2.75}$$

Since ϕ is constant, Eq. (2.75) may be integrated from the inlet value of x_i to the point $z = z$ to give

$$x = x_i + \frac{\phi}{Gh_{fg}} \frac{4z}{D} \tag{2.76}$$

Substituting for Eq. (2.76) in Eq. (2.73), denoting the density at inlet by ρ_i, and integrating over the tube length L, it is eventually found that

$$\Delta p = \frac{2L}{D} \frac{C_f G^2}{\rho_i} + \left(\frac{2L}{D}\right)^2 C_f G \phi \frac{v_{fg}}{h_{fg}} + G\phi \frac{v_{fg}}{h_{fg}} \frac{4L}{D} + g \cos \theta \frac{DGh_{fg}}{4\phi v_{fg}} \ln \left(1 + \frac{\phi v_{fg} \rho_i}{Gh_{fg}} \frac{4L}{D}\right) \tag{2.77}$$

In the special case where $x_i = 0$ an alternative form of Eq. (2.77) in terms of the quality at $z = L$ is

$$\Delta p = \frac{2C_f LG^2 v_f}{D}\left(1 + \frac{v_{fg}x}{v_f 2}\right) + G^2 v_{fg} x + \frac{Lg\cos\theta}{x v_{fg}} \ln\left(1 + \frac{v_{fg}}{v_f}x\right) \tag{2.78}$$

This problem is discussed in greater detail by Owens.[8] In applying Eq. (2.77) it is most advisable to check the assumptions since compressibility and kinetic-energy effects rapidly become important as mass velocities are increased and flashing must also be considered whenever the relative change in pressure is appreciable.

Example 2.4 Saturated water at a rate of $G_f = 2 \times 10^5$ lb/(hr)(ft²) enters the bottom of a vertical evaporator tube ½ in. in diameter and 5 ft long. The tube receives a heat flux of 2×10^5 Btu/(hr)(ft²), and there are no heat losses. Calculate the pressure drop through the evaporator for inlet pressures of (*a*) 350 psia and (*b*) 1000 psia. Assume a constant friction factor of 0.005.

Solution Assuming homogeneous flow theory this is simply an application of Eq. (2.77).

(*a*) From steam tables, at 350 psia, $v_{fg} = 1.3064$ ft³/lb, $v_f = 0.01912$ ft³/lb, and $h_{fg} = 794.7$ Btu/lb.

From Eq. (2.77)

$$\begin{aligned}\Delta p &= \frac{(2)(120)(0.005)(2\times10^5)^2(0.01912)}{(32.2)(3600)^2(144)} + \frac{(240)^2(0.005)(2\times10^5)^2(1.3064)}{(3600)^2(794.7)(32.2)(144)} \\ &\quad + \frac{(2\times10^5)^2(1.3064)(4)(120)}{794.7(32.2)(3600)^2(144)} \\ &\quad + \frac{(32.2)(2\times10^5)(794.7)(1/24)}{(4)(2\times10^5)(1.3064)(32.2)(144)} \ln\left[1 + \frac{(2\times10^5)(1.3064)(480)}{(2\times10^5)(794.7)(0.01912)}\right] \\ &= 0.015 + 0.315 + 0.505 + 0.165 \text{ psi} \\ &= 1.02 \text{ psi}\end{aligned}$$

Since this value is small compared with 350 psia, the assumption of constant fluid properties is justified.

We should also check for compressibility effects and confirm that the exit steam is not superheated.

The exit quality, from Eq. (2.76), is

$$x_e = \frac{\phi}{Gh_{fg}}\frac{4L}{D} = \frac{(2\times10^5)(480)}{(2\times10^5)(794.7)} = 0.605$$

The compressibility and flashing effects are checked by evaluating the terms in the denominator of Eq. (2.55). From steam tables we find

$$\frac{dv_g}{dp} = -3.8\times10^{-3} \text{ ft}^3/(\text{lb})(\text{psi})$$

$$\frac{dv_f}{dp} \approx 0$$

$$\begin{aligned}\left(\frac{\partial x}{\partial p}\right)_h &\approx 0 &&\text{at } x = 1 \\ &= -3.9\times10^{-4}\ \text{psi}^{-1} &&\text{at } x = 0 \\ &= -2\times10^{-4}\ \text{psi}^{-1} &&\text{at } x = 0.5\end{aligned}$$

The largest numerical value of the bracket in the denominator therefore occurs at $x = 0.6$. The value of the whole denominator is

$$1 - \frac{(2 \times 10^5)^2(0.6)(3.8 \times 10^{-3})}{(144)(32.2)(3600)^2} = 0.9985$$

The assumptions are therefore justified. However, if both G and ϕ were increased by a factor of 10, the denominator would become 0.85 and could not be put equal to unity without introducing significant error.

(b) At 1000 psia the relevant property values are $v_{fg} = 0.424$ ft³/lb, $v_f = 0.0216$ ft³/lb, $h_{fg} = 649.5$ Btu/lb. Putting in the numbers as above we get

$$\Delta p_A = 0.525 \frac{0.424}{1.3064} \frac{794.7}{649.5} = 0.208 \text{ psi}$$

$$\Delta p_F = 0.015 \frac{0.0216}{0.01912} + 0.315 \frac{0.424}{1.3064} \frac{794.7}{649.5} = 0.143 \text{ psi}$$

$$\Delta p_G = 0.111 \ln 17.3 = 0.317 \text{ psi}$$

The total pressure drop is therefore $0.208 + 0.143 + 0.317 = 0.668$ psi.

Example 2.5 Air and water at 70°F flow in a vertical pipe of diameter 0.98 in. at rates $W_f = 31.3 \times 10^{-3}$ slug/sec and $W_g = 0.583 \times 10^{-3}$ slug/sec. The exit pressure is 14.7 psia and the pressure 18 in. upstream is found to be 15.0 psia. How does this compare with the predictions of homogeneous flow theory with $C_f = 0.005$?

Solution Since the pressure drop is low, it is a reasonable approximation to assume constant property values in calculating the gravitational and frictional pressure drops. The acceleration component is due to the expansion of the gas. Since $W_f \gg W_g$, the thermodynamic path is probably isothermal (see Example 2.2).

Equation (2.44) is first put into a convenient form by utilizing Eqs. (2.31) and (2.32) to eliminate x, using Eqs. (2.11) and (2.12). Since the area change, quality change, and compressibility of the liquid are essentially zero, we obtain

$$-\frac{dp}{dz} = \frac{2C_f jG/D + g \cos \theta[\alpha\rho_g + (1 - \alpha)\rho_f]}{1 + Gj_g\rho_g(dv_g/dp)} \tag{2.79}$$

Since the gas expansion is isothermal, we have

$$\frac{dv_g}{dp} = -\frac{1}{p\rho_g} \tag{2.80}$$

The variables of interest are evaluated as follows:

$$j = \frac{Q_f + Q_g}{A} = 50.3 \text{ ft/sec}$$

$$G = \frac{W_f + W_g}{A} = 6.08 \text{ slugs/(sec)(ft}^2\text{)}$$

$$\alpha = \frac{Q_g}{Q_f + Q_g} = 0.939$$

Making the requisite substitutions in Eq. (2.79) we find

$$-\frac{dp}{dz} = \frac{(2.17 \times 10^{-3})(2.22 \times 10^{-3})}{1 - 0.136} = 5.08 \times 10^{-3} \text{ psi/in.}$$

Since the pressure gradient is so low, the velocities hardly change down the duct and the overall pressure drop is found by multiplying the gradient by the length, as follows:

$$\Delta p = (5.08 \times 10^{-3})18 = 0.092 \text{ psi}$$

The error in pressure-drop prediction is mostly due to the poor estimate of liquid fraction which is given by homogeneous theory at these values of the flow rates. The flow pattern is, in fact, slug or annular and the gas flows much faster than the liquid. The experimental liquid fraction was actually 0.23, which is much greater than the value of 0.061 which homogeneous theory predicts. Errors of this order of magnitude are to be expected when using a theory for a flow regime to which it does not apply. The separated-flow model is far more appropriate in the present case. Slug-flow theory or a theory which is valid in the transition region from slug flow to annular flow is even more accurate, as will be shown in Chaps. 10 and 11.

In many cases the solution to a problem cannot be found in closed form and it is necessary to put equations such as (2.55) or (2.79) in terms of finite differences and to use numerical methods of integration.

Note that the denominator of Eq. (2.79) together with Eq. (2.80) can be combined to show that the square of the Mach number in an isothermal homogeneous gas-liquid system is

$$M^2 = \frac{Gj_g}{p} \tag{2.81}$$

2.4 PRESSURE DROP IN BENDS, TEES, ORIFICES, VALVES, ETC.

The usual way of calculating pressure drop through pipe fittings in single-phase flow is to replace the fitting by an equivalent length of pipe. The same procedure can be applied to two-phase flows. The equivalent pipe lengths tend to be somewhat longer in the two-phase flow case.

Several problems relating to pressure-drop prediction in nozzles and orifices will be found at the end of this chapter.

2.5 UNSTEADY FLOW

Homogeneous theory can be extended to unsteady flows by including time-dependent terms in the equations of continuity, momentum, and energy. For one-dimensional flow in a constant area duct these equations are, in differential form.

Continuity $$\frac{\partial \rho_m}{\partial t} + \frac{\partial}{\partial z}(\rho_m v) = 0 \tag{2.82}$$

Momentum $$\rho_m\left(\frac{\partial v}{\partial t} + v\frac{\partial v}{\partial z}\right) = -\frac{\partial p}{\partial z} - \rho_m g \cos\theta - \frac{P}{A}\tau_w \tag{2.83}$$

Energy $$\frac{\partial}{\partial t}\left[\rho_m\left(e + \frac{v^2}{2}\right)\right] + \frac{\partial}{\partial z}\left[\rho_m v\left(h + \frac{v^2}{2}\right)\right] = \frac{1}{A}\left(\frac{\partial q_e}{\partial z} - \frac{\partial w}{\partial z}\right) - \rho_m v g \cos\theta \tag{2.84}$$

These equations can be combined in numerous ways, depending upon the conditions of interest. A particularly useful development is to substitute the identity

$$\rho_m e = \rho_m h - p \tag{2.85}$$

into Eq. (2.84) and expand the differentials to get

$$\left(h + \frac{v^2}{2}\right)\left[\frac{\partial \rho_m}{\partial t} + \frac{\partial(\rho_m v)}{\partial z}\right] + \rho_m\left(\frac{\partial h}{\partial t} + v\frac{\partial h}{\partial z}\right) + \rho_m v\left(\frac{\partial v}{\partial t} + v\frac{\partial v}{\partial z} + g\cos\theta\right) = \frac{\partial p}{\partial t} + \frac{1}{A}\left(\frac{\partial q_e}{\partial z} - \frac{\partial w}{\partial z}\right) \tag{2.86}$$

Making use of Eqs. (2.82) and (2.83) then gives the result

$$\frac{\partial h}{\partial t} + v\frac{\partial h}{\partial z} = \frac{1}{\rho_m}\left(\frac{\partial p}{\partial t} + v\frac{\partial p}{\partial z}\right) + v\frac{P}{A}\frac{\tau_w}{\rho_m} + \frac{1}{A\rho_m}\left(\frac{\partial q_e}{\partial z} - \frac{\partial w}{\partial z}\right) \tag{2.87}$$

The use of this equation may be illustrated by considering the case of flow in a high-pressure straight-tube evaporator in which pressure changes and viscous dissipation are small compared with the other energy terms during a transient. Let the tube diameter be D, the wall heat flux ϕ, and let there be no shaft work. Equation (2.87) then becomes

$$\frac{\partial h}{\partial t} + v\frac{\partial h}{\partial z} = \frac{1}{\rho_m}\frac{4\phi}{D} \tag{2.88}$$

If pressure changes and their effect on properties are negligible, the enthalpy and density may be expressed as

$$h = h_f + xh_{fg} \tag{2.89}$$

$$\frac{1}{\rho_m} = v_f + xv_{fg} \tag{2.90}$$

Substituting Eqs. (2.89) and (2.90) in Eq. (2.88) eventually yields

$$\frac{\partial x}{\partial t} + v\frac{\partial x}{\partial z} = \left(\frac{v_f}{v_{fg}} + x\right)\Omega \tag{2.91}$$

where

$$\Omega = \frac{4v_{fg}\phi}{Dh_{fg}} \tag{2.92}$$

Equation (2.91) expresses the *propagation equation* for quality changes and can be integrated to obtain the dynamic response of a boiler channel. The left-hand side of the equation is the *substantial time derivative* of the quality and represents the time rate of change of quality for a given fluid particle (lagrangian viewpoint). The quantity Ω has the dimensions of inverse time and can be called a *reaction frequency*. If a given particle is identified by the time t_0 at which it starts to evaporate,

Eq. (2.91) can be integrated, if properties are constant, to give

$$x(t,t_0) = \frac{v_f}{v_{fg}} (e^{\Omega(t-t_0)} - 1) \qquad (2.93)$$

Therefore the quality of a given fluid particle increases exponentially with time. Further developments along these lines lead to prediction of the transient response of evaporators and condensers.[11]

PROBLEMS

2.1. Deduce Eq. (2.16) from Eq. (2.10) and state what conditions are necessary for an error of less than 5% in Eq. (2.16).

2.2. Solve Example 2.2 if the required exit velocity is (*a*) 1000 fps; (*b*) 1500 fps; and (*c*) 2500 fps.

2.3. Suppose that a homogeneous two-phase flow between parallel plates is assumed to consist of a large number of parallel sheets of the two phases oriented in the direction of the plates. Show that the equivalent viscosity is given by

$$\frac{1}{\mu} = \frac{\alpha}{\mu_2} + \frac{1-\alpha}{\mu_1}$$

2.4. Solve Prob. 2.3 if the sheets of fluid are oriented perpendicular to the plates and show that

$$\mu = \alpha\mu_2 + (1-\alpha)\mu_1$$

2.5. Compare the equivalent viscosities predicted by Eqs. (2.59), (2.60), and (2.61) and the results of Probs. 2.3 and 2.4 for steam-water mixtures of 0.1, 1, and 10% quality at 10, 100, and 1000 psia.

2.6. For air-water flow in a 2-in.-diam pipe at 70°F and 30 psia, calculate the friction factor by using the various techniques in the text for the following conditions:

j_f, fps	1	10	10	100	10	200
j_g, fps	10	10	1	10	100	200

2.7. Still dusty air is sampled by a stagnation pitot tube mounted on a jeep traveling at 50 mph over the desert. The air contains its own weight of sand per unit weight. What is the measured stagnation pressure? State clearly all the assumptions which are made.

2.8. A vertical tubular test section is installed in an experimental high-pressure water loop. The tube is 0.4 in. ID and 7 ft long and is uniformly heated with 100 kw of power. Saturated water enters at the base at 1000 psia and with a flow rate of 1000 lb/hr. Calculate the frictional, gravitational, accelerational, and total pressure drops using the various equations in the text. Compare the prediction for total pressure drop with the measured value of 8 psi.

2.9. Estimate the critical flow rate per unit area for a steam-water mixture, 26.9% quality, at a pressure of 125 psia. The measured value is 1265 $lb_m/(ft^2)(sec)$.

2.10. Air and water flow from a large tank through a converging nozzle with exit area of 1 $in.^2$. The air flow rate is 10 lb/hr, the water flow rate is 10 lb/sec, and the fluids are intimately mixed. The temperature is 70°F and the external pressure is 14.7 psia.

What is the pressure in the tank? (Use Bernoulli's equation for the homogeneous mixture and neglect wall shear and gravity.)

2.11. Consider the isentropic, adiabatic homogeneous equilibrium flashing discharge of carbon dioxide from a large storage cylinder at 1000 psia and 80°F. What are the quality, velocity, density, and flow per unit area as a function of pressure? What is the critical pressure at which choking occurs? Assume that choking corresponds to the maximum flow per unit area.

2.12. Consider low-velocity horizontal laminar flow of a gas-liquid mixture in a pipe. Let the gas volumetric flow rate be Q_g and the liquid volumetric flow rate Q_f. The pressure drop for the liquid alone in the pipe is Δp_f. If the two-phase pressure drop is Δp_{TP}, use equation (2.58) to show that, if $\rho_f \gg \rho_g$

$$\frac{\Delta p_{TP}}{\Delta p_f} - 1 = \frac{2Q_g}{Q_f}$$

2.13. If the liquid is dirty, small gas bubbles tend to behave like solid spheres. In this case, solve Prob. 2.12 using Eq. (2.57) to show that

$$\frac{\Delta p_{TP}}{\Delta p_f} - 1 = 3.5\frac{Q_g}{Q_f}$$

2.14. Solve Prob. 2.12 for turbulent flow assuming that the friction factor is the same as for the liquid alone. Do the solutions to Probs. 2.12, 2.13, and 2.14 suggest a method of plotting two-phase flow data? (See Ref. 12.)

2.15. A coal-water slurry is pumped in a horizontal 2-in.-diam pipe over a distance of 100 ft at an average velocity of 10 fps. The slurry is nonnewtonian with a limiting viscosity at high shear rates of 0.01 lb/(ft)(sec). It consists of 45% by volume of coal with a density of 85 lb/ft^3. Single-phase flow tests in the same pipe gave the following expression for the turbulent flow friction factor

$$C_f = 0.026\text{Re}^{-0.12}$$

What is the pressure drop? What difference would it make if the viscosity of water at 70°F were used in the Reynolds number instead of the limiting viscosity of the slurry?

2.16. It has been suggested that the intensity of "water hammer" in hydraulic lines could be significantly decreased by suspending small air bubbles in the fluid. Discuss the possibilities of this idea.

2.17. Develop equations for describing the adiabatic compressible flow of a dusty gas in a long horizontal pipeline, assuming a constant value of the friction factor (Fanno line).

2.18. Develop equations for describing the flow of a dusty gas in a long pipeline if there is no friction but a constant wall heat flux (Rayleigh line).

2.19. Combine the solutions to Probs. 2.17 and 2.18 to generate the normal shock relationships for a homogeneous dusty gas.

2.20. Consider two-phase flow through the nozzle shown in Fig. 2.4. Assuming no phase change and incompressible homogeneous flow show that

$$p_3 - p_2 = \frac{G_1^2}{\rho_f}\left(\frac{A_1}{A_2} - 1\right)\left[1 + x\left(\frac{\rho_f}{\rho_g} - 1\right)\right]$$

and

$$p_1 - p_3 = \frac{G_1^2}{2\rho_f}\left(\frac{A_1}{A_2} - 1\right)^2\left[1 + x\left(\frac{\rho_f}{\rho_g} - 1\right)\right]$$

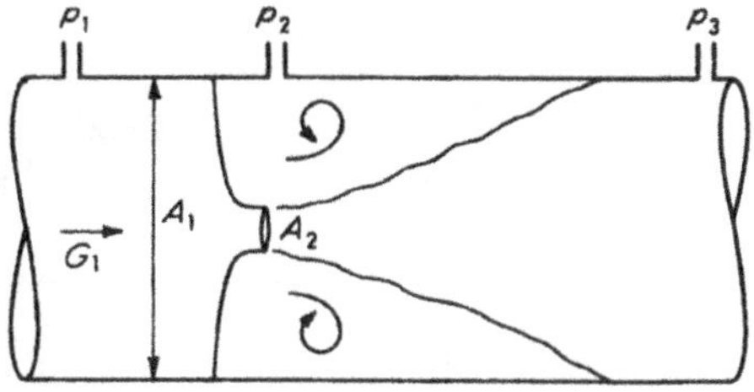

Fig. 2.4 Flow of a gas-liquid mixture through a nozzle. (*Prob.* 2.20.)

How are these results altered if the nozzle is sharp-edged and produces a *vena contracta*?

2.21. Figure 2.5 shows a proposed method for controlling the flow of oil through a nozzle valve by varying the air flow supplied to the set of hypodermic needles. What is the relationship between the air and oil flow rates W_g and W_f?

2.22. Diehl has measured the pressure drop Δp for two-phase flow over tube banks. His results were correlated by plotting $\Delta p/\Delta p_g^*$ versus $Q_f\rho_f/(Q_f + Q_g)\rho_g$, where Δp_g^* was the pressure drop for the same mass flow $(W_f + W_g)$, of gas alone. Show that homogeneous flow theory, assuming a friction factor equal to the all gas friction factor, is quite close to the data presented in Fig. 2.6.

2.23. Show, by differentiating Eq. (1.30) for homogeneous flow, that the void fraction propagation equation analogous to Eq. (2.91) is

$$\frac{\partial\alpha}{\partial t} + v\frac{\partial\alpha}{\partial z} = \left[\frac{v_f}{v_{fg}} + (1 - \alpha)\right]\Omega$$

Deduce that, if t_0 is the time at which an element of fluid starts to evaporate and Ω is constant, the density of the same fluid element after time t will be

$$\rho(t,t_0) = \rho_f e^{-\Omega(t-t_0)}$$

2.24. Derive the normal shock-wave relationships for an isothermal, homogeneous bubbly mixture. (See Chap. 9.)

2.25. Equation (2.55) is not adequate for predicting the pressure gradient in evaporators if the kinetic energy of the fluid is appreciable, since in this case the enthalpy cannot be determined directly from a heat balance. Use the energy equation to show that,

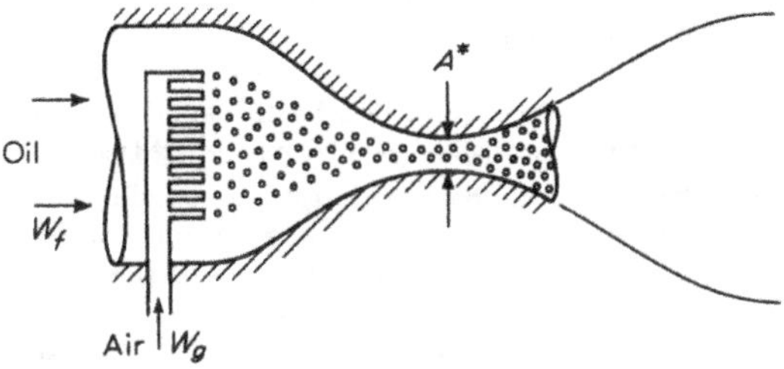

Fig. 2.5 A two-phase flow control valve. The air and oil stagnation temperatures and stagnation pressures are 100°F and 400 psia. Assume choking at the throat where $A^* = 0.1$ in.^{2}.

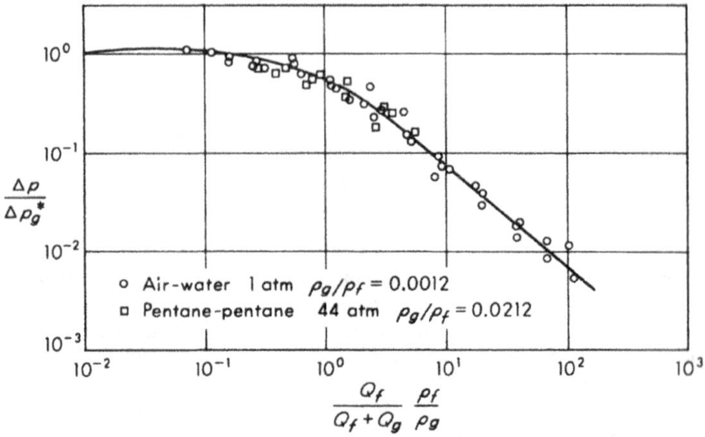

Fig. 2.6 Diehl's correlation[13] of two-phase pressure drop for turbulent horizontal crossflow through tube bank with 45° layout. (*Prob.* 2.22.)

in the absence of shaft work effects, Eq. (2.55) becomes

$$-\frac{dp}{dz} = \frac{\dfrac{2C_f G^2}{D\rho_m} + \left(\dfrac{Gv_{21}}{h_{21}A}\dfrac{dq_e}{dz} - \dfrac{G^2}{A\rho_m}\dfrac{dA}{dz} + g\rho_m \cos\theta\right)\Phi}{1 + G^2\left[x\dfrac{\partial v_2}{\partial p} + (1-x)\dfrac{\partial v_1}{\partial p} + v_{12}\left(\dfrac{\partial x}{\partial p}\right)_h\right]\Phi}$$

where

$$\Phi = \frac{1}{1 + G^2 v_{21}/h_{21}\rho_m}$$

What is the choking condition in this case?

2.26. Compare the order of magnitude of the various terms in Eqs. (2.44) and (2.55) for vertical upward steam-water flow at 10% quality with $G = 10^6$ lb/(hr)(ft²) and a wall heat flux of 10^6 Btu/(hr)(ft²) in a 1-in.-diam pipe with a taper of $1/100$ in./in. at pressures of 1, 10, 100, and 1000 psia. Is the correction introduced in Prob. 2.25 important under any of these conditions?

2.27. Investigate the effect on the thrust of a jet engine of injecting small, solid particles into the hot gases before expanding them through a nozzle.

2.28. What is the discharge rate from a round-edged 1-in.-diam hole in the pressure vessel of a nuclear reactor containing saturated water at 2000 psia? Assume choked isentropic, adiabatic homogeneous flow and the following various circumstances: (*a*) Equilibrium flow; (*b*) ½ of the amount of vapor predicted by the equilibrium assumption everywhere; (*c*) $1/10$ of the amount of vapor predicted by the equilibrium assumption; (*d*) no vapor formation.

2.29. Air and water flow upward in a 10-ft-long, 2-in.-diam vertical pipe and discharge into the atmosphere at 14.7 psi. Assuming homogeneous isothermal flow, at a temperature of 70°F calculate the inlet pressure for the following volumetric fluxes meas-

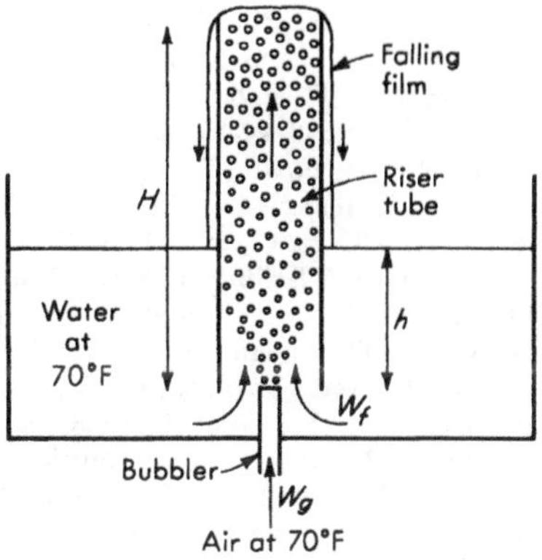

Fig. 2.7 Air-lift pump demonstration. (*Prob.* 2.30.)

ured at atmospheric pressure:

j_f, fps	0.5	1	10	15	30	35	40	40
j_g, fps	1	2	2	10	30	35	40	60

Under what conditions is the exit flow choked? What is the exit pressure under choking conditions? Use numerical techniques.

2.30. Figure 2.7 shows an air-lift pump demonstration for use in a laboratory. Predict pressure drop in the riser as a function of the air and water flow rates W_g and W_f. If $h = 1$ ft and $H = 2$ ft, what is the relationship between W_g and W_f? Under what conditions is the energy expended to run the air compressor per pound of water pumped a minimum?

2.31. Solve Prob. 2.30 for the case where $h = 50$ ft and $H = 100$ ft and the pipe diameter is increased to 6 in. Use numerical techniques.

2.32. Under conditions of rapid phase change, thermodynamic equilibrium as represented by Eq. (2.52) is not obtained. Assume that enthalpy is constant in a rapid expansion but that the quality is given as a function of time, thus

$$x = x(p,t)$$

Explore the one-dimensional flow relationships, assuming some simple expressions for this function such as power laws, exponentials, series expansions, etc.

REFERENCES

1. Shapiro, A. H.: "The Dynamics and Thermodynamics of Compressible Fluid Flow," The Ronald Press Company, New York, 1953.
2. Taylor, G. I.: *Proc. Roy. Soc.* (*London*), vol. A148, p. 141, 1932.
3. Einstein, A.: *Ann. Phys.*, vol. 4, p. 289, 1906.
4. McAdams, W. H., *et al.*: Vaporisation Inside Horizontal Tubes. II. Benzene-Oil Mixtures, *Trans. ASME*, vol. 64, p. 193, 1942.
5. Cicchitti, A., *et al.*: Two-phase Cooling Experiments—Pressure Drop, Heat Transfer and Burnout Measurements, *Energi Nucl.*, vol. 7, no. 6, pp. 407–425, 1960.

6. Dukler, A. E., *et al.*: Pressure Drop and Hold-up in Two-phase Flow, *A. I. Ch. E. J.*, vol. 10, no. 1, pp. 38–51, 1964.
7. Meyer, P. E., and G. B. Wallis: Co-current Upwards Annular-mist Flow, AEC Rept. NYO-3114-10, Sept., 1965.
8. Owens, W. L., Jr.: Two-phase Pressure Gradient, International Developments in Heat Transfer, *Am. Soc. Mech. Engrs.*, paper 41, vol. 2, pp. 363–368, 1961.
9. Martinelli, R. C., and D. B. Nelson: Prediction of Pressure Drop during Forced Circulation Boiling of Water, *Trans. ASME*, vol. 70, p. 695, 1948.
10. Muscettola, M.: Two-phase Pressure Drop, UKAEA Rept. AEEW-R284, 1963.
11. Wallis, G. B., and J. H. Heasley: Oscillations in Two-phase Systems, *Trans. ASME*, ser. C, vol. 83, p. 363, 1961.
12. Wallis, G. B.: Some Hydrodynamic Aspects of Two-phase Flow and Boiling. Part III. International Developments in Heat Transfer, *Am. Soc. Mech. Engrs.*, paper 38, vol. 2, pp. 330–340, 1961.
13. Diehl, J. E., and C. H. Unruh: Two-phase Pressure Drop for Horizontal Crossflow through Tube Banks, *Am. Soc. Mech. Engrs.*, paper 58-HT-20, 1958.
14. Simpson, H. C., and R. S. Silver: Theory of One-dimensional Two-phase, Homogeneous, Non-equilibrium Flow, *Symp. Two-phase Fluid Flow*, London, Inst. Mech. Engrs., pp. 45–56, 1962.

3
Separated Flow

3.1 INTRODUCTION

The separated flow model takes account of the fact that the two phases can have differing properties and different velocities. It may be developed with various degrees of complexity. In the most sophisticated version, separate equations of continuity, momentum, and energy are written for each phase and these six equations are solved simultaneously, together with rate equations which describe how the phases interact with each other and with the walls of the duct. In the simplest version, only one parameter, such as velocity, is allowed to differ for the two phases while conservation equations are only written for the combined flow. When the number of variables to be determined exceeds the available number of equations, correlations or simplifying assumptions are introduced.

3.2 STEADY FLOW IN WHICH THE PHASES ARE CONSIDERED TOGETHER BUT THEIR VELOCITIES ARE ALLOWED TO DIFFER

Suppose that one of the assumptions of homogeneous equilibrium flow is relaxed in order to allow different velocities for the two phases. The conservation laws for mass, momentum, and energy in steady one-dimensional flow can then be derived in terms of the two velocities v_1 and v_2. Alternatively, the extra degree of freedom of the system can be accounted for by introducing both the void fraction and the quality into the equations. Some of the most useful forms of these equations will now be developed.

CONTINUITY

Normally, no mass is added to the flow from outside the duct and the overall mass flow rate is constant. Therefore

$$W = W_1 + W_2 = \text{const} \tag{3.1}$$

In the absence of phase change both W_1 and W_2 are constant individually.

The mass flow rates can be expressed in terms of other variables in many ways. For example, if the areas of cross section of the two streams are A_1 and A_2, we have

$$W_1 = \rho_1 v_1 A_1 \tag{3.2}$$
$$W_2 = \rho_2 v_2 A_2 \tag{3.3}$$

The mass flux of each stream is then, from the relationships derived in Chap. 1,

$$G_1 = \rho_1 v_1 (1 - \alpha) \tag{3.4}$$
$$G_2 = \rho_2 v_2 \alpha \tag{3.5}$$

By using the definition of x, the above two equations can be used to give two alternative expressions for the overall mass flux as follows:

$$G = \rho_1 v_1 \frac{1 - \alpha}{1 - x} \tag{3.6}$$

$$G = \rho_2 v_2 \frac{\alpha}{x} \tag{3.7}$$

MOMENTUM

Many alternative forms of the momentum equation can be derived by manipulating relationships among α, x, G, v_1, v_2, and other variables. For steady flow in a round pipe, for example, one version is

$$-\frac{dp}{dz} = \frac{4\tau_w}{D} + G\frac{d}{dz}[xv_2 + (1 - x)v_1] + [\alpha\rho_2 + (1 - \alpha)\rho_1]g \cos\theta \tag{3.8}$$

ENERGY

The energy equation is conveniently written in terms of the quality. Thus

$$\frac{1}{W}\left(\frac{dq_e}{dz} - \frac{dw}{dz}\right) = \frac{d}{dz}\left[xh_2 + (1-x)h_1\right] + \frac{d}{dz}\left[x\frac{v_2^2}{2} + (1-x)\frac{v_1^2}{2}\right] + g\cos\theta \tag{3.9}$$

In order to solve the above set of equations, two further relationships are required besides knowledge about the relationships between thermodynamic properties. While it is more satisfactory to derive these relationships by analyzing each component separately, a common technique is to use empirical correlations for τ_w and α in terms of the flow rates, fluid properties, and geometry. The method of solution is then mostly determined by the form of these correlations.

When using correlations to fill the gaps in the theory, it should be remembered that they are usually established for adiabatic flow with low-pressure gradients. Under conditions of rapid phase change and acceleration they may be in serious error.

A common way of correlating the wall shear stress in gas-liquid flow was found by Martinelli and coworkers.[1,2] The actual two-phase shear stress is expressed as a factor ϕ^2 times the wall shear stress which would occur in a related single-phase flow. For example, one method that was introduced in the previous chapter relates the wall shear stress to the stress which would act if all the mass flow were composed of liquid. This is evidently motivated by problems of boiling and condensation. Defining $(C_f)_{fo}$ as the appropriate friction factor for the all-liquid flow, we have

$$-\left(\frac{dp}{dz}\right)_F = \frac{2(C_f)_{fo}G^2\mathcal{v}_f\phi_{fo}^2}{D} \tag{3.10}$$

The void fraction correlation is often expressed, for a vapor-liquid mixture of a given substance, in the form

$$\alpha = \alpha(p,x) \tag{3.11}$$

In this case it is an exercise to show that the equivalent of Eq. (2.44) is

$$-\frac{dp}{dz} = \left(\frac{2(C_f)_{fo}G^2\mathcal{v}_f\phi_{fo}^2}{D} + G^2\frac{dx}{dz}\left\{\left(\frac{2x\mathcal{v}_g}{\alpha} - 2\frac{1-x}{1-\alpha}\mathcal{v}_f\right) + \left(\frac{\partial\alpha}{\partial x}\right)_p\left[\frac{(1-x)^2}{(1-\alpha)^2}\mathcal{v}_f - \frac{x^2}{\alpha^2}\mathcal{v}_g\right]\right\} - \frac{G^2}{A}\frac{dA}{dz}\left[\frac{x^2}{\alpha}\mathcal{v}_g + \frac{(1-x)^2}{1-\alpha}\mathcal{v}_f\right] + g\cos\theta\left(\frac{1-\alpha}{\mathcal{v}_f} + \frac{\alpha}{\mathcal{v}_g}\right)\right)$$
$$\left(1 + G^2\left\{\frac{x^2}{\alpha}\frac{\partial\mathcal{v}_g}{\partial p} + \frac{(1-x)^2}{1-\alpha}\frac{\partial\mathcal{v}_f}{\partial p} + \left(\frac{\partial\alpha}{\partial p}\right)_x\left[\frac{(1-x)^2}{(1-\alpha)^2}\mathcal{v}_f - \frac{x^2}{\alpha^2}\mathcal{v}_g\right]\right\}\right)^{-1} \tag{3.12}$$

For the solution of equations such as this, one is virtually forced to use numerical methods.

The denominator of Eq. (3.12) may be equated to zero to obtain the condition for "choking" in the absence of flashing, using Eqs. (3.6) and (3.7) as follows:

$$\alpha \frac{v_g^2}{c_g^2} + (1 - \alpha) \frac{v_f^2}{c_f^2} - \left(\frac{\partial \alpha}{\partial p}\right)_x (\rho_f v_f^2 - \rho_g v_g^2) = 1 \tag{3.13}$$

However, this equation is misleading since the term $(\partial\alpha/\partial p)_x$ is usually derived from a correlation which was obtained at moderate values of the pressure gradient when frictional forces dominated the inertia terms. A more rigorous treatment that considers the separate equations of motion of the two streams will be considered later (p. 68). If inertia effects dominate, the choking condition is found to be

$$\frac{\alpha}{\rho_g}\left(\frac{1}{c_g^2} - \frac{1}{v_g^2}\right) + \frac{1 - \alpha}{\rho_f}\left(\frac{1}{c_f^2} - \frac{1}{v_f^2}\right) = 0 \tag{3.14}$$

Equations (3.13) and (3.14) are compatible only for the particular case in which

$$\rho_g v_g^2 = \rho_f v_f^2 \tag{3.15}$$

Example 3.1 Consider the discharge of flashing liquid through a nozzle. Let the initial state be saturated liquid at a high pressure and possessing negligible kinetic energy. At a much lower pressure downstream let the quality be x. Assume adiabatic equilibrium flow and neglect gravitational effects. What is the maximum possible flow rate per unit area at the downstream point?

Solution Let the enthalpy change from the initial to the final condition be Δh. From Eq. (3.9), in view of the assumptions, we have

$$2\,\Delta h = x v_g^2 + (1 - x) v_f^2 \tag{3.16}$$

The mass flow rate per unit area can be expressed in terms of the quality by eliminating α between Eqs. (3.6) and (3.7) to obtain

$$G = \left(\frac{x}{\rho_g v_g} + \frac{1 - x}{\rho_f v_f}\right)^{-1} \tag{3.17}$$

Denoting v_g/v_f by the symbol n and eliminating v_f^2 from Eqs. (3.16) and (3.17), we get

$$2\,\Delta h G^{-2} = \left(\frac{x}{n\rho_g} + \frac{1 - x}{\rho_f}\right)^2 (x n^2 + 1 - x) \tag{3.18}$$

If the thermodynamic path is known, Δh and x are fixed and Eq. (3.18) can be used to predict the influence of n, the velocity ratio, (or "slip ratio") on the mass flux G. In particular, G will be a maximum when

$$0 = \frac{d}{dn}(2\,\Delta h G^{-2}) = 2nx\left(\frac{x}{n\rho_g} + \frac{1 - x}{\rho_f}\right)^2 - 2(xn^2 + 1 - x)\frac{x}{n^2\rho_g}\left(\frac{x}{n\rho_g} + \frac{1 - x}{\rho_f}\right) \tag{3.19}$$

that is, when

$$\left(\frac{x}{n\rho_g} + \frac{1-x}{\rho_f}\right)\left(\frac{n}{\rho_f} - \frac{1}{n^2\rho_g}\right)x(1-x) = 0 \tag{3.20}$$

The first term in parentheses can never be zero, and the only nontrivial solution of Eq. (3.20) is

$$n^3 = \frac{\rho_f}{\rho_g} \quad \text{or} \quad \frac{v_g}{v_f} = \left(\frac{\rho_f}{\rho_g}\right)^{1/3} \tag{3.21}$$

and the maximum value of G is then, from Eq. (3.18),

$$G_{max} = \frac{(2\,\Delta h)^{1/2}}{[(1-x)/\rho_f^{2/3} + x/\rho_g^{2/3}]^{3/2}} \tag{3.22}$$

Variations in n close to the value given in Eq. (3.21) have small effects on the value of G.

If choking is regarded as a condition of maximum possible discharge through a given nozzle exit area, then it can be argued that this condition corresponds to the value of G given by Eq. (3.22). Moody[3] has had some success in plotting Eq. (3.22) versus available data for steam-water flows as is shown in Fig. 3.1. Since x does not vary much with the thermodynamic path and Δh is a maximum for an isentropic process, the maximum value of G should occur if the flashing is reversible and isentropic.

In the graphs shown in Fig. 3.1 the abscissa is the measured pressure p_m at the exit of the nozzle. Since the flow was choked, this pressure was not the same as the receiver pressure. Figure 3.2 gives value of p_m as a function of the upstream stagnation condition, p_0 and h_0, based on the additional criterion that G in Eq. (3.22) is to be maximized as a function of the pressure p_2. Thus p_m corresponds to the further condition

$$\frac{\partial G}{\partial p_2} = 0 \qquad \text{when } p_2 = p_m \tag{3.23}$$

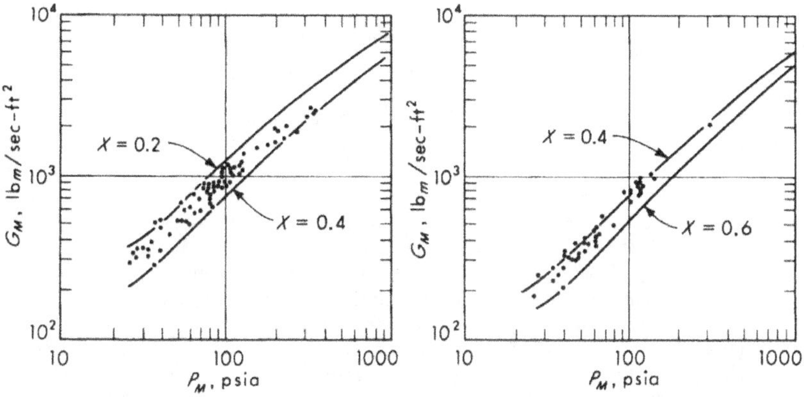

Fig. 3.1 Comparison between measured choked flow rates of steam-water mixtures and Eq. (3.22) for two ranges of steam quality, x. (*Moody.*[3])

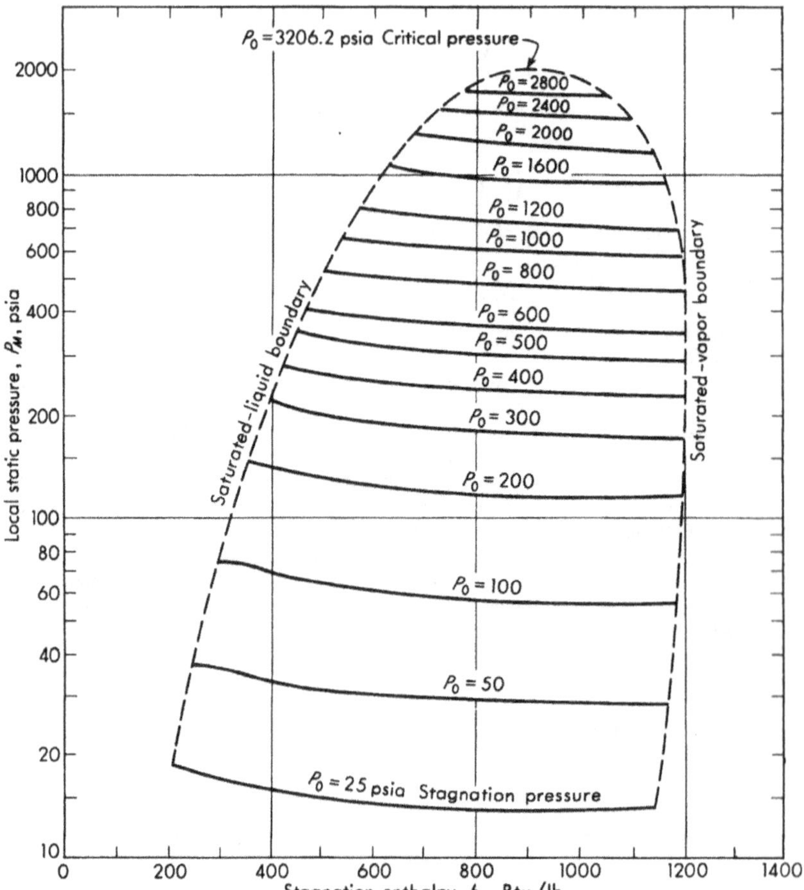

Fig. 3.2 Local static pressure and stagnation properties at maximum steam-water flow rate in terms of upstream stagnation conditions. (*Moody.*[3])

If the receiver pressure is less than p_m, the flow is assumed to be choked and G is calculated from Eq. (3.22) using thermodynamic properties appropriate to p_m. In general, the theoretical values given in Fig. 3.2 tend to be above the values observed by Fauske[4] who found that, to a good approximation, $p_m/p_0 \approx 0.55$.

In spite of its impressive correlation of data, this theory of maximum flow at a given kinetic energy (or the equivalent minimum kinetic energy at a given flow rate) is by no means correct in detail. Much less vapor is produced than is predicted from the equilibrium assumption. The velocity ratio is not given accurately by Eq. (3.21). Part of the success of Eqs. (3.22) and (3.23) is caused by the fact that they predict an extremum of extremes which is not too sensitive to the initial assumptions.

EVALUATION OF WALL SHEAR STRESS AND VOID FRACTION

Equations (3.1), (3.8), and (3.9) cannot be solved, in general, without additional expressions for the wall shear stress (or *frictional pressure drop*) and void fraction. One possibility is to use the homogeneous flow assumptions, in which case the equations reduce to those which were given in the previous chapter. An alternative usually associated with the separated flow model is the correlation scheme which was developed by Martinelli and others[1,2] in the period of 1944–1949.

Perhaps the simplest way to approach the correlation is via the frictional pressure drop. For given flow rates of the gas and the liquid it is assumed that we know how to calculate the pressure gradient which would occur if either fluid were flowing alone in the pipe. Denote these pressure gradients by $(dp/dz)_g$ and $(dp/dz)_f$. Now define new variables in terms of the ratio between the observed pressure gradient dp/dz without phase change, acceleration, or body force effects, and the values for each component alone. Thus

$$\phi_g{}^2 = \frac{dp/dz}{(dp/dz)_g} \tag{3.24}$$

$$\phi_f{}^2 = \frac{dp/dz}{(dp/dz)_f} \tag{3.25}$$

When there is no gas flow, the following results apply.

$$\frac{1}{\phi_f{}^2} = 1 \qquad \frac{1}{\phi_g{}^2} = 0 \tag{3.26}$$

Whereas when there is no liquid flow we have, similarly,

$$\frac{1}{\phi_g{}^2} = 1 \qquad \frac{1}{\phi_f{}^2} = 0 \tag{3.27}$$

Furthermore, at the critical point where the phases are indistinguishable the relationships between ϕ_f and ϕ_g are

For laminar flow,

$$\frac{1}{\phi_f{}^2} + \frac{1}{\phi_g{}^2} = 1 \tag{3.28}$$

For turbulent flow obeying Blasius' smooth-pipe law,

$$\left(\frac{1}{\phi_f}\right)^{8/7} + \left(\frac{1}{\phi_g}\right)^{8/7} = 1 \tag{3.29}$$

A very simple model of separated flow can also be developed by assuming that the two phases flow, without interaction, in two horizontal separate cylinders and that the areas of the cross sections of these cylinders add up to the cross-sectional area of the actual pipe. The pressure drop

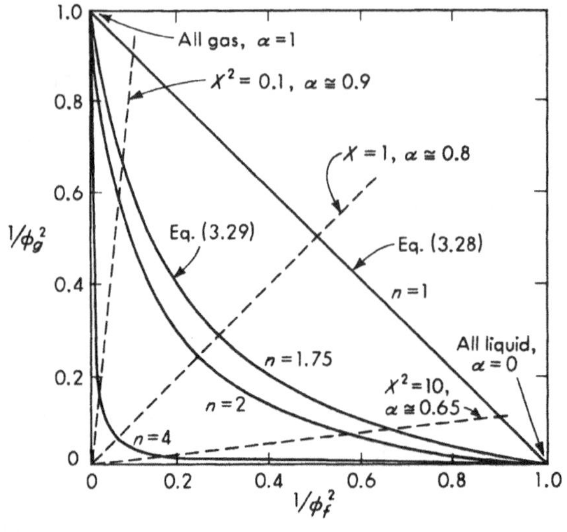

Fig. 3.3 Martinelli's correlation scheme represented on the basis of Eq. (3.30).

in each of the imagined cylinders is the same as in the actual flow, is due to frictional effects only, and is calculated from single-phase flow theory. This *separate-cylinders* model[5] resembles Martinelli's original formulation but has the virtue that it can be pursued to an analytical conclusion, whereas Martinelli was content with a correlation. The results of this analysis are (see Probs. 3.3, 3.4, and 5.12 and Example 3.2):

$$\left(\frac{1}{\phi_f^2}\right)^{1/n} + \left(\frac{1}{\phi_g^2}\right)^{1/n} = 1 \tag{3.30}$$

where $n = 2$ for laminar flow, 2.375 to 2.5 for turbulent flows analyzed on a basis of friction factor, and 2.5 to 3.5 for turbulent flows calculated on a mixing-length basis. Equation (3.30) is compatible with Eqs. (3.26) to (3.29), which are shown in Fig. 3.3.

Equation (3.30) gives a one-parameter family of curves which can be used to fit experimental data.

Example 3.2 Develop the separate-cylinders model for turbulent flow, assuming a constant friction factor for both phases.

Solution Let the gas flow in a cylinder of radius r_g and the liquid in one with radius r_f. The gas and liquid volumetric fractions will then be

$$\alpha = \frac{r_g^2}{r_0^2} \qquad 1 - \alpha = \frac{r_f^2}{r_0^2}$$

where r_0 is the radius of the actual pipe.

For the gas the pressure gradient in the imagined cylinder is

$$-\frac{dp}{dz} = \frac{C_f}{r_g}\rho_g\left(\frac{j_g}{\alpha}\right)^2 = \frac{C_f\rho_g j_g^2}{r_0}\frac{1}{\alpha^{5/2}}$$

and is the same as the actual pressure gradient.

The first factor on the right-hand side is identical with $(-dp/dz)_g$; therefore, from Eq. (3.24)

$$\phi_g^2 = \frac{1}{\alpha^{5/2}}$$

Similarly for the liquid

$$\phi_f^2 = \frac{1}{(1-\alpha)^{5/2}}$$

Eliminating α between these expressions, we get

$$\left(\frac{1}{\phi_g^2}\right)^{2/5} + \left(\frac{1}{\phi_f^2}\right)^{2/5} = 1$$

which is Eq. (3.30) with $n = 2.5$.

In order to avoid having the unknown two-phase pressure drop in both the correlating parameters it is convenient to divide them to yield a new variable. Thus

$$X^2 = \frac{\phi_g^2}{\phi_f^2} = \frac{(dp/dz)_f}{(dp/dz)_g} \tag{3.31}$$

X^2 gives a measure of the degree to which the two-phase mixture behaves as the liquid rather than as the gas. Thus, one might expect some relationship between X^2 and the void fraction as indicated in Fig. 3.3. Because of the much greater specific volume of the gas, the void-fraction variation is not symmetrical about $X = 1$ and tends to be lopsided at low pressures, as shown in the figure.

Since the pressure drop can be calculated from laminar- or turbulent-flow theory for both the liquid and the gas, four different combinations are possible. In annular flow the gas is seldom viscous if the liquid is turbulent, and therefore this combination has not been studied extensively. Martinelli's data for low pressures for the other three permutations are shown in Figs. 3.4, 3.5, and 3.6 together with the correlating line which was drawn by the original authors.

A reasonable correlation for all flow regimes is given by Eq. (3.30) with $n = 3.5$, as shown in Fig. 3.7; $n = 4$ gives a better correlation in the turbulent-turbulent regime.

Martinelli's empirical correlation for void fraction at low pressures is shown in Fig. 3.8. The line can be represented very well by the equation

$$\alpha = (1 + X^{0.8})^{-0.378} \tag{3.32}$$

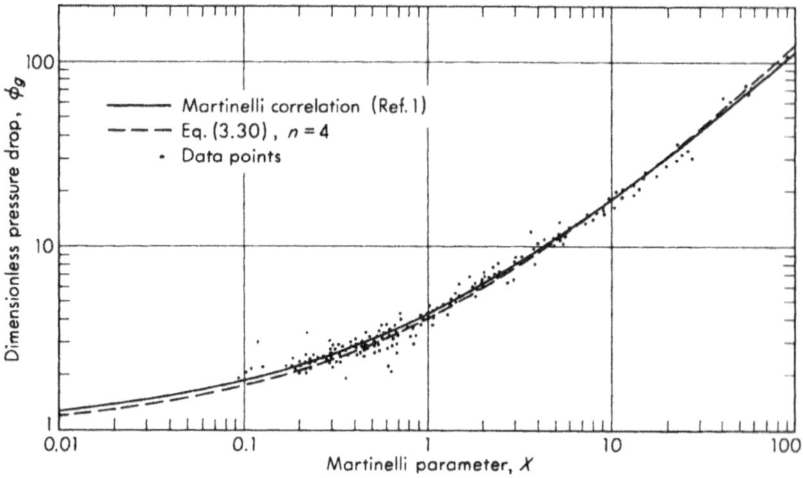

Fig. 3.4 Comparison between pressure drop predicted by Eq. (3.30) with $n = 4$ and empirical results of Martinelli for turbulent-turbulent flow.[5]

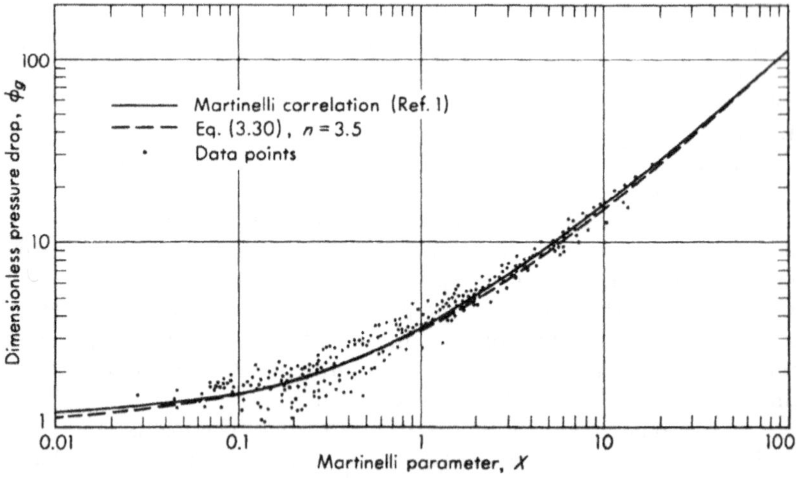

Fig. 3.5 Comparison between pressure drop predicted by Eq. (3.30) with $n = 3.5$ and empirical results of Martinelli for viscous liquid-turbulent gas flow.[5]

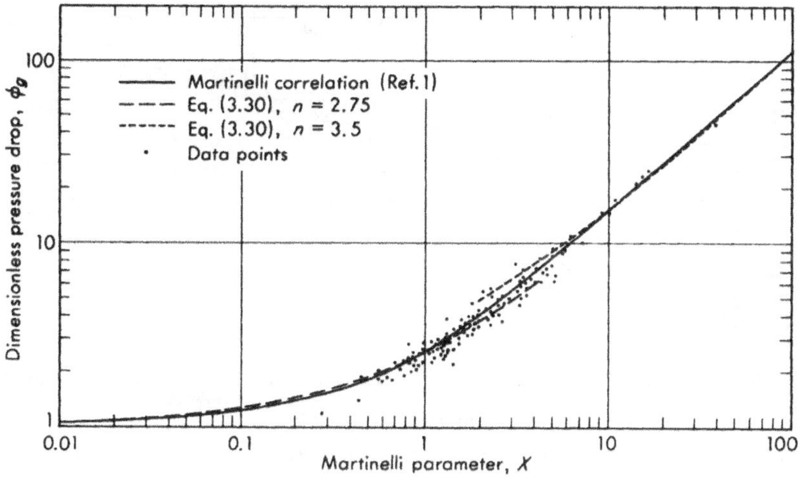

Fig. 3.6 Comparison between pressure drop predicted by Eq. (3.30) and empirical results of Martinelli for viscous-viscous flow.[5]

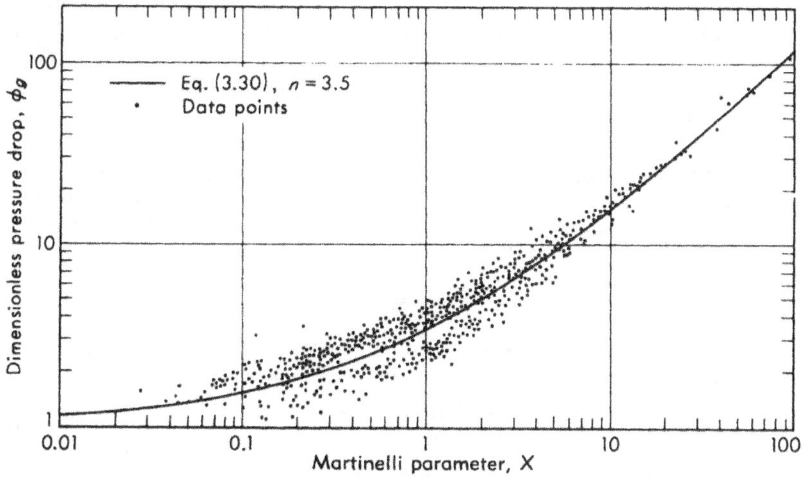

Fig. 3.7 Comparison between pressure drop predicted by Eq. (3.30) with $n = 3.5$ and empirical results of Martinelli for all flow regimes.[5]

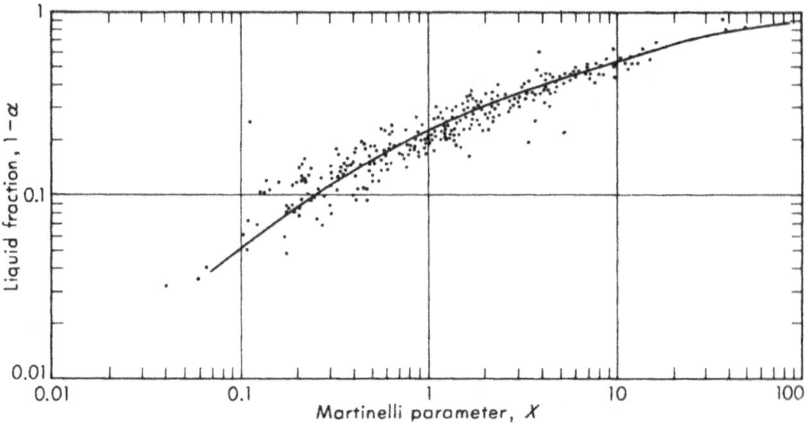

Fig. 3.8 Martinelli's correlation for void fraction. (*Ref.* 1.)

Although Martinelli's correlation was specifically derived for horizontal flow without phase change or significant acceleration, it is often used to calculate both the void fraction and frictional pressure drop for insertion into Eq. (3.12), even when these other effects are not negligible. This procedure leads to progressively increasing errors as the frictional component of pressure drop decreases in proportion to the other terms. The Martinelli correlation in essence balances frictional shear stresses versus pressure drop. More elaborate methods are required when body forces and inertia forces are significant.

Example 3.3 Solve Example 2.5 using the Martinelli-Lockhart correlations.

Solution (working in cgs units) The Reynolds number for the liquid flowing alone in the pipe is found to be 23,000; therefore the flow is turbulent and the friction factor is 0.005. The pressure drop is calculated to be

$$\left(-\frac{dp}{dz}\right)_f = \frac{(2)(0.005)(94.5)^2(1.0)}{2.48} = 36 \text{ dynes/cm}^3 \qquad (a)$$

Similarly for the gas alone the Reynolds number is found to be 24,000, giving a turbulent friction factor of 0.005 and a pressure drop

$$\left(-\frac{dp}{dz}\right)_g = \frac{(2)(0.005)(1.4 \times 10^3)^2(1.25 \times 10^{-3})}{2.48} = 9.4 \text{ dynes/cm}^3 \qquad (b)$$

The value of the Martinelli parameter X_{tt} is

$$X_{tt} = \left[\frac{(dp/dz)_f}{(dp/dz)_g}\right]^{1/2} = \left(\frac{36}{9.4}\right)^{1/2} = 1.96 \qquad (c)$$

From Figs. 3.4 and 3.8,

$$\phi_g = 6.3 \qquad (1 - \alpha) = 0.31 \qquad (d)$$

The friction pressure drop is therefore

$$\Delta p_F = \phi_g^2 \left(-\frac{dp}{dz}\right)_g L = (40)(9.4)(45.6)(1.4 \times 10^{-5}) = 0.24 \text{ psi} \qquad (e)$$

The gravitational pressure drop is

$$\Delta p_G = \left[\frac{(0.31)(62.5) + (0.67)(0.076)}{144}\right] 1.5 = 0.20 \text{ psi} \qquad (f)$$

There is a problem in calculating the acceleration pressure drop because it is very sensitive to the accuracy with which small changes in α can be calculated. However, it is unlikely to be greater than the homogeneous flow acceleration pressure drop. Therefore we use Eq. (2.79) of Example 2.5 to get an expected overestimate of the pressure drop. Thus

$$\Delta p = \frac{0.24 + 0.20}{0.866} = 0.51 \text{ psi}$$

These predictions for liquid fraction and pressure drop both exceed the measured values (0.23 and 0.3 psi). This is because the actual flow regime is slug-annular, liquid bridges move with the gas velocity, and the liquid fraction is therefore decreased. Example 10.5 shows how an accurate solution can be obtained.

FLOW OF BOILING WATER IN STRAIGHT PIPES

Because of its practical importance, the flow of boiling water has perhaps been studied more than any other two-phase system. Correlations based on the Martinelli method are highly developed and tables and curves have been generated for convenient use. The values for ϕ_{f0}^2 which result from the Martinelli-Nelson[2] correlation were already presented in Table 2.1*b*. A graph which allows a rapid prediction of void fraction as a function of quality and pressure is shown in Fig. 3.9. These empirical relationships, together with Eq. (3.12) and the assumption of thermal equilibrium, enable pressure-drop predictions to be made by numerical integration.

In many cases considerable simplification is possible. For example, in straight pipes with no area change (constant G), Eq. (3.8) can be integrated to give

$$\Delta p = \frac{4}{D}\int_0^L \tau_w \, dz + G[xv_g + (1 - x)v_f]_0^L + g \cos\theta \int_0^L [\alpha\rho_g + (1 - \alpha)\rho_f] \, dz \qquad (3.33)$$

For constant heat flux and negligible work, kinetic- or potential-energy changes, or property variations, Eq. (3.9) gives in the two-phase region

$$\frac{dx}{dz} = \frac{4\phi}{DGh_{fg}} \qquad (3.34)$$

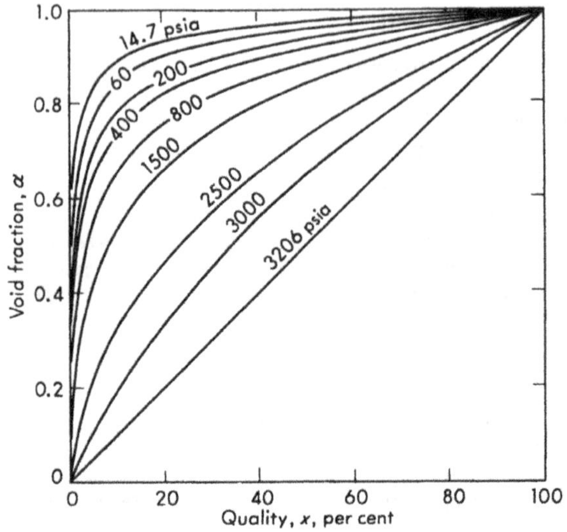

Fig. 3.9 Void fraction versus quality for steam-water flow at various pressures. (*Martinelli and Nelson.*[2])

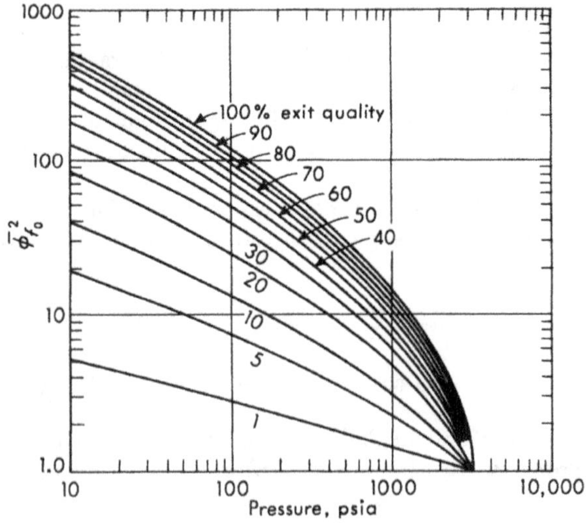

Fig. 3.10 $\bar{\phi}_{fo}^2$, from Eq. (3.36), as a function of pressure and quality for water. (*Martinelli and Nelson.*[2])

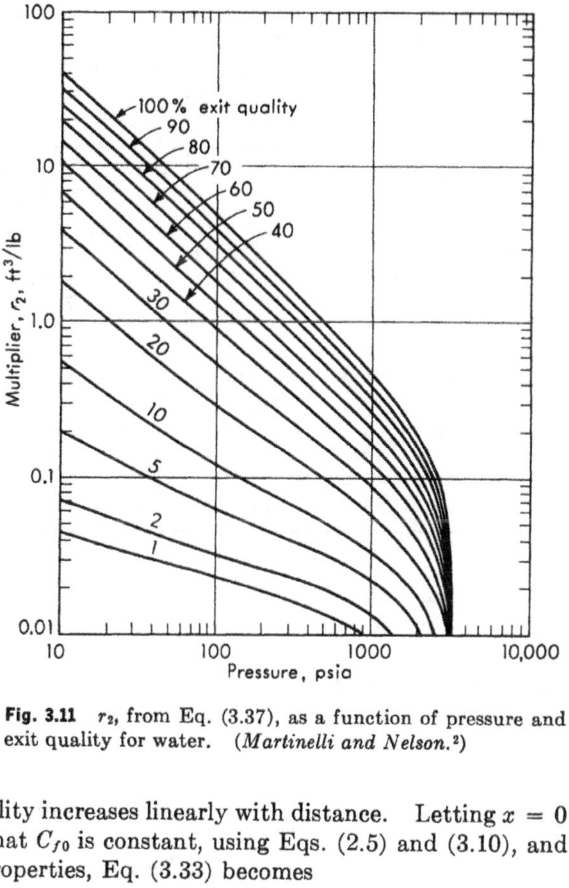

Fig. 3.11 r_2, from Eq. (3.37), as a function of pressure and exit quality for water. (*Martinelli and Nelson.*[2])

showing that the quality increases linearly with distance. Letting $x = 0$ at $z = 0$, assuming that C_{f0} is constant, using Eqs. (2.5) and (3.10), and assuming constant properties, Eq. (3.33) becomes

$$\Delta p = \frac{2C_{f0}G^2L}{D\rho_f}\left(\frac{1}{x}\int_0^x \phi_{f0}^2\,dx\right) + G^2\left\{\frac{x^2}{\alpha\rho_g} + \frac{1}{\rho_f}\left[\frac{(1-x)^2}{1-\alpha} - 1\right]\right\}$$
$$+ g\cos\theta \int_0^L [(1-\alpha)\rho_f + \alpha\rho_g]\,dz \quad (3.35)$$

The various terms in parentheses can be evaluated from the Martinelli-Nelson correlation. This has been done for water for the first two terms and the results are shown in Figs. 3.10 and 3.11 in terms of the parameters

$$\bar{\phi}_{f0}^2 = \frac{1}{x}\int_0^x \phi_{f0}^2\,dx \quad (3.36)$$

$$r_2 = \frac{1}{\rho_f}\left[\frac{x^2}{\alpha}\frac{\rho_f}{\rho_g} + \frac{(1-x)^2}{1-\alpha} - 1\right] \quad (3.37)$$

The final term in Eq. (3.35) has to be calculated by integrating the results

Table 3.1 Revised values of Martinelli factors as given by Thom[6]

Steam quality, wt %	*Pressure, psia* 250		600		1250		2100		3000	
	ϕ_{fo}^2	$\bar{\phi}_{fo}^2$	ϕ_{fo}^2	$\bar{\phi}_{fo}^2$	ϕ_{fo}^2	$\bar{\phi}_{fo}^2$	ϕ_{fo}^2	$\bar{\phi}_{fo}^2$	ϕ_{fo}^2	$\bar{\phi}_{fo}^2$
1	2.12	1.49	1.46	1.11	1.10	1.03	.....	.....	.....	.....
5	6.29	3.71	2.86	2.09	1.62	1.31	1.21	1.10	1.02	.....
10	11.1	6.30	4.78	3.11	2.39	1.71	1.48	1.21	1.08	1.06
20	20.6	11.4	8.42	5.08	3.77	2.47	2.02	1.46	1.24	1.12
30	30.2	16.2	12.1	7.00	5.17	3.20	2.57	1.72	1.40	1.18
40	39.8	21.0	15.8	8.80	6.59	3.89	3.12	2.01	1.57	1.26
50	49.4	25.9	19.5	10.6	8.03	4.55	3.69	2.32	1.73	1.33
60	59.1	30.5	23.2	12.4	9.49	5.25	4.27	2.62	1.88	1.41
70	68.8	35.2	26.9	14.2	10.19	6.00	4.86	2.93	2.03	1.50
80	78.7	40.1	30.7	16.0	12.4	6.75	5.45	3.23	2.18	1.58
90	88.6	45.0	34.5	17.8	13.8	7.50	6.05	3.53	2.33	1.66
100	98.86	49.93	38.30	19.65	15.33	8.165	6.664	3.832	2.480	1.740

from Fig. 3.9. These graphs provide a very rapid method for estimating pressure drops for boiling water.

Alternative (generally better) values for ϕ_{fo}^2, $\bar{\phi}_{fo}^2$, α, and $\rho_f r_2$ for high-pressure water were deduced by Thom,[6] using much more data than the original Martinelli-Nelson correlation, and are shown in Tables 3.1 and 3.2.

Table 3.2 Revised values of Martinelli factors as given by Thom[6]

Steam quality, wt %	*Pressure, psia* 250		600		1250		2100		3000	
	α	$r_2\rho_f$	α	$r_2\rho_f$	α	$r_2\rho_f$	α	$r_2\rho_f$	α	$r_2\rho_f$
1	0.288	0.4125	0.168	0.2007	0.090	0.0955	0.0476	0.0431	0.0213	0.0132
5	0.678	2.169	0.512	1.040	0.340	0.4892	0.207	0.2182	0.102	0.0657
10	0.816	4.620	0.690	2.165	0.521	1.001	0.355	0.4431	0.193	0.1319
20	0.910	10.39	0.833	4.678	0.710	2.100	0.553	0.9139	0.350	0.2676
30	0.945	17.30	0.895	7.539	0.808	3.292	0.679	1.412	0.480	0.4067
40	0.964	25.37	0.930	10.75	0.866	4.584	0.767	1.937	0.589	0.5495
50	0.975	34.58	0.952	14.30	0.908	5.958	0.832	2.490	0.682	0.6957
60	0.984	44.93	0.967	18.21	0.936	7.448	0.881	3.070	0.763	0.8455
70	0.990	56.44	0.979	22.46	0.959	9.030	0.920	3.678	0.834	0.9988
80	0.994	69.09	0.988	27.06	0.976	10.79	0.952	4.512	0.895	1.156
90	0.997	82.90	0.995	32.01	0.989	12.48	0.978	5.067	0.951	1.316
100	1	98.10	1	37.30	1	14.34	1	5.664	1	1.480

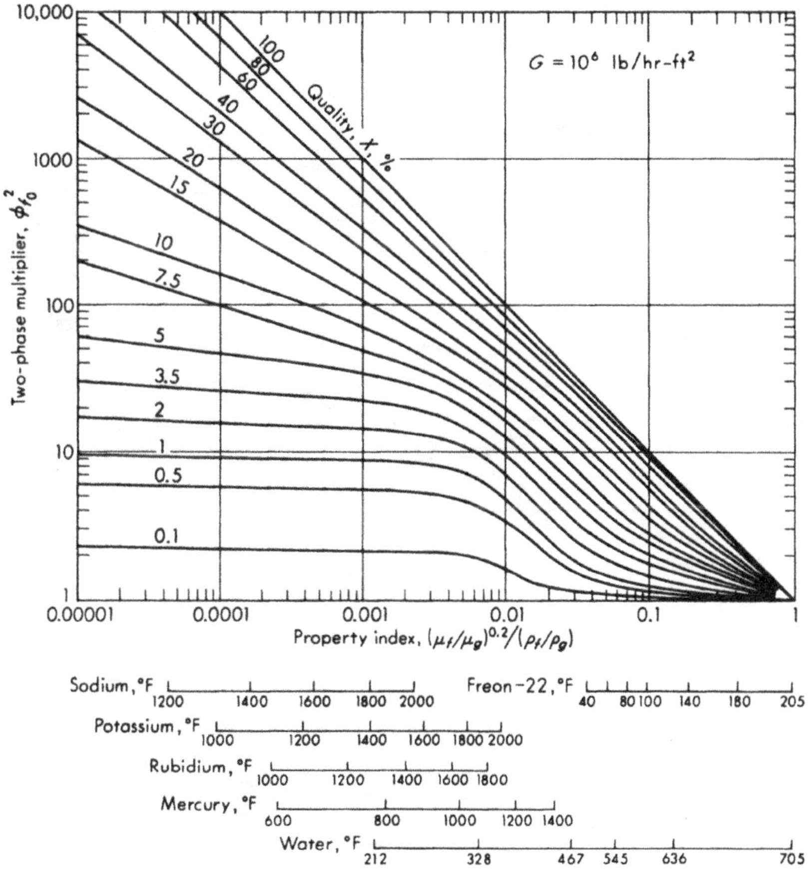

Fig. 3.12 Baroczy's correlation[7] of ϕ_{fo}^2 for $G = 10^6$ lb/(hr)(ft^2).

A further extension of the method by Baroczy[7] enables it to be used for fluids other than water over a wide range of mass fluxes. The parameter ϕ_{fo}^2 is expressed as a function of a *property index* $(\mu_f/\mu_g)^{0.2}(\rho_g/\rho_f)$ for a given value of $G = 10^6$ lb/(hr)(ft^2) (Fig. 3.12). For other mass fluxes, this value of ϕ_{fo}^2 is multiplied by a correction factor Ω shown in Fig. 3.13 in order to get the appropriate value of ϕ_{fo}^2. The whimsical fluctuations in the curves in Fig. 3.13 provide an excellent stimulus for mirth among those skeptics who are unimpressed by grand correlation schemes. However, they do enable a design engineer to get on with the job of making performance predictions in the absence of any better information.

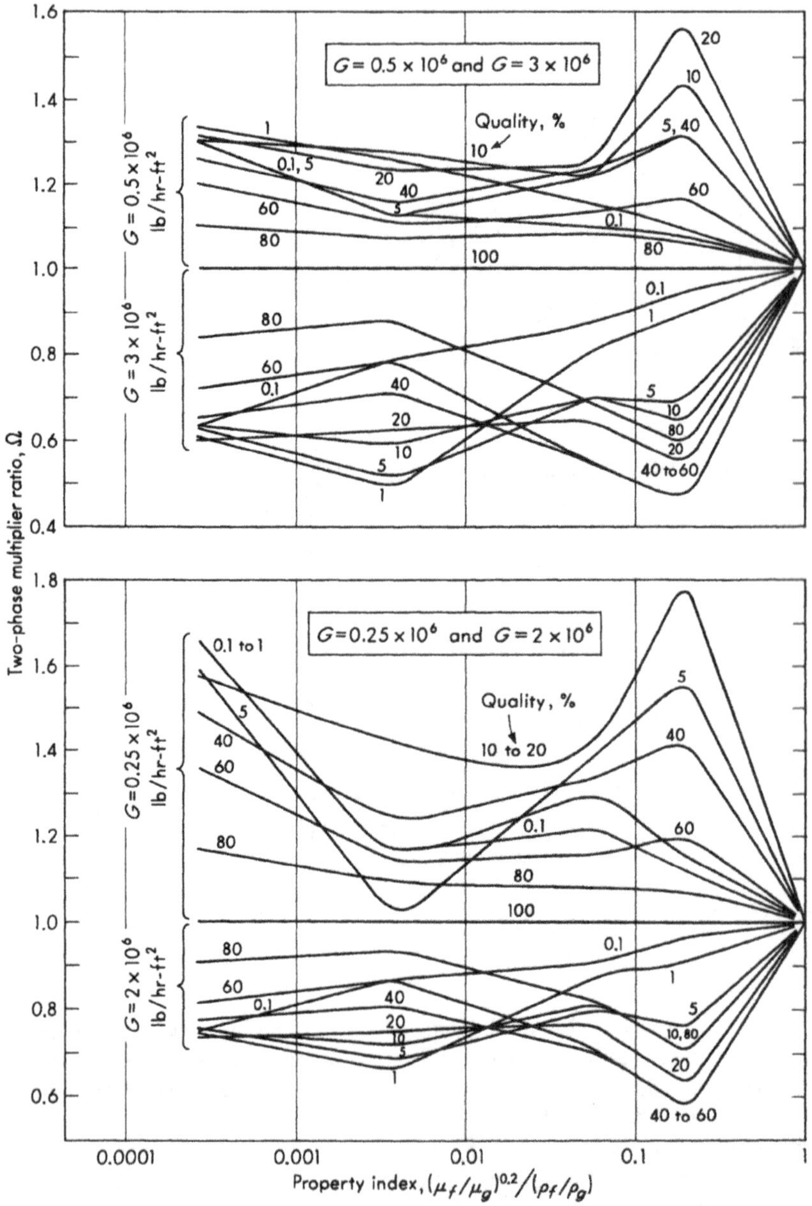

Fig. 3.13 Mass flux correction factor versus property index for Baroczy correlation.[7]

3.3 ONE-DIMENSIONAL SEPARATED FLOW IN WHICH THE PHASES ARE CONSIDERED SEPARATELY

In order to avoid the extensive use of correlations to make predictions for separated flows, it is necessary to use more equations to relate the variables. This can be done by writing separate conservation laws for the components, rather than by using only the equations for the entire mixture. Thus increased accuracy is bought at the price of increased complexity.

At this stage it is convenient to keep these equations in a very general form which will provide the basis for a unified treatment in later chapters. This will involve the introduction of symbols and concepts which will probably appear rather abstract, with a meaning which is difficult to grasp until more specific cases are treated. Examples will be included in order to help to make this meaning clear.

CONTINUITY EQUATIONS

The equations of conservation of mass, or continuity equations, can be written quite generally in differential form:

$$\frac{\partial}{\partial t}[\rho_1(1-\alpha)] + \nabla \cdot [\rho_1(1-\alpha)\mathbf{v}_1] = S_{12} + S_1 \qquad (3.38a)$$

$$\frac{\partial}{\partial t}(\rho_2\alpha) + \nabla \cdot (\rho_2\alpha\mathbf{v}_2) = -S_{12} + S_2 \qquad (3.38b)$$

S_{12} is a source term which represents the mass rate of phase change per unit volume. S_1 and S_2 are external sources of matter and are almost invariably zero.

In one-dimensional form, Eqs. (3.38) become, after integration across the duct,

$$\frac{\partial}{\partial t}[\rho_1(1-\alpha)A] + \frac{\partial}{\partial z}[\rho_1(1-\alpha)v_1A] = \int (S_{12} + S_1)\, dA \qquad (3.39a)$$

$$\frac{\partial}{\partial t}(\rho_2\alpha A) + \frac{\partial}{\partial z}[\rho_2\alpha v_2A] = \int (-S_{12} + S_2)\, dA \qquad (3.39b)$$

Corresponding integral versions of these equations can be derived in exactly the same way as they are in the usual single-phase flow cases.

MOMENTUM EQUATIONS

The momentum equations, equations of motion, or Newton's law for the two phases can be written quite generally in three-dimensional vector form:

$$\rho_1\left(\frac{\partial\mathbf{v}_1}{\partial t} + \mathbf{v}_1 \cdot \nabla\mathbf{v}_1\right) = \mathbf{b}_1 + \mathbf{f}_1 - \nabla p \qquad (3.40a)$$

$$\rho_2\left(\frac{\partial\mathbf{v}_2}{\partial t} + \mathbf{v}_2 \cdot \nabla\mathbf{v}_2\right) = \mathbf{b}_2 + \mathbf{f}_2 - \nabla p \qquad (3.40b)$$

$\mathbf{b}_1$ and $\mathbf{b}_2$ are the body forces, per unit volume of that component, which act on each component. ∇p is some average gradient of pressure or bulk stress which is suitably defined and is usually the thermodynamic pressure of one or both of the fluid phases. $\mathbf{f}_1$ and $\mathbf{f}_2$ are simply "what is left over." Since Eqs. (3.38) and (3.39) are merely force (or momentum) balances, the f's are introduced to keep the accounts straight and the way in which they are evaluated will depend on the particular flow regime and conditions of the problem.

For example, if Eqs. (3.38) and (3.39) refer to elements of incompressible newtonian fluid containing only one component which is not undergoing phase change, the usual results of viscous flow are obtained, thus

$$\mathbf{f}_1 = \mu_1 \nabla^2 \mathbf{v}_1 \tag{3.41a}$$
$$\mathbf{f}_2 = \mu_2 \nabla^2 \mathbf{v}_2 \tag{3.41b}$$

To obtain a quasicontinuum model of two-phase flow, one usually calculates the f's using an element of the flow which is larger than the particles, drops, or bubbles that occupy the flow field. In this case the f's represent the average total surface force, per unit volume, that is not contained in the pressure gradient. The f's contain components due to hydrodynamic drag, apparent mass effects during relative acceleration, particle-particle forces, forces due to momentum changes during evaporation or condensation, and so on. The evaluation of the f's in specific cases often requires considerable care. For example, most "drag" forces on particles suspended in fluids are determined experimentally and contain both the effects of ∇p and f; in addition, phase change modifies hydrodynamic drag, while bubble or droplet shapes, and hence drag forces, are altered by the surrounding force field.

F_1 and F_2 are defined as the equivalent of the f's per unit volume of the whole flow field. Thus

$$F_1 = f_1 (1 - \alpha) \tag{3.42}$$
$$F_2 = f_2 \alpha \tag{3.43}$$

If these forces are entirely due to mutual hydrodynamic drag, action and reaction are equal and we have

$$F_1 = -F_2 = F_{12} \tag{3.44}$$

In the case of one-dimensional flow, Eqs. (3.40*a*) and (3.40*b*) are resolved in the direction of motion to give

$$\rho_1 \left(\frac{\partial v_1}{\partial t} + v_1 \frac{\partial v_1}{\partial z} \right) = b_1 + f_1 - \frac{\partial p}{\partial z} \tag{3.45}$$

$$\rho_2 \left(\frac{\partial v_2}{\partial t} + v_2 \frac{\partial v_2}{\partial z} \right) = b_2 + f_2 - \frac{\partial p}{\partial z} \tag{3.46}$$

Example 3.4 In *annular* flow, liquid flows as a film on the wall of a pipe while gas flows in a central cylindrical core. For a pipe of diameter D let the interfacial and wall shear stresses be τ_i and τ_w. Assuming symmetrical vertical flow with the positive direction measured upward, derive the values of F_f, F_g, f_f, f_g, b_f, and b_g and hence the equations of motion.

Solution The diameter of the gas core is $D\sqrt{\alpha}$, where α is the void fraction. Therefore

$$F_g = \frac{-\tau_i \pi D\sqrt{\alpha}}{\pi D^2/4} = -\frac{4\tau_i\sqrt{\alpha}}{D}$$

and

$$F_f = \frac{\tau_i \pi D\sqrt{\alpha} - \tau_w \pi D}{\pi D^2/4} = \frac{4(\tau_i\sqrt{\alpha} - \tau_w)}{D}$$

The corresponding values of f_f and f_g are

$$f_f = \frac{F_f}{1-\alpha} = \frac{4(\tau_i\sqrt{\alpha} - \tau_w)}{D(1-\alpha)}$$

$$f_g = \frac{F_g}{\alpha} = -\frac{4\tau_i}{D\sqrt{\alpha}}$$

Also

$$b_f = -\rho_f g$$

and

$$b_g = -\rho_g g$$

The equations of motion are then, from Eqs. (3.45) and (3.46),

$$\rho_f\left(\frac{\partial v_f}{\partial t} + v_f\frac{\partial v_f}{\partial z}\right) = -\rho_f g + \frac{4}{D(1-\alpha)}(\tau_i\sqrt{\alpha} - \tau_w) - \frac{\partial p}{\partial z}$$

$$\rho_g\left(\frac{\partial v_g}{\partial t} + v_g\frac{\partial v_g}{\partial z}\right) = -\rho_g g - \frac{4\tau_i}{D\sqrt{\alpha}} - \frac{\partial p}{\partial z}$$

Example 3.5 The Carman-Kozeny equation for the frictional pressure drop during viscous flow through a packed bed of spheres with diameter d and void fraction ϵ is

$$-\left(\frac{\partial p}{\partial z}\right)_F = 180\,\frac{\mu_f j_{f0}}{d^2}\,\frac{(1-\epsilon)^2}{\epsilon^3}$$

where j_{f0} is the fluid flux relative to the particles. ϵ has been chosen to represent the liquid fraction because α is more commonly used to denote the volumetric fraction of the discontinuous phase. Deduce the values of f_f and f_s. The subscripts f and s refer to the fluid and the solid, respectively. What are the components of f_s due to the fluid and the particles?

Solution Since we are only concerned with frictional pressure drop, the inertia and gravitational terms are dropped from Eqs. (3.45) and (3.46) to give

$$f_f = f_s = \left(\frac{\partial p}{\partial z}\right)_F$$

The force f_s on the particles is made up of two parts, one due to the fluid, $(f_s)_f$, and the other due to the particles, $(f_s)_s$. Thus

$$f_s = (f_s)_f + (f_s)_s$$

The mutual force between the fluid and the particles obeys Eq. (3.44). Therefore

$$F_{fs} = f_f\epsilon = -(f_s)_f(1 - \epsilon)$$

Combining these relationships we find

$$f_f = -\frac{180\mu_f j_{f0}}{d^2}\frac{(1-\epsilon)^2}{\epsilon^3}$$

$$(f_s)_f = 180\frac{\mu_f j_{f0}}{d^2}\frac{1-\epsilon}{\epsilon^2}$$

$$(f_s)_s = -180\frac{\mu_f j_{f0}}{d^2}\frac{1-\epsilon}{\epsilon^3}$$

Once the values of the f's have been determined, they can be used to solve more complex problems (such as those encountered in fluidization and soil compaction) in which other terms in Eqs. (3.45) and (3.43) are not negligible.

Example 3.6 In a *fluidized bed* particles are supported by an upward flow of fluid around them and interparticle forces are negligible. Use the results of the previous example to deduce the fluid flux necessary to cause fluidization in a bed with void fraction ϵ. What is the pressure gradient through the bed in this case?

Solution The condition for fluidization is

$$(f_s)_s = 0$$

Use Eqs. (3.45) and (3.46) without the inertia terms to obtain

$$-\frac{dp}{dz} - \rho_f g - 180\frac{\mu_f j_{f0}}{d^2}\frac{(1-\epsilon)^2}{\epsilon^3} = 0$$

$$-\frac{dp}{dz} - \rho_s g + 180\frac{\mu_f j_{f0}}{d^2}\frac{1-\epsilon}{\epsilon^2} = 0$$

Solving these equations simultaneously we get

$$j_{f0} = \frac{d^2 g(\rho_s - \rho_f)}{180\mu_f}\frac{\epsilon^3}{1-\epsilon}$$

$$-\frac{dp}{dz} = g[\epsilon\rho_f + (1-\epsilon)\rho_s]$$

The pressure gradient is simply equal to the weight of the bed, solids plus fluid, per unit depth of the bed.

3.4 FLOW WITH PHASE CHANGE

The general momentum equations for one-dimensional separated flow in a duct in the presence of phase change will now be developed. To be more specific about the f's, let the drag forces from the duct walls on components 1 and 2 be F_{w1} and F_{w2}, per unit volume of the flow, respectively. Let the drag force between the components be F_{12} acting on component 1 in the direction of motion and in the opposite direction on component 2. Since the two components have differing velocities, any phase change will result in a change of momentum. The mass rate of phase change per unit length is $W\,dx/dz$ and the velocity change is $v_2 - v_1$. Therefore the force that is necessary to account for momentum increase due to phase change per unit volume of the duct is

$$(v_2 - v_1)\frac{W}{A}\frac{dx}{dz} = (v_2 - v_1)G\frac{dx}{dz} \tag{3.47}$$

It is not clear at this juncture how much of this force is to be charged to stream 1 and how much to stream 2. In fact, this assignment will depend on the process (e.g., boiling or condensation) and on the interaction between the phase change and the hydrodynamic mechanism which gives rise to F_{12}. For example, the drag force on an evaporating droplet will depend on its rate of vaporization.

To be quite general, therefore, we ascribe a fraction η, of the force represented by Eq. (3.47), to stream 2 and a fraction $1 - \eta$ to stream 1. The choice of η may be different for different systems.

The two momentum equations of steady flow are now, if gravity is the only body force,

$$\rho_1 v_1 \frac{dv_1}{dz} = -\frac{dp}{dz} - \rho_1 g \cos\theta + \frac{F_{12} - F_{w1}}{1-\alpha} - \frac{1-\eta}{1-\alpha}(v_2 - v_1)G\frac{dx}{dz} \tag{3.48}$$

$$\rho_2 v_2 \frac{dv_2}{dz} = -\frac{dp}{dz} - \rho_2 g \cos\theta - \frac{F_{12} + F_{w2}}{\alpha} - \frac{\eta}{\alpha}(v_2 - v_1)G\frac{dx}{dz} \tag{3.49}$$

Equations (3.48) and (3.49) can be combined in several ways. The following two results are particularly convenient.

Multiply (3.48) by $1 - \alpha$ and (3.49) by α and add. The result is the *equation of motion for the mixture* and could have been obtained by considering the momentum balance for both components taken together, thus

$$(1-\alpha)\rho_1 v_1 \frac{dv_1}{dz} + \alpha\rho_2 v_2 \frac{dv_2}{dz} = -\frac{dp}{dz} - g\cos\theta\,[\rho_1(1-\alpha) + \rho_2\alpha] - (F_{w1} + F_{w2}) - (v_2 - v_1)G\frac{dx}{dz} \tag{3.50}$$

Furthermore, the last term on the right-hand side can be combined with the left-hand side, using Eqs. (3.2) through (3.5) and (3.47) to give

$$\frac{1}{A}\frac{d}{dz}(W_1 v_1 + W_2 v_2) = -\frac{dp}{dz} - g\cos\theta\,[\rho_1(1-\alpha) + \rho_2\alpha] - (F_{w1} + F_{w2}) \tag{3.51}$$

which is plainly the momentum balance for the mixture.

If, on the other hand, Eq. (3.49) is subtracted from Eq. (3.48), the result is

$$\rho_1 v_1 \frac{dv_1}{dz} - \rho_2 v_2 \frac{dv_2}{dz} = g \cos\theta\,(\rho_2 - \rho_1) - \frac{F_{w1}}{1-\alpha} + \frac{F_{w2}}{\alpha} + \frac{F_{12}}{\alpha(1-\alpha)} - G\frac{dx}{dz}\left(\frac{1-\eta}{1-\alpha} - \frac{\eta}{\alpha}\right)(v_2 - v_1) \tag{3.52}$$

This equation does not include the pressure gradient and could be thought of as the *relative motion equation* since it describes the difference between the rates at which the two phases are gaining kinetic energy.

Usually only one component is in contact with the wall and the appropriate wall shear stress can be obtained from a correlation. The drag force between the components is a function of the relative velocity and can also be estimated. However, the detailed solution of the resulting equations is often quite a formidable problem.

Example 3.7 Use the Martinelli correlation scheme for gas-liquid flow to deduce the values of F_{fg} and F_{wf}, if F_{wg} is assumed to be zero due to the liquid wetting the entire wall of the duct (annular flow).

Solution Martinelli's correlation represents a balance between the F's and the pressure gradient, since it does not include the effects of inertia, gravity, or phase change. From Eqs. (3.48) and (3.49), therefore,

$$F_{wf} = \frac{F_{fg}}{\alpha} = -\frac{dp}{dz}$$

It appears most reasonable to express the force on the liquid in terms of liquid properties and the force on the gas in terms of gas properties. Using Eqs. (3.24) and (3.25) we therefore have

$$F_{wf} = -\phi_f^2 \left(\frac{dp}{dz}\right)_f$$

$$F_{fg} = -\alpha\phi_g^2 \left(\frac{dp}{dz}\right)_g$$

If ϕ_f^2 and ϕ_g^2 are given as functions of α by Martinelli's correlation, F_{wf} and F_{fg} can be expressed in terms of known parameters, leaving α to be determined by solution of Eqs. (3.48) and (3.49), even when gravity and inertia are not negligible.

The momentum equations can be combined with the continuity equations to give results which parallel corresponding developments which are common in gas dynamics. Consider first component 2. Write the identity

$$\rho_2 v_2 \frac{dv_2}{dz} = v_2 \frac{d(\rho_2 v_2)}{dz} - v_2^2 \frac{d\rho_2}{dz} \tag{3.53}$$

Now, the change in the density of phase 2 will be related to the change of pressure and the particular thermodynamic path which the components are following. For example, in the two-phase region, for a pure substance in equilibrium, the densities are only a function of pressure. For particle-gas systems a "virtual equation of state" may be appropriate, as discussed in Example 2.2.

In any case we may write

$$\frac{d\rho_2}{dz} = \frac{dp/dz}{\partial p/\partial \rho_2} \tag{3.54}$$

where it is understood that $\partial p/\partial \rho_2$ is evaluated for the prevailing conditions.

Define a pseudo-sonic velocity c_2 for component 2 by the equation

$$c_2{}^2 = \frac{\partial p}{\partial \rho_2} \tag{3.55}$$

Combining Eqs. (3.53), (3.54), and (3.55) and using the result in Eq. (3.49), we get, after rearrangement,

$$-\frac{dp}{dz}\left(1 - \frac{v_2{}^2}{c_2{}^2}\right) = v_2 \frac{d}{dz}(\rho_2 v_2) + \rho_2 g \cos\theta + \frac{F_{12} + F_{w2}}{\alpha} + \frac{\eta}{\alpha}(v_2 - v_1)G\frac{dx}{dz} \tag{3.56}$$

Equation (3.3) is now rewritten as

$$Wx = A\alpha\rho_2 v_2 \tag{3.57}$$

and is differentiated to yield

$$\frac{1}{x}\frac{dx}{dz} = \frac{1}{A}\frac{dA}{dz} + \frac{1}{\alpha}\frac{d\alpha}{dz} + \frac{1}{\rho_2 v_2}\frac{d(\rho_2 v_2)}{dz} \tag{3.58}$$

Eliminating $d(\rho_2 v_2)/dz$ between Eqs. (3.56) and (3.58) then gives the result

$$-\frac{dp}{dz}\frac{1}{\rho_2 v_2{}^2}\left(1 - \frac{v_2{}^2}{c_2{}^2}\right) = \frac{1}{x}\frac{dx}{dz} - \frac{1}{\alpha}\frac{d\alpha}{dz} - \frac{1}{A}\frac{dA}{dz} + \frac{1}{\rho_2 v_2{}^2}\left[\rho_2 g \cos\theta + \frac{F_{12} + F_{w2}}{\alpha} + \frac{\eta}{\alpha}(v_2 - v_1)G\frac{dx}{dz}\right] \tag{3.59a}$$

The similar equation for the other phase is

$$-\frac{dp}{dz}\frac{1}{\rho_1 v_1{}^2}\left(1 - \frac{v_1{}^2}{c_1{}^2}\right) = -\frac{1}{1-x}\frac{dx}{dz} + \frac{1}{1-\alpha}\frac{d\alpha}{dz} - \frac{1}{A}\frac{dA}{dz} + \frac{1}{\rho_1 v_1{}^2}\left[\rho_1 g \cos\theta - \frac{F_{12} - F_{w1}}{1-\alpha} + \frac{1-\eta}{1-\alpha}(v_2 - v_1)G\frac{dx}{dz}\right] \tag{3.59b}$$

The above equations resemble the general one-dimensional steady flow equations of single-component flow, except for the effects of phase change and the additional degree of freedom which is introduced by the variable α. "Choking" occurs in the individual components when $v_1 = c_1$ or

$v_2 = c_2$, but this does not necessarily correspond to the "compound choking" of the combined flows since α is free to adjust to local conditions.

In order to investigate compound choking we eliminate $d\alpha/dz$ between Eqs. (3.57) and (3.58). The result is

$$-\frac{dp}{dz}\left[\frac{\alpha}{\rho_2 v_2^2}\left(1 - \frac{v_2^2}{c_2^2}\right) + \frac{1-\alpha}{\rho_1 v_1^2}\left(1 - \frac{v_1^2}{c_1^2}\right)\right] = -\frac{1}{A}\frac{dA}{dz}$$
$$+ g\cos\theta\left(\frac{\alpha}{v_2^2} + \frac{1-\alpha}{v_1^2}\right) + F_{12}\left(\frac{1}{\rho_2 v_2^2} - \frac{1}{\rho_1 v_1^2}\right) + \frac{F_{w2}}{\rho_2 v_2^2} + \frac{F_{w1}}{\rho_1 v_1^2}$$
$$+ \frac{dx}{dz}\left[\frac{\alpha}{x} - \frac{1-\alpha}{1-x} + G(v_2 - v_1)\left(\frac{1-\eta}{\rho_1 v_1^2} + \frac{\eta}{\rho_2 v_2^2}\right)\right] \quad (3.60)$$

As long as F_{12} and dx/dz are independent of the pressure gradient, the choking condition is

$$\frac{\alpha}{\rho_2 v_2^2}\left(1 - \frac{v_2^2}{c_2^2}\right) + \frac{1-\alpha}{\rho_1 v_1^2}\left(1 - \frac{v_1^2}{c_1^2}\right) = 0 \quad (3.61)$$

Since all the factors except those in parentheses are positive, it is evident that one of the ratios $M_2^2 = v_2^2/c_2^2$ and $M_1^2 = v_1^2/c_1^2$ must be less than unity while the other is greater than unity. Thus one stream is supersonic while the other is subsonic.

The choking condition is considerably modified if *flashing* (phase change as a result of pressure change) is significant. This effect introduces a dependence on pressure gradient in the terms on the right-hand side of Eq. (3.60) which involve dx/dz.

In general, if the thermodynamic path is known (or assumed), then $\partial x/\partial p$ can be evaluated and we may write

$$\frac{dx}{dz} = \frac{\partial x}{\partial p}\frac{dp}{dz} \quad (3.62)$$

Combining Eq. (3.62) with Eq. (3.60) the choking condition in the presence of flashing is found to be

$$\frac{\alpha}{\rho_2 v_2^2}(1 - M_2^2) + \frac{1-\alpha}{\rho_1 v_1^2}(1 - M_1^2)$$
$$+ \frac{\partial x}{\partial p}\left[\frac{\alpha}{x} - \frac{1-\alpha}{1-x} + G(v_2 - v_1)\left(\frac{1-\eta}{\rho_1 v_1^2} + \frac{\eta}{\rho_2 v_2^2}\right)\right] = 0 \quad (3.63)$$

This result depends on the value of η except in the particular case where

$$\rho_1 v_1^2 = \rho_2 v_2^2 \quad (3.64)$$

3.5 FLOW IN WHICH INERTIA EFFECTS DOMINATE

In certain cases in which a two-component separated flow is accelerated rapidly through a nozzle, the inertia and pressure-drop terms dominate

Eqs. (3.48) and (3.49) entirely. In this case the velocity changes of the components can be related. In steady one-dimensional flow without phase change, for example, if the f's are zero, we have

$$-\frac{dp}{dz} = \rho_1 v_1 \frac{dv_1}{dz} = \rho_2 v_2 \frac{dv_2}{dz} \tag{3.65}$$

In particular, if both components start with a low velocity and density changes are small, their final velocities after expansion are related by the simple expression

$$\frac{v_1}{v_2} = \sqrt{\frac{\rho_2}{\rho_1}} \tag{3.66}$$

This expression has no universal validity but is approximately true under conditions of rapid expansion at low Mach numbers.

Example 3.8 Air and water flow from a large tank through a converging nozzle having an exit area of 1 in.². The flow rates measured at atmospheric pressure are 2000 in.³/sec of air and 27 in.³/sec of water. What are the pressure in the tank and the exit velocities of the components if the external pressure is atmospheric (14.7 psia) and the temperature is 70°F? Neglect wall shear stresses.

Solution The problem as stated above is indeterminate because of the lack of detail about the method of mixing the components and the resulting flow pattern. The best that can be done is to perform a limiting analysis considering the two extreme cases: (1) no forces acting between the components (i.e., $f_1 = f_2 = 0$); (2) large forces suppressing relative motion (i.e., homogeneous flow).

Making assumption 1, we use Eq. (3.65) and first assume negligible density change for the air (this can be taken care of later if necessary). Integrating Eq. (3.65), we get

$$\Delta p = \frac{\rho_1 v_1^2}{2} = \frac{\rho_2 v_2^2}{2} \tag{3.67}$$

Let component 2 be the air and let its volumetric concentration in the exit of the nozzle be α. In the usual way we have

$$v_1 = \frac{j_1}{1 - \alpha} \tag{3.68a}$$

$$v_2 = \frac{j_2}{\alpha} \tag{3.68b}$$

Substituting Eqs. (3.68a) and (3.68b) into Eq. (3.67) gives

$$\frac{\rho_1 j_1^2}{(1 - \alpha)^2} = \frac{\rho_2 j_2^2}{\alpha^2} \tag{3.69}$$

whence

$$\frac{1 - \alpha}{\alpha} = \frac{j_1}{j_2}\left(\frac{\rho_1}{\rho_2}\right)^{1/2} \tag{3.70}$$

Therefore

$$\alpha = \left(1 + \frac{j_1}{j_2}\sqrt{\frac{\rho_1}{\rho_2}}\right)^{-1} \tag{3.71}$$

For this particular problem

$j_1 = 27$ in./sec $\quad j_2 = 2000$ in./sec
$\rho_1 = 62.5$ lb/ft^3 $\quad \rho_2 = 0.076$ lb/ft^3

Therefore

$$\alpha = \frac{1}{1 + {}^{27}\!/_{2000}(62.5/0.076)^{1/2}} = \frac{1}{1.388} = 0.72$$

and $1 - \alpha = 0.28$.

Substituting these values in Eqs. (3.68*a*) and (3.68*b*) we get

$$v_1 = \frac{27}{0.28} = 96.5 \text{ in./sec} \qquad v_2 = \frac{2000}{0.72} = 2780 \text{ in./sec}$$

The pressure drop, from Eq. (3.67), is then

$$\Delta p = \frac{62.5}{(32.2)(2)} \left(\frac{96.5}{12}\right)^2 \frac{1}{144} = 0.455 \text{ psi},$$

or, alternatively (as a check),

$$\frac{0.076}{(2)(32.2)} \left(\frac{2780}{144}\right)^2 = 0.439 \text{ psi}$$

Using assumption 2 we have $v_2 = v_1 = j = 2027$ in./sec. Since there exist forces between the components in this case, we must use the momentum equation for both components taken together, i.e., Eq. (3.51), which, for the homogeneous flow case with no area change, reduces to

$$G \frac{dv}{dz} = -\frac{dp}{dz} \tag{3.72}$$

or, using Eq. (2.32),

$$\rho_m v \frac{dv}{dz} = -\frac{dp}{dz} \tag{3.73}$$

Integrating, and assuming incompressible flow for the moment, we get

$$\frac{1}{2} \rho_m v^2 = \Delta p \tag{3.74}$$

which is simply Bernoulli's equation applied to the homogeneous mixture. Expressing Eq. (3.74) in terms of j and G we find

$$\Delta p = \frac{jG}{2} = \frac{(2027)[(2000)(0.076) + (27)(62.5)]}{(32.2)(144)^2(2)}$$
$$= 2.8 \text{ psi}$$

Note the large difference between the answers which are predicted by the two methods. Since it is possible to achieve almost any value of pressure drop in this range by suitable design of the way in which the fluids mix, it is evident that there is often great latitude for design of two-phase flow apparatus. For instance, if one were interested in achieving minimum pressure drop, for economic reasons, it would be advantageous to allow the air and water to separate in the tank and ensure that the interface was on the level of the nozzle so that a stratified flow could take place with little friction between the components. On the other hand, if a large pressure drop is desired (for instance, to restrict leakage from the container) it is better to mix the components thoroughly and form a mixture which would be essentially homogeneous. It is unlikely that bubbly flow could be stable at such high void fractions. A droplet dispersion would be needed. The air velocities derived in this example are actually barely sufficient to produce effective atomization.

Example 3.9 Show how to analyze the characteristics of an isentropic, adiabatic convergent-divergent nozzle which is to carry a stratified flow of air and water. Neglect friction, phase change, and droplet entrainment. In particular, show how the maximum possible air flow rate is related to the rate of water discharge for given upstream conditions and a fixed throat geometry.

Solution The maximum rate of discharge will be governed by compound choking at the throat.

Let the upstream stagnation conditions be p_{0f} for the liquid and p_{0g}, T_{0g} for the gas.

Since the water is incompressible, it will obey Bernoulli's equation and its velocity at the place where the pressure is p will be

$$v_f = \left[\frac{2(p_{0f} - p)}{\rho_f}\right]^{1/2} \tag{3.75}$$

The corresponding flow rate per unit area will be

$$\frac{W_f}{A_f} = [2(p_{0f} - p)\rho_f]^{1/2} \tag{3.76}$$

Differentiating Eq. (3.76) we obtain, since W_f is constant,

$$\frac{1}{A_f}\frac{dA_f}{dz} = \left[2\left(\frac{p_{0f}}{p} - 1\right)\right]^{-1}\frac{1}{p}\frac{dp}{dz} \tag{3.77}$$

The gas will expand isentropically in a nozzle which has an area variation determined by the total area minus the liquid area. In terms of the local pressure this area is given by the well-known equation[8]

$$\frac{A_g}{W_g} = \frac{\sqrt{T_{0g}}}{p_{0g}}\left(\frac{p_{0g}}{p}\right)^{1/\gamma}\left\{\frac{2\gamma}{R_g(\gamma - 1)}\left[1 - \left(\frac{p}{p_{0g}}\right)^{(\gamma-1)/\gamma}\right]\right\}^{-1/2} \tag{3.78}$$

where γ is the isentropic exponent for the gas.

In principle, Eqs. (3.76) and (3.78) contain all that is necessary to predict the relationships between flow rates and overall area for the passage. At the throat, under choked conditions, the gas is supersonic and the liquid subsonic. The overall qualitative appearance of the area variation with length is shown in Fig. 3.14. Note that the stagnation pressures of the two streams are not necessarily identical, although the local pressures are assumed to be the same.

The equation corresponding to Eq. (3.77) for the gas is, from Eq. (3.59*a*),

$$\frac{1}{A_g}\frac{dA_g}{dz} = \frac{1}{A}\frac{dA}{dz} + \frac{1}{\alpha}\frac{d\alpha}{dz} = \frac{1}{\rho_g v_g^2}(1 - M_g^2)\frac{dp}{dz} \tag{3.79}$$

Assuming air to be a perfect gas we can replace $\rho_g v_g^2$ by $\gamma p M_g^2$ in the usual way to get

$$\frac{1}{A_g}\frac{dA_g}{dz} = \left(\frac{1}{M_g^2} - 1\right)\frac{1}{\gamma p}\frac{dp}{dz} \tag{3.80}$$

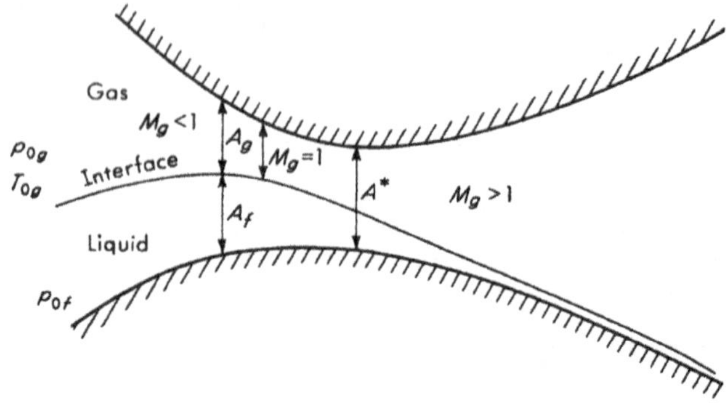

Fig. 3.14 Separated gas-liquid flow through a converging-diverging nozzle under choked conditions.

Now, at the throat dA/dz is zero. Therefore

$$\frac{dA_f}{dz} + \frac{dA_g}{dz} = 0 \tag{3.81}$$

From Eqs. (3.77) and (3.80), therefore

$$A_g\left(\frac{1}{M_g^2} - 1\right) + \frac{\gamma A_f}{2(p_{0f}/p - 1)} = 0 \tag{3.82}$$

which is another form of Eq. (3.61).

Moreover, if A^* is the throat area we also have

$$A_f + A_g = A^* \tag{3.83}$$

Eliminating A_g from Eqs. (3.82) and (3.83) gives

$$A_f = A^*\left[1 + \frac{\gamma}{2(1 - 1/M_g^2)(p_{0f}/p - 1)}\right]^{-1} \tag{3.84}$$

In addition we have, from Eq. (3.76),

$$A_f = W_f[2\rho_f(p_{0f} - p)]^{-1/2} \tag{3.85}$$

Equations (3.84) and (3.85) can now be solved simultaneously to obtain the pressure at the throat and the area of the liquid stream. The other variables can then be deduced by substitution in the appropriate equations.

A method for visualizing the result graphically is shown in Fig. 3.15. Equations (3.84) and (3.85) are plotted to give A_f as a function of p. Equation (3.84) is independent of the liquid flow rate. M_g is only a function of the pressure ratio for the gas stream, as follows:

$$M_g^2 = \frac{2}{\gamma - 1}\left[\left(\frac{p_{0g}}{p}\right)^{(\gamma-1)/\gamma} - 1\right] \tag{3.86}$$

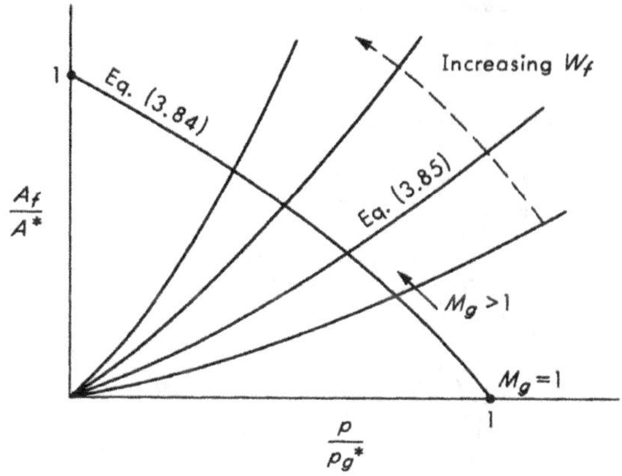

Fig. 3.15 Simultaneous solution of Eqs. (3.84) and (3.85) of Example 3.9.

A_f is equal to zero in Eq. (3.84) where $M_g^2 = 1$, and this corresponds to choking of a flow of gas alone. As M_g is increased above unity, p decreases and A_f increases as shown in the figure.

Equation (3.85), on the other hand, shows A_f increasing with p. The intersection then moves to the left as W_f is increased. The limit occurs when liquid fills the duct entirely, the throat pressure is zero, and there is no gas flow.

Knowing p and A_g from Fig. 3.15, we can calculate the gas flow rate from Eq. (3.78).

The simplest procedure for deriving the locus of maximum flow rates is to use the throat pressure as a parameter. First calculate A_f from Eq. (3.84). Then use Eq. (3.85) to get W_f. Use Eq. (3.83) to find A_g and Eq. (3.78) to compute W_g.

The allowable flow rates will then lie to the left of the lines shown in Fig. 3.16, where the boundary is calculated as indicated above.

The method of solution can be compared with similar methods of treating separated compound gas flows.[9]

3.6 USE OF THE CONCEPT OF ENTROPY GENERATION TO EVALUATE THE COEFFICIENT η

Up to now we have deliberately left the quantity η, which apportions the effects of phase change between the components, as a variable which can be selected according to the conditions of interest. A particularly intriguing development is based on the requirement that the flow should be isentropic.[10] We shall see that this enables us to assign a definite value to η. For convenience in description we shall refer to the phases as "liquid" and "vapor" when it is necessary to give physical significance to the results.

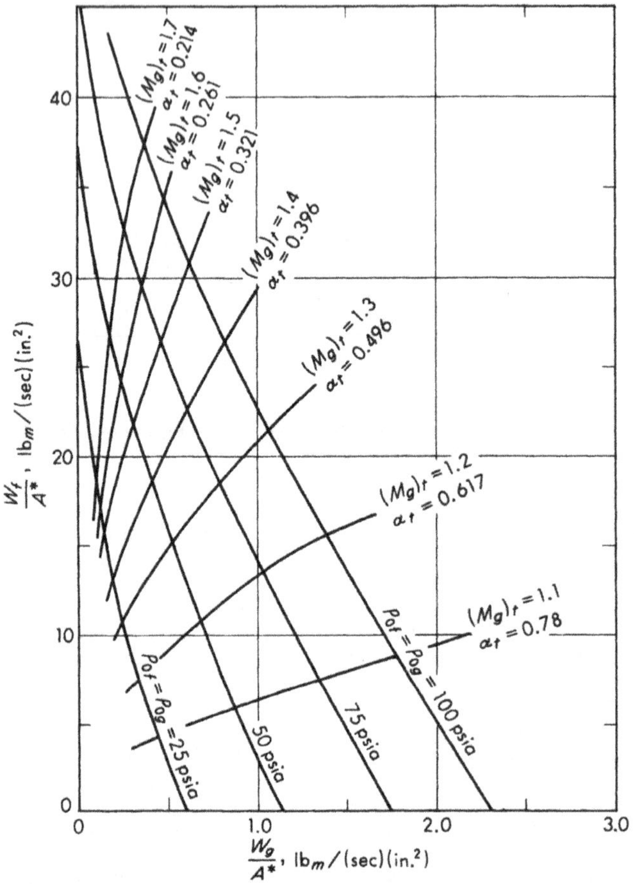

Fig. 3.16 Relationship between air and water flow rates for stratified critical flow with equal stagnation pressures ($p_{0f} = p_{0g}$). (*D. A. Sullivan, Dartmouth College.*) $k = 1.4$, $R_g = 53.3$ (ft)(lb)$_f$/(lb$_m$(°R)), $T_0 = 530$°R, $\rho_f = 62.4$ lb$_m$/ft^3. $(M_g)_t$, α_t, and A^* are the values of the gas-phase Mach number, void fraction, and total area at the throat under conditions of compound choking.

First, write the energy equation for the two phases taken together, neglecting any shaft or shear work, as follows:

$$\frac{1}{W}\frac{dq_e}{dz} = \frac{d}{dz}\left[xh_2 + (1-x)h_1 + \tfrac{1}{2}xv_2^2 + \tfrac{1}{2}(1-x)v_1^2\right] + g\cos\theta \tag{3.87}$$

If S_s denotes a source of entropy per unit length caused by irreversible processes and heat transfer, an entropy balance for steady one-dimensional flow gives

$$S_s = W \frac{d}{dz} [xs_2 + (1 - x)s_1] \tag{3.88}$$

Equation (3.51) is rewritten by using the definition of x to yield

$$G \frac{d}{dz} [xv_2 + (1 - x)v_1] = -\frac{dp}{dz} - g \cos \theta \, [\alpha\rho_2 + (1 - \alpha)\rho_1] - (F_{w1} + F_{w2}) \tag{3.89}$$

Now, from thermodynamic theory it is known that, for a homogeneous pure substance in equilibrium,

$$T\, ds = dh - \frac{1}{\rho} dp \tag{3.90}$$

For a two-phase mixture, moreover,

$$T\, d[xs_2 + (1 - x)s_1] = d[xh_2 + (1 - x)h_1] - \left(\frac{x}{\rho_2} + \frac{1 - x}{\rho_1}\right) dp \tag{3.91}$$

Although Eq. (3.91) is usually deduced for the particular case of a homogeneous mixture, it is an algebraic relationship between thermodynamic properties which is valid for any change of state for which the phases remain in equilibrium. Therefore, we may divide Eq. (3.91) by dz, substitute from Eqs. (3.87), (3.88), and (3.89), and make use of the relationships between x and α to eventually obtain

$$\begin{aligned} TS_s = \frac{dq_e}{dz} + \alpha(1 - \alpha)(v_2 - v_1)A \Bigg\{ g \cos \theta(\rho_1 - \rho_2) \\ + \frac{1}{2}\left[G(v_2 - v_1)\left(\frac{1}{1 - \alpha} - \frac{1}{\alpha}\right)\frac{dx}{dz} - \rho_2 \frac{dv_2^2}{dz} + \rho_1 \frac{dv_1^2}{dz}\right]\Bigg\} \\ + GA(F_{w1} + F_{w2})\left(\frac{x}{\rho_2} + \frac{1 - x}{\rho_1}\right) \end{aligned} \tag{3.92}$$

Furthermore, the final term in Eq. (3.92) can be rearranged as follows:

$$\begin{aligned} GA(F_{w1} + F_{w2})\left(\frac{x}{\rho_2} + \frac{1 - x}{\rho_1}\right) &= A[\alpha v_2 + (1 - \alpha)v_1](F_{w1} + F_{w2}) \\ &= [v_1 + \alpha(v_2 - v_1)]AF_{w1} + [v_2 - (1 - \alpha)(v_2 - v_1)]AF_{w2} \\ &= v_1 AF_{w1} + v_2 AF_{w2} + A\alpha\,(1 - \alpha)(v_2 - v_1)\left(\frac{F_{w1}}{1 - \alpha} - \frac{F_{w2}}{\alpha}\right) \end{aligned} \tag{3.93}$$

Combining Eqs. (3.92) and (3.93) yields

$$TS_s = \frac{dq_e}{dz} + v_1AF_{w1} + v_2AF_{w2}$$
$$+ \alpha(1-\alpha)(v_2 - v_1)A\left\{g \cos\theta(\rho_1 - \rho_2) + \frac{F_{w1}}{1-\alpha} - \frac{F_{w2}}{\alpha}\right.$$
$$\left. + \frac{1}{2}\left[G(v_2 - v_1)\left(\frac{1}{1-\alpha} - \frac{1}{\alpha}\right)\frac{dx}{dz} - \rho_2\frac{dv_2^2}{dz} + \rho_1\frac{dv_1^2}{dz}\right]\right\} \quad (3.94)$$

Equation (3.94) expresses four mechanisms of entropy generation:

1. Heat transfer
2. Wall shear on the liquid
3. Wall shear on the gas
4. Relative motion

The entropy generation due to relative motion is a new phenomenon to those who are used to working with single-phase flows and is often overlooked. It is represented by the final term in Eq. (3.94) and is zero only if one of the factors in that term is zero. Taking the factors in order, we have the following conditions for zero entropy generation due to relative motion:

$\alpha = 0$	i.e., single-phase liquid
$(1-\alpha) = 0$	i.e., single-phase gas
$v_2 = v_1$	i.e., homogeneous flow

The only nontrivial result is found if the curly bracket is equated to zero and rearranged to give

$$\rho_1 v_1 \frac{dv_1}{dz} + g\rho_1 \cos\theta + \frac{1}{2}\frac{G(v_2 - v_1)}{1-\alpha}\frac{dx}{dz} + \frac{F_{w1}}{1-\alpha}$$
$$= \rho_2 v_2 \frac{dv_2}{dz} + g\rho_2 \cos\theta + \frac{1}{2}\frac{G(v_2 - v_1)}{\alpha}\frac{dx}{dz} + \frac{F_{w2}}{\alpha} \quad (3.95)$$

To discover the physical significance of Eq. (3.95), we let both sides equal λ, take $(1-\alpha)$ times the left-hand side and add it to α times the right-hand side. The result is

$$\lambda = (1-\alpha)\rho_1 v_1 \frac{dv_1}{dz} + \alpha\rho_2 v_2 \frac{dv_2}{dz} + (v_2 - v_1)G\frac{dx}{dz}$$
$$+ g\cos\theta\,[(1-\alpha)\rho_1 + \alpha\rho_2] + F_{w1} + F_{w2} \quad (3.96)$$

Comparison with Eq. (3.50) reveals the simple result that

$$\lambda = -\frac{dp}{dz} \quad (3.97)$$

Equation (3.95) is therefore equivalent to the two equations, which are equations of motion for each phase, in the absence of interfacial shear, as follows:

$$\rho_1 v_1 \frac{dv_1}{dz} = -\frac{dp}{dz} - \rho_1 g \cos\theta - \frac{F_{w1}}{1-\alpha} - \frac{1}{2}\frac{G(v_2 - v_1)}{1-\alpha}\frac{dx}{dz} \tag{3.98}$$

$$\rho_2 v_2 \frac{dv_2}{dz} = -\frac{dp}{dz} - \rho_2 g \cos\theta - \frac{F_{w2}}{\alpha} - \frac{1}{2}\frac{G(v_2 - v_1)}{\alpha}\frac{dx}{dz} \tag{3.99}$$

Thus it is seen that, in order to conserve entropy, the force represented by Eq. (3.47) has to be shared equally by the components, each of which experiences a reaction $-\frac{1}{2}G(v_2 - v_1)(dx/dz)$ per unit volume of the duct. Since the terms in the equations of motion represent forces per unit volume of the specific component, this force must be divided by the volumetric concentration of each, as appropriate.

For isentropic flow, therefore, the value of η should be chosen to be $\frac{1}{2}$. The above equations are equally valid for evaporation or condensation and are perfectly symmetrical. At first sight it may appear very strange that the liquid (or solid) and the vapor should share the reaction which is due to vaporization. Most authors are wont to attribute all of this reaction to the vapor stream during evaporation and to the liquid stream during condensation. The explanation, however, should be sought in the two-dimensional effects which have been obscured by the one-dimensional idealizations. In fact, the local velocities of the liquid and vapor streams tangential to the interface will be equal and there will be velocity variations over the flow field. If solutions could be obtained to the detailed three-dimensional flow problem, this would give a more complete representation of reversible evaporation and condensation.

The conclusion that the force represented by Eq. (3.47) should be shared equally between the components could also have been derived from consideration of reversibility. Unless the force is shared equally, the equations differ in the cases of evaporation and condensation and the fluid cannot be caused to return to its initial state by reversing either process in a symmetrical way.

Colloquially this result could be expressed as follows: "If you are going to use a one-dimensional flow model, and you are going to require that there be no entropy generation, then you must include the final terms in Eqs. (3.98) and (3.99) in order to keep all the relevant equations consistent."

Now, suppose that there is irreversible momentum transfer taking place between the vapor and liquid streams. This could be imagined to resemble the process proposed by Osborne Reynolds[17] to describe turbulent transfer phenomena involving M units of mass per unit volume per unit time, which alternately contact the vapor and liquid and share momentum with each. The resulting force on the liquid will be

$$F_{12} = (v_2 - v_1)M \tag{3.100}$$

in the direction of motion. The corresponding reaction on the gas is

$$-F_{12} = -(v_2 - v_1)M \tag{3.101}$$

The total force on the liquid per unit volume of the flow field, resulting from the relative motion, assuming that the two processes of reversible and irreversible momentum transfer can be superposed linearly, is

$$(v_2 - v_1)\left(M - \tfrac{1}{2}G\frac{dx}{dz}\right)$$

and on the gas, similarly,

$$(v_2 - v_1)\left(-M - \tfrac{1}{2}G\frac{dx}{dz}\right)$$

An alternative description can be developed in terms of mass and momentum fluxes at the two-phase interface. If the interfacial perimeter is P_i in a pipe of area A, then the surface area per unit volume is P_i/A. If ϵ_0 is defined to be the irreversible *Reynolds flux*† per unit area of the interface, then we have

$$\epsilon_0 = \frac{MA}{P_i} \tag{3.102}$$

The rate of vaporization per unit area of the interface is

$$m = \frac{GA}{P_i}\frac{dx}{dz} = m_{12} = -m_{21} \tag{3.103}$$

The net forces on the liquid and vapor per unit volume can then be rewritten as

$$F_1 = (v_2 - v_1)\frac{P_i}{A}\left(\epsilon_0 - \frac{m}{2}\right) \tag{3.104}$$

$$F_2 = -(v_2 - v_1)\frac{P_i}{A}\left(\epsilon_0 + \frac{m}{2}\right) \tag{3.105}$$

Dividing Eqs. (3.104) and (3.105) by $1 - \alpha$ and α respectively, to obtain the force per unit volume of each component, and incorporating

† The concept of *Reynolds flux* was first proposed by Osborne Reynolds[17] in 1874. It has been used by Nusselt,[18] Silver,[19–21] and extensively by Spalding[11] to analyze problems of combustion and simultaneous heat and mass transfer. Applications to two-phase flows are comparatively recent.[10,22–24] The subject is sadly neglected in many textbooks.

the result in the equations of motion (3.98) and (3.99), we obtain

$$\rho_1 v_1 \frac{dv_1}{dz} = -\frac{dp}{dz} - \rho_1 g \cos\theta - \frac{F_{w1}}{1-\alpha} - \frac{P_i}{A(1-\alpha)}(v_1 - v_2)\left(\epsilon_0 - \frac{m}{2}\right) \tag{3.106}$$

$$\rho_2 v_2 \frac{dv_2}{dz} = -\frac{dp}{dz} - \rho_2 g \cos\theta - \frac{F_{w2}}{\alpha} - \frac{P_i}{A\alpha}(v_2 - v_1)\left(\epsilon_0 + \frac{m}{2}\right) \tag{3.107}$$

These equations are quite symmetrical and are valid for phase change in either direction, depending on the sign of m. The last term in each could be regarded as representing a modified interfacial shear stress (and momentum transfer) as a result of phase changes. An alternative derivation of this same result will be given in Chap. 7.

The interpretation of these results is interesting from the point of view of irreversible thermodynamics. The term involving m involves orderly mass transfer and does not contribute to the entropy production. The Reynolds flux ϵ_0, however, is a measure of inherent disorderly mass exchange, irreversibility, and entropy production. Use of Eqs. (3.106) and (3.107) in Eq. (3.94), for example, yields the result

$$TS_s = \frac{dq_e}{dz} + A(v_1 F_{w1} + v_2 F_{w2}) + (v_2 - v_1)^2 \epsilon_0 P_i \tag{3.108}$$

Note that the terms involving wall shear stress in Eq. (3.108) are not necessarily positive. In vertical slug flow, for example, the second term can be negative while the third term is zero. The second law of thermodynamics, however, requires that the final term should be large enough to ensure that the net dissipation is positive.

It is likely that a similar analysis could be performed to account for entropy production due to energy and mass transfer between the phases across finite differences of temperature and Planck potential. Such a treatment would extend the methods of Spalding[11] to two-phase flow and could provide a very powerful technique for dealing with multicomponent chemical reaction, combustion, and other practically important problems. The Reynolds flux concept would provide the key to a unified treatment.

The last term in Eq. (3.108) could be regarded as a force (per unit length), $(v_2 - v_1)\epsilon_0 P_i$, times a finite velocity difference, $v_2 - v_1$, which agrees with the usual ideas of entropy production in linear systems.

Since entropy production due to interphase friction (in a flow with initially equal velocities) is zero in the extremes of infinite ϵ_0 (homogeneous flow) and zero ϵ_0, it will evidently be a maximum somewhere in between. For example, in the case of the expansion from stagnation conditions of two incompressible fluids with no wall shear stress and negligible phase change or gravitational effects, entropy production is zero when either $v_1/v_2 = 1$ or $v_1/v_2 = (\rho_2/\rho_1)^{1/2}$, from Eq. (3.66).

3.7 ENERGY EQUATIONS

General one-dimensional energy conservation equations can be written for the two phases by considering a control volume of length dz in the duct. For component 2, for example, we have

$$\frac{\partial}{\partial t}\left[A\alpha\rho_2\left(e + \frac{v_2^2}{2}\right)\right] + \frac{\partial}{\partial z}\left[A\alpha\rho_2 v_2\left(h + \frac{v_2^2}{2}\right)\right] = (q_e + q_m - w_s - w_\tau)A - A\rho_2\alpha v_2 g \cos\theta \quad (3.109)$$

The symbols q_e, q_m, $-w_s$, and $-w_\tau$ represent average rates of energy addition per unit volume of the duct due to heat transfer, mass transfer, shaft work, and shear work. Since these processes may all occur both between the components and between each component and the duct wall, the general formulation becomes extremely cumbersome. Since this text is concerned primarily with fluid mechanics, rather than with heat and mass transfer, we shall avoid a general discussion and development of the energy equations. Simplified versions will be introduced when they are necessary and appropriate.

Example 3.10 What is the one-dimensional equation governing internal energy changes for steady flow of particles in a gas stream? Neglect mass transfer, heat transfer, and friction with the walls and shaft work.

Solution Let the particles be component 2. Since the flow is steady and there is no mass transfer

$$A\rho_2\alpha v_2 = W_2 = \text{const} \quad (3.110)$$

Equation (3.109) then reduces to

$$\rho_2\alpha v_2\left[\frac{d}{dz}\left(h_2 + \frac{v_2^2}{2}\right) + g\cos\theta\right] = q_e - w_\tau \quad (3.111)$$

Since the particles are incompressible and gravity is the only body force, Eq. (3.46) can be put into the form

$$\rho_2\left[\frac{d}{dz}\frac{v_2^2}{2} + g\cos\theta + \frac{d}{dz}(p v_2)\right] = f_2 \quad (3.112)$$

Combining Eqs. (3.111) and (3.112) and using the thermodynamic relation $h_2 = e_2 + p v_2$, we get

$$\rho_2\alpha v_2\left(\frac{de_2}{dz} + \frac{f_2}{\rho_2}\right) = q_e - w_\tau \quad (3.113)$$

or

$$\frac{de_2}{dz} = \frac{q_e - w_\tau - F_2 v_2}{\rho_2\alpha v_2} \quad (3.114)$$

in view of Eq. (3.43).

If the mechanical work terms, w_τ and F_2v_2, are small compared with the heat-transfer term q_e, there is no difficulty in evaluating de_2/dz as long as q_e can be determined from heat-transfer theory. In particular, if the heat-transfer coefficient between the particles and the fluid is h, the temperatures are T_1 and T_2, and the particles are spherical with diameter d and a constant specific heat c_2, we have

$$q_e = h(T_1 - T_2)\pi d^2 \frac{\alpha}{\frac{1}{6}\pi d^3} \tag{3.115}$$

and, therefore

$$\frac{dT_2}{dz} = \frac{6h(T_1 - T_2)}{c_2\rho_2 v_2 d} \tag{3.116}$$

In general, w_τ and F_2v_2 are not necessarily small, nor are they equal and opposite, as the following example shows, and the proper evaluation of the energy relationships can be quite tricky.

Example 3.11 Consider one-dimensional steady flow with no mass transfer. Assume no shaft work or heat transfer, but let there be a force F_{12} acting between the components per unit volume of the flow field (conventionally acting on component 1 in the direction of its motion). Discuss the energy equations for the components, neglecting gravity.

Solution The only term surviving on the right-hand side of Eq. (3.109) and its equivalent for the other phase is the shear-work term and we obtain, under the specified conditions, apparently

$$\rho_1(1 - \alpha)v_1 \frac{d}{dz}\left(h_1 + \frac{v_1^2}{2}\right) = -w_{\tau 1} \tag{3.117}$$

$$\rho_2\alpha v_2 \frac{d}{dz}\left(h_2 + \frac{v_2^2}{2}\right) = w_{\tau 2} \tag{3.118}$$

Now, the rate at which component 2 is doing work is $F_{12}v_2$. However, component 1 only receives shear work amounting to $F_{12}v_1$. Therefore, due to the relative motion there is a "dissipation" of work, per unit volume, equal to

$$\dot{\mathfrak{D}} = F_{12}(v_2 - v_1) \tag{3.119}$$

This dissipation is equal to a source of entropy times the temperature and appears as "heat," but there is no way of telling, at this level of sophistication, how to apportion this energy between Eqs. (3.117) and (3.118) in order to balance accounts.

One way around this difficulty is to increase the level of sophistication and to consider a differential analysis which does not allow discontinuities in velocity. This leads to a study of three-dimensional temperature and velocity variations, "recovery factors," and other considerations which are beyond the scope of a one-dimensional analysis.

The simpler solution is to apportion the dissipation between the components as if it were a heat source. Thus we write

$$\rho_1(1 - \alpha)v_1 \frac{d}{dz}\left(h_1 + \frac{v_1^2}{2}\right) = F_{12}v_1 + \zeta F_{12}(v_2 - v_1) \tag{3.120}$$

$$\rho_2\alpha v_2 \frac{d}{dz}\left(h_2 + \frac{v_2^2}{2}\right) = -F_{12}v_2 + (1 - \zeta)F_{12}(v_2 - v_1) \tag{3.121}$$

ζ is merely a convenient parameter which fulfills much the same general function as η in Eqs. (3.48) and (3.49).

Note that the right-hand side of both Eqs. (3.120) and (3.121) could be written as a force times a weighted mean velocity, $\zeta v_2 + (1 - \zeta)v_1$, which could be regarded as an effective interface velocity.

In cases where heat, mass, and momentum transfer occur simultaneously, a rigorous treatment of the energy exchange processes becomes very difficult. A convenient concept for simplification purposes is the Reynolds flux which will be discussed in Chap. 7.

PROBLEMS

3.1. Use a one-dimensional flow control volume approach to derive the momentum equations for component 2 in the form

$$\frac{\partial}{\partial t}(v_2\alpha\rho_2 A) + \frac{\partial}{\partial z}(\rho_2\alpha v_2^2 A) = AF_2' + \alpha A b_2$$

where F_2' is the total resultant surface force per unit volume of the duct acting on component 2.

Combine this equation with the continuity equation in order to get Eq. (3.46). How do you account for the emergence of an apparent component of force due to phase change?

3.2. Prove that, for the separate cylinders model in laminar flow, $n = 2$ in Eq. (3.30).

3.3. Prove that, for the separate cylinders model in turbulent flow, $n = 2.5$ if the friction factor is assumed to be constant, whereas $n = 2.375$ if the Blasius equation is valid.

3.4. Solve Prob. 2.10 using separated-flow theory.

3.5. Consider homogeneous flow theory with a constant friction factor. Rearrange the frictional pressure-drop prediction in terms of the Martinelli parameters. For various values of the density ratio, plot ϕ_g versus X and compare with Fig. 3.4.

3.6. Solve Prob. 3.5 using other assumptions about the friction factor in both laminar and turbulent flow.

3.7. Using the results of Example 3.6 and Eqs. (3.48) and (3.49), show how to extend Martinelli's correlation to vertical flow when inertia and phase-change effects are negligible. What will the curves of constant α or constant dp/dz look like on a graph of $(dp/dz)_g$ versus $(dp/dz)_f$?

3.8. In a straight pipe, instead of using the energy equation in Example 3.1, it is possible to start from the momentum equation. If the total wall friction force divided by the pipe cross-sectional area is F, show that, if inlet momentum is ignored,

$$\Delta p - F = G[xv_g + (1 - x)v_f]$$

Using the continuity equation, deduce that G is a maximum at a given value of F if $v_g/v_f = (\rho_f/\rho_g)^{1/2}$. Furthermore, show that the greatest maximum value of G occurs when $F = 0$ and the flow is isentropic. What is this absolute maximum value of G in terms of Δp and x?

Compare the predicted values for this maximum with the predictions of Example 3.1 for specific cases of the discharge of initially saturated water. The results should not be very different.

This model of minimum momentum flux was initially proposed by Fauske.[4]

3.9. Rather than maximizing the discharge in terms of the slip ratio for given momentum or energy fluxes, it is perhaps more reasonable to solve the energy and momentum equations simultaneously. Do this by plotting the mass flux G versus v_g/v_f for both the energy and momentum methods (Example 3.1 and Prob. 3.8). For a typical case (e.g., saturated water at 2000 psia expanded to 1200 psia) does it make much difference whether one uses the intersection of the curves or their maxima to make predictions?

3.10 What is the "best" value of n as a function of pressure in Eq. (3.30) in order to fit the empirical curves in Fig. 3.17. Express the result as a correlation of n versus the reduced pressure (pressure divided by critical pressure), a representation which might justify extrapolation to fluids other than water.

3.11. A horizontal pipeline, 7.75 in. ID, 11,317 ft long, with an inlet pressure of 1022 psia, was passing 5484 barrels per day of oil and 50,000 lb/hr of gas. The line temperature was 80°F. The density and viscosity of the oil were 6.499 lb/gal and 0.574 cp, respectively. The density and viscosity of the gas were 3.48 lb/ft^3 and 0.014 cp, respectively. Determine the two-phase pressure drop in this line. The measured value was 32 psi.

3.12. (*a*) Estimate the critical flow rate for a steam-water mixture, quality 26.9% by weight, issuing from a horizontal pipe with 0.269 in. ID at a pressure of 125 psia. Compare the value calculated with that determined by Fauske[12] [1265.4 lb/(ft^2)(sec)].

(*b*) Calculate the pressure distribution over the last 4 ft of the tube for a flow of 1265.4 lb/(ft^2)(sec) and compare this with that measured by Fauske[12] and shown in Fig. 3.18. The inlet pressure is 350 psi and the inlet quality 19.04% by weight. Assume a fluid enthalpy decrease due to heat losses of 2.5 Btu/(lb)(ft) run of tube. Use the Baroczy correlation for the frictional pressure-drop component.

3.13. Solve Prob. 2.29 using the Martinelli correlations.

3.14. Solve Prob. 2.20 using the separated-flow model and suitable assumptions.

3.15. Solve Prob. 2.30 using the Baroczy correlations.

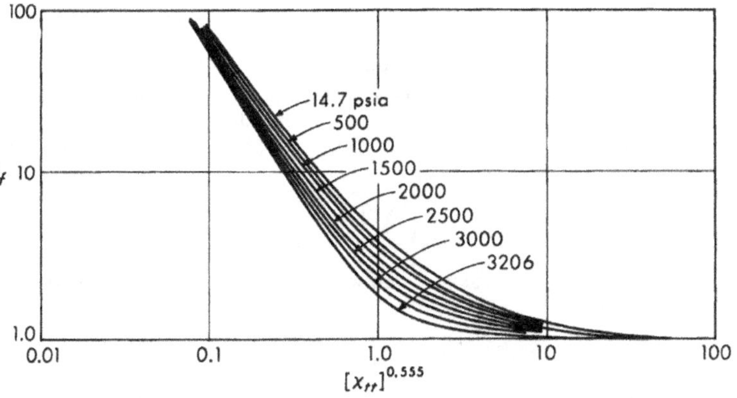

Fig. 3.17 Martinelli and Nelson's empirical dependence of ϕ_f on pressure and X for water.[2]

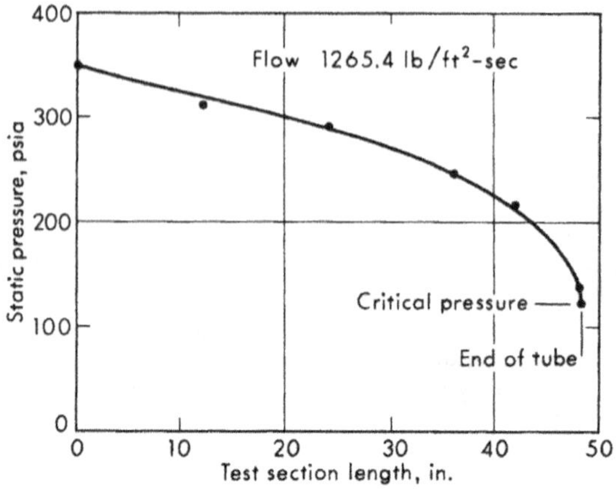

Fig. 3.18 Pressure variation during flashing in a tube. (*Fauske.*[12])

3.16. Consider condensation in a round duct of diameter D. Let there be no direct contact between the vapor and the wall and let the condensate film be so thin that $v_f \ll v_g$ and $\alpha \approx 1$. From Eq. (3.59) show that the pressure gradient is

$$-\frac{dp}{dz} = \frac{\rho_g g \cos\theta + \rho_g v_g^2 \left[-\frac{1}{A}\frac{dA}{dz} + (1+\eta)\frac{1}{x}\frac{dx}{dz} + \frac{2C_{fi}}{D} \right]}{1 - M_g^2}$$

where C_{fi} is a suitably defined interfacial friction factor.

Evaluate the pressure gradient for steam at 90°F traveling upward at 500 fps in a 1-in.-diam straight vertical pipe if $C_{fi} = 0.005$ and the condensing rate is 10 lb/(hr)(ft²). Consider the two limiting cases $\eta = 0$ and $\eta = \frac{1}{2}$.

Suppose that the pipe is tapered so that $dD/dz = 0.05$; what is the pressure gradient?

Solve the same problem for steam velocities of 1000 and 1500 fps.

3.17. Consider uniform condensation in a straight horizontal tube at a low value of Mach number. Assume constant properties, an inlet vapor velocity of V_1 and an outlet velocity of V_2 in a pipe of length L. Using the results of Prob. 3.16 show that the pressure recovery is[23]

$$\frac{p_2 - p_1}{\frac{1}{2}\rho_g V_1^2} = (1+\eta)\left[1 - \left(\frac{V_2}{V_1}\right)^2\right] - \frac{4}{3}C_{fi}\frac{L}{D}\left[1 + \frac{V_2}{V_1} + \left(\frac{V_2}{V_1}\right)^2\right]$$

If the pipe is short and condensation is complete, show that the predicted pressure recovery is

$$p_2 - p_1 = (1+\eta)\frac{1}{2}\rho_g V_1^2$$

By applying Bernoulli's equation to the central streamline for the vapor, show that this result is only valid if the centerline velocity is $(1+\eta)^{1/2}$ times the average. Is this compatible with the usual turbulent flow velocity profiles if $\eta \approx \frac{1}{2}$?

3.18. The final term in Eq. (3.106) could be taken to represent the effect of modified interfacial shear stress as a result of phase change. Show that this shear stress is given by

$$\tau_i = (v_2 - v_1)\left(\epsilon_0 - \frac{m}{2}\right)$$

If evaporation is rapid enough, $m/2$ can be greater than ϵ_0. One may argue that negative shear stress would be unreasonable so that for high rates of vaporization we must have $(\epsilon_0 - m/2) = 0$. At these rapid rates of vaporization, inertia and phase-change terms are usually much larger than the friction and body-force terms. Show that in this case, for gas-liquid flow, Eqs. (3.106) and (3.107) become

$$\rho_f v_f \frac{dv_f}{dz} = -\frac{dp}{dz}$$

$$\rho_g v_g \frac{dv_g}{dz} = -\frac{dp}{dz} - \frac{v_g - v_f}{\alpha} G \frac{dx}{dz}$$

For evaporation in a straight duct (G = const) deduce that, if ρ_f is constant

$$\frac{d}{dz}\left[\frac{\rho_f x^2}{\rho_g \alpha} - \frac{1}{2}\frac{(1-x)^2}{(1-\alpha)^2} + \frac{(1-x)^2}{1-\alpha}\right] = 0 \tag{3.122}$$

This is the result obtained by Levy[13] in his momentum exchange model which is reasonably valid for rapid evaporation in pipes. (It is equivalent to assuming $\eta = 1$.)

For small values of x during boiling or flashing from initially pure liquid, deduce that the slip ratio is approximately

$$\frac{v_g}{v_f} \approx \left(\frac{\alpha}{2}\frac{\rho_f}{\rho_g}\right)^{1/2}$$

Integrate Eq. (3.122) from $x = 0$ and deduce a relationship between x and α at any point in the duct.

3.19. The choking condition in the presence of flashing is given by Eq. (3.63). Assume that Eq. (3.64) is approximately valid, so that the actual value of η is irrelevant. Furthermore, assume $\rho_f \gg \rho_g$ and $c_f^2 \gg c_g^2$ in gas-liquid flow. Show that the mass flux is then given by[10]

$$G^2 = \left(\frac{x^2}{\rho_g^2 \alpha c_g^2} - 2\frac{\partial x}{\partial p}\frac{x}{\alpha \rho_g}\right)^{-1}$$

while α is given by Eq. (3.71).

3.20. From the results of Prob. 3.19 deduce the maximum possible values for the mass flux of water as a function of pressure for x = 0.03, 0.1, 0.5, and 0.8. The results should be virtually indistinguishable from those presented by Levy[14] and shown in Fig. 3.19.

3.21. For expansion from stagnation conditions with constant properties and no gravitational effects, entropy production is zero if $v_1/v_2 = 1$ (homogeneous flow) or $v_1/v_2 = (\rho_2/\rho_1)^{1/2}$ (frictionless separated flow) throughout. Zivi[15] claims that the maximum (actually he called it a minimum) rate of entropy production occurs if $v_1/v_2 = (\rho_2/\rho_1)^{1/3}$ everywhere. Do you agree?

3.22. Show that in separated flow:

(*a*) The entropy generation per unit volume due to interfacial effects alone is

$$S_s = \frac{v_2 - v_1}{T}\left\{F_{12} + G\frac{dx}{dz}\left[(1-\eta)v_1 + \eta v_2 - \frac{v_1 + v_2}{2}\right]\right\}$$

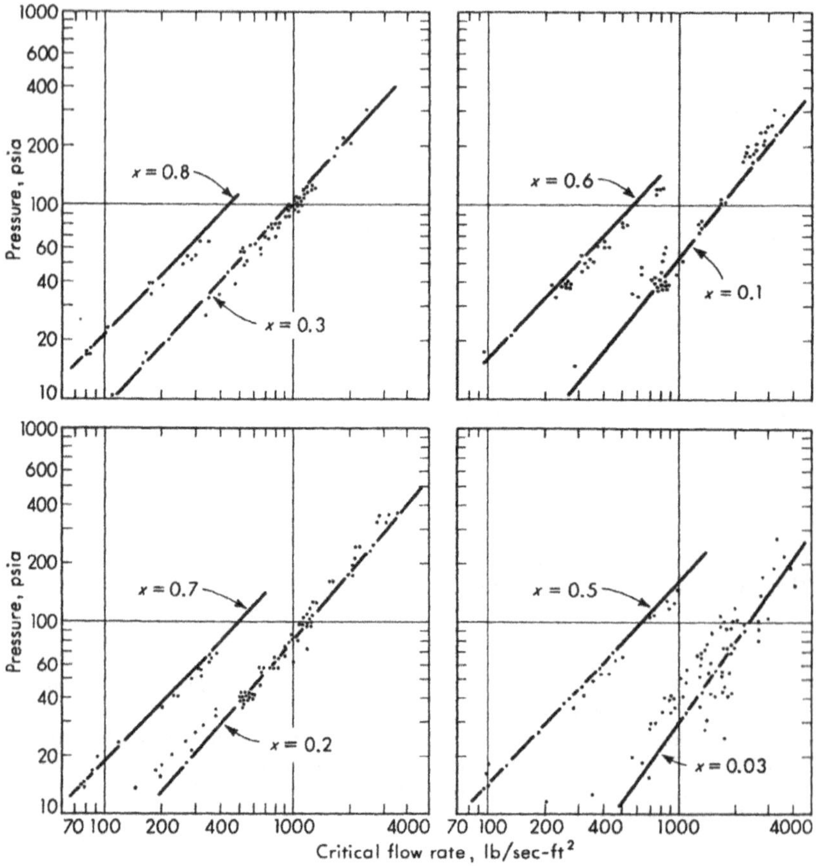

Fig. 3.19 Critical flow rates of steam-water mixtures. Dashes are Levy's predictions.[14] Dots are data points. These results should agree with the solutions to Probs. 3.19 and 3.20.

Deduce that the entropy increase due to phase change depends on the difference between an "interface velocity," $(1 - \eta)v_1 + \eta v_2$, and the arithmetic mean velocity $(v_1 + v_2)/2$.

(*b*) the above equation reduces to

$$S_s = \frac{v_2 - v_1}{T}\left[F_{12} + G\frac{dx}{dz}(v_2 - v_1)\left(\eta - \frac{1}{2}\right)\right]$$

Deduce that the entropy generation due to phase change is zero if $\eta = \frac{1}{2}$ and that for evaporation we must have $\eta \geq \frac{1}{2}$ and for condensation, $\eta \leq \frac{1}{2}$.

3.23. Show that the interfacial forces per unit duct volume in Eqs. (3.48) and (3.49) can be written as

$$F_1 = \left[F_{12} + \left(\eta - \frac{1}{2}\right)(v_2 - v_1)G\frac{dx}{dz}\right] - \frac{1}{2}(v_2 - v_1)G\frac{dx}{dz}$$

$$F_2 = -\left[F_{12} + \left(\eta - \frac{1}{2}\right)(v_2 - v_1)G\frac{dx}{dz}\right] - \frac{1}{2}(v_2 - v_1)G\frac{dx}{dz}$$

Deduce that entropy generation is dependent on the "odd" forces, which act in opposite directions on components 1 and 2 and not on the "even" forces. Show that ϵ_0 is proportional to the odd-force component, whereas m is proportional to the even component. Relate to general ideas of reversibility in thermodynamics.

3.24. How is η related to the quantity ζ in Example 3.10? Are the two interface velocities, $\zeta v_2 + (1 - \zeta)v_1$ and $\eta v_2 + (1 - \eta)v_1$, the same? How are these related to boundary layer or mixing phenomena on each side of the interface?

3.25. A vertical tubular test section is to be installed in an experimental high-pressure water loop. The tube is 0.400-in. ID and 12 ft long heated uniformly over its length. An estimate of the pressure drop across the test section is required as a function of the flow rate of water entering the test section at 400°F and 1000 psia.

(*a*) Calculate the pressure drop over the test section for a water flow of 2 gpm with a power of 100 kw applied to the tube using (*i*) the homogeneous model; (*ii*) the Martinelli-Nelson model; (*iii*) the Thom correlation; and (*iv*) the Baroczy correlation.

(*b*) Estimate the pressure drop versus flow-rate relationship over the range of 2 to 15 gpm for a power of 100 and 200 kw applied to the tube using (*i*) the Martinelli-Nelson correlation; (*ii*) the Baroczy correlation.

3.26. Water and air flow together through a 350-ft section of horizontal 1-in.-ID smooth pipe. The flow rates are 1000 and 15 lb/hr of water and air, respectively. The discharge end of the pipe is at a pressure of 1 atm and the system is isothermal at a temperature of 68°F. Calculate the pressure drop over the pipe work using the Lockhart-Martinelli correlation.

3.27. For separated frictionless incompressible flow of two fluids through a nozzle from a large container, show that

$$\Delta p^{1/2} = \Delta p_1^{1/2} + \Delta p_2^{1/2}$$

where Δp_1 and Δp_2 are the pressure drops for each component flowing alone through the nozzle. Compare with Eq. (3.30) and the Martinelli correlation.

(Murdoch[16] found that $\Delta p^{1/2} = 1.26\Delta p_f^{1/2} + \Delta p_g^{1/2}$ for air-water and steam-water mixtures.)

3.28. The one-dimensional flow model ignores variations across the channel. As a result, errors are introduced due to the various different "averages" which can be derived by integration over the flow cross section. Show how correction factors can be applied to the various equations in the text to allow for these effects. Explore the possibility for improving the theories in this way.

3.29. Solve Prob. 2.9 using (*a*) the minimum kinetic-energy flux theory, from Example 3.1, (*b*) the minimum momentum flux theory from Prob. 3.8, (c) the momentum exchange model from Prob. 3.18, and (*d*) the results of Prob. 3.19.

3.30. Rework Example 3.9 for the case of (*a*) two incompressible liquids, (*b*) two compressible gases, and (*c*) multiphase flow of n compressible fluids.[9]

3.31. Solve Example 3.5 for the case of vertical flow. Show that when gravitational effects are considered, the expressions for f_f and $(f_s)_f$ are unchanged but

$$(f_s)_s = (\rho_s - \rho_f)g - \frac{180\mu_f j_{f0}}{d^2}\frac{1-\epsilon}{\epsilon^3}$$

Show that this result is compatible with the solution to Example 3.6.

REFERENCES

1. Lockhart, R. W., and R. C. Martinelli: *Chem. Eng. Progr.*, vol. 45, p. 39, 1949.
2. Martinelli, R. C., and D. B. Nelson: *Trans. ASME*, vol. 70, p. 695, 1948.
3. Moody, F. J.: *Trans. ASME J. Heat Transfer*, vol. 87, No. 1, pp. 134–142, 1965.
4. Fauske, H. K.: ANL Rept. 6633, 1962.
5. Turner, J. M.: Ph.D. thesis, Dartmouth College, Hanover, N.H., 1966, also AEC Repts., J. M. Turner and G. B. Wallis, Rept. NYO-3114-6, 1965, and G. B. Wallis, Rept. NYO-3114-14, 1966.
6. Thom, J. R. S.: *Intern. J. Heat Mass Transfer*, vol. 7, pp. 709–724, 1964.
7. Baroczy, C. J.: A. I. Ch. E. J. preprint no. 37, *Eighth National Heat Transfer Conference*, Los Angeles, Calif., 1965.
8. Shapiro, A. H.: "The Dynamics and Thermodynamics of Compressible Fluid Flow," The Ronald Press Company, New York, 1953.
9. Bernstein, A., W. H. Heiser, and C. Hevenor: *Trans. ASME J. Appl. Mech.*, vol. 34, pp. 548–554, 1967.
10. Wallis, G. B.: *Intern. J. Heat Mass Transfer*, vol. 11, pp. 445–472, 1968.
11. Spalding, D. B.: "Convective Mass Transfer," McGraw-Hill Book Company, New York, 1963.
12. Fauske, H. K.: *Inst. Mech. Eng. Symp. Two-phase Flow*, paper 10, Feb., 1962.
13. Levy, S.: *Trans. ASME J. Heat Transfer*, vol. 82, pp. 113–124, 1960.
14. Levy, S.: *Trans. ASME J. Heat Transfer*, vol. 87, pp. 53–58, 1965.
15. Zivi, S. M.: *Trans. ASME J. Heat Transfer*, vol. 86, pp. 247–252, 1964.
16. Murdoch, J. W.: *Trans. ASME J. Basic Eng.*, vol. 84, pp. 419–433, 1962.
17. Reynolds, O.: *Proc. Lit. Phil. Soc.*, Manchester, England, vol. 14, pp. 7–12, 1874.
18. Nusselt, W.: *Z. Ver. Deut. Ing.*, vol. 60, pp. 102–107, 1916.
19. Silver, R. S.: *Nature*, vol. 165, p. 725, 1950.
20. Silver, R. S.: *Fuel*, vol. 32, pp. 121–150, 1953.
21. Silver, R. S.: *Proc. 3d Intern. Heat Trans. Conf.*, A. I. Ch. E., New York, 1966.
22. Silver, R. S.: *Proc. Inst. Mech. Engrs.*, vol. 178, pp. 339–376, 1963.
23. Wallis, G. B.: *Proc. Inst. Mech. Engrs.*, vol. 180, pp. 27–35, 1965–1966.
24. Silver, R. S., and G. B. Wallis: *Proc. Inst. Mech. Engrs.*, vol. 180, pp. 36–40, 1965–1966.

4
The Drift-flux Model

4.1 INTRODUCTION

The *drift-flux* model is essentially a separated-flow model in which attention is focused on the relative motion rather than on the motion of the individual phases. Although the theory can be developed in a way which is quite general, it is particularly useful if the relative motion is determined by a few key parameters and is independent of the flow rate of each phase. For example, in bubbly flow at low velocities in large vertical pipes, the relative motion between the bubbles and the liquid is governed by a balance between buoyancy and drag forces; it is a function of the void fraction but not of the flow rate. Drift-flux theory has widespread application to the bubbly, slug, and drop regimes of gas-liquid flow as well as to fluid-particle systems such as fluidized beds. It provides a starting point for extension of the theory to flows in which two- and three-dimensional effects, such as density and velocity variations across a channel, are significant. It is also the key to a rapid solution of unsteady-flow problems of sedimentation and foam drainage.

4.2 GENERAL THEORY

The drift flux j_{21} was introduced in Chap. 1 where it was shown to represent the volumetric flux of either component relative to a surface moving at the volumetric average velocity j. It can be expressed in terms of the relative velocity by using Eq. (1.39),

$$j_{21} = v_{21}\alpha(1 - \alpha) \tag{4.1}$$

or in terms of the component fluxes by using Eq. (1.36),

$$j_{21} = (1 - \alpha)j_2 - \alpha j_1 \tag{4.2}$$

Since $j = j_1 + j_2$, Eq. (4.2) can be expressed in the alternative forms

$$j_1 = (1 - \alpha)j - j_{21} \tag{4.3}$$

$$j_2 = \alpha j + j_{21} \tag{4.4}$$

Equation (4.3) shows that the volumetric flux of component 1 is the sum of the volumetric concentration times the average volumetric flux and a flux $-j_{21} = j_{12}$ due to the relative motion. Equation (4.4) is a similar statement for component 2. The drift flux is therefore analogous to the diffusion flux in the molecular diffusion of gases and provides a convenient way of modifying homogeneous theory to account for the relative motion. Indeed, all of the properties of the flow, such as void fraction, mean density, and momentum flux can be expressed as the homogeneous flow value together with a correction factor or an additional term which is a function of the ratios of j_{21} to the component fluxes. For example, the void fraction is, from Eq. (4.2),

$$\alpha = \frac{j_2}{j}\left(1 - \frac{j_{21}}{j_2}\right) \tag{4.5}$$

The mean density is

$$\rho_m = \frac{j_1\rho_1 + j_2\rho_2}{j} + (\rho_1 - \rho_2)\frac{j_{21}}{j} \tag{4.6}$$

When j_{21} is zero these results reduce to the homogeneous flow values.

4.3 GRAVITY-DOMINATED FLOW REGIMES WITH NO WALL SHEAR

Drift-flux theory is particularly convenient for analyzing flow regimes in which gravity (or some other body force) is balanced by the pressure gradient and the forces between the components. For vertical flow,

Eqs. (3.45) and (3.46) then reduce to

$$0 = -\frac{dp}{dz} - g\rho_1 + \frac{F_{12}}{1-\alpha} \tag{4.7}$$

$$0 = -\frac{dp}{dz} - g\rho_2 - \frac{F_{12}}{\alpha} \tag{4.8}$$

Subtracting Eq. (4.8) from Eq. (4.7) we get

$$F_{12} = \alpha(1-\alpha)g(\rho_1 - \rho_2) \tag{4.9}$$

In the absence of wall effects, F_{12}, the mutual drag force per unit volume, is a function of the properties of the components, their geometry, the void fraction, and the relative motion. For a given system, therefore, F_{12} is a function only of α and j_{21}. Thus, in view of Eq. (4.9), j_{21} must be only a function of α and the system properties and we can write

$$j_{21} = j_{21}\,[\alpha, \text{properties of system}] \tag{4.10}$$

Evidently, in the absence of infinite relative velocities, j_{21} must go to zero at $\alpha = 0$ and $\alpha = 1$, in view of Eq. (4.1).

Example 4.1 Using the results of Example 3.5 determine the relationship between j_{sf} and ϵ for a vertically moving fluid-solid system to which the Carman-Kozeny equation applies.

Solution Let component 1 be the fluid and let $\epsilon = 1 - \alpha$. From Example 3.5

$$F_{fs} = -\frac{180\mu_f j_{f0}}{d^2}\left(\frac{1-\epsilon}{\epsilon}\right)^2 \tag{4.11}$$

j_{f0} denotes the fluid flux when $j_s = 0$. Therefore, from Eq. (4.2),

$$j_{sf} = -(1-\epsilon)j_{f0} \tag{4.12}$$

Combining the above relationships with Eq. (4.9) we obtain

$$j_{sf} = -\frac{d^2 g(\rho_s - \rho_f)\epsilon^3}{180\mu_f} \tag{4.13}$$

This equation is reasonably valid for packed beds with values of ϵ below 0.6. It is not a reasonable expression for j_{sf} at high values of ϵ since it does not go to zero at $\epsilon = 1$. Since the sign convention has been adopted that the upward direction is positive, the negative value of j_{sf} indicates that the particles "drift" downward relative to the average motion if $\rho_s > \rho_f$. Other sign conventions are possible as long as consistency is maintained.

For a given system j_{21} can be plotted as a function of α, using Eq. (4.10), on a graph such as Fig. 4.1. Then, if the overall flow rates Q_1 and Q_2 are specified in a given situation, j_1 and j_2 can be calculated from Eqs. (1.19) and (1.20), and Eq. (4.2) predicts a linear relationship between j_{21} and α. In fact, Eq. (4.2) represents a line joining the points $\alpha = 0$, $j_{21} = j_2$ and $\alpha = 1$, $j_{21} = -j_1$ in Fig. 4.1. The intersections between

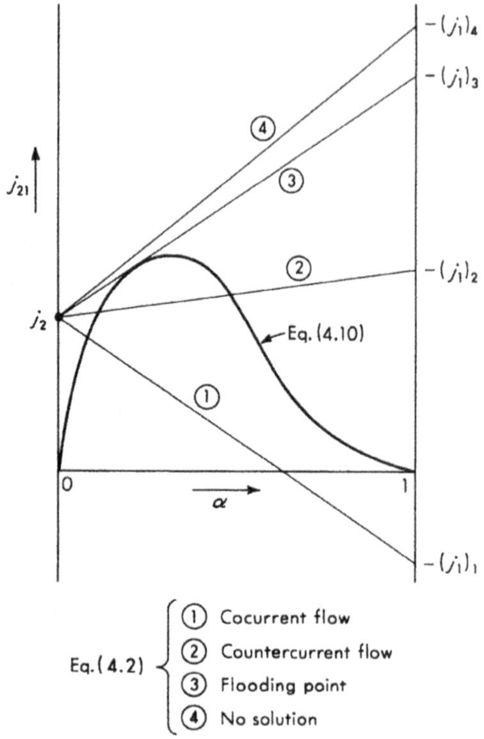

Fig. 4.1 Graphical solution of Eqs. (4.2) and (4.10).

this line and the curve determine the values of α which are obtained in practice.[1–3]

This graphical method of solving Eqs. (4.2) and (4.10) is particularly convenient as a means of visualizing the effects of changing the flow rates Q_1 and Q_2 since behavior in cocurrent and countercurrent flow in either direction can be predicted simply by moving a straight edge.

Figure 4.1 has been drawn for the particular case of small bubbles suspended in a liquid flowing in a vertical pipe. It shows that for cocurrent upward or downward flow there is always a possible solution; for countercurrent flow with the gas flowing downward there are no solutions; while for countercurrent flow with the gas flowing upward there are either two solutions or none depending on the magnitude of the flow rates. In the case of a dispersion of solid particles in a fluid the j_{21} versus α curve passes through a sudden discontinuity at the point where the particles pack together randomly and form a compact bed. If the particle concentration is denoted by α, then there is a maximum allowable value of α as shown in Fig. 4.2. In this figure the drift flux has been rendered dimensionless by dividing by the terminal velocity v_∞ of a single particle in an infinite fluid. The curve shown represents a typical equation which has been found to correlate a wide variety of data.

$$\frac{j_{21}}{v_\infty} = \alpha(1 - \alpha)^n \tag{4.14}$$

n is a function of a suitably defined Reynolds number, and $n = 3$ is an intermediate value for fluid-particle systems. α is the volumetric concentration of the dispersed component in general.

The actual value of α_{max} at which the particles pack together depends on the particle shape and the nature of the interparticle forces. In the extreme case of flocculated suspensions α_{max} can be as low as 0.1. For hard spheres the range is usually $0.58 < \alpha_{max} < 0.62$, depending on the way in which the packing is achieved. Tapping or shaking the bed

Material	Diameter	Material	Diameter
× Socony beads	0.129"	▽ Sea sand	0.0147"
◻ Socony beads	0.174"	▷ Sea sand	0.0219"
○ Glass beads	5 mm	◁ Sea sand	0.0393"
◇ Glass beads	0.0201"	+ Crushed rock	0.0557"
△ Lead shot	0.0505"	— Eq. (4.14) {n = 3}	

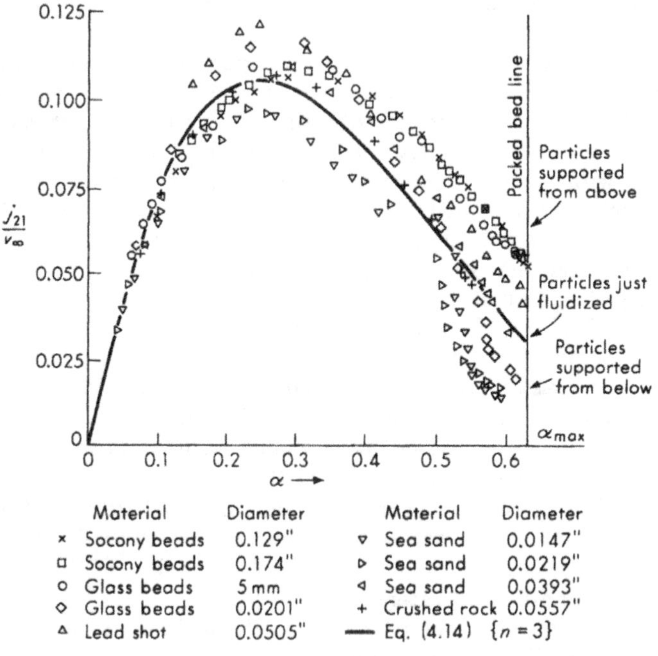

Fig. 4.2 Dimensionless drift flux plotted versus particle volumetric concentration for fluidization of various materials with water. (*Data of Wilhelm and Kwauk.*[4])

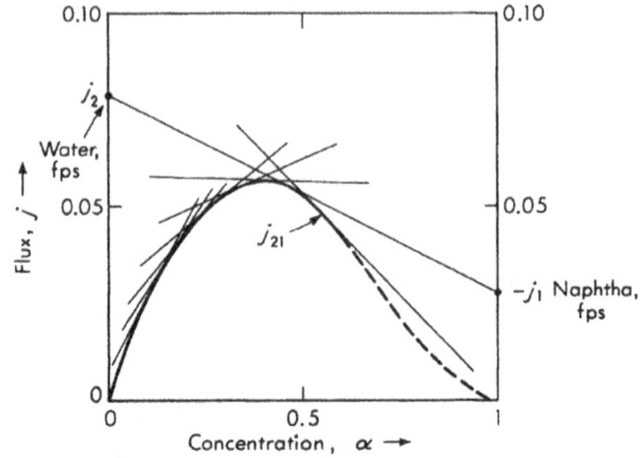

Fig. 4.3 Derivation of the drift-flux-concentration curve from Blanding and Elgin's flooding data.[5]

will promote a higher value of α_{max}. The curve of j_{21}/v_∞ versus α shown in Fig. 4.2 represents a balance between fluid dynamic drag and buoyancy. α_{max} is the point at which particle-particle forces become significant. If α exceeds α_{max} then it is necessary to return to Eqs. (4.7) and (4.8) and include terms to describe the particle-particle interaction. If the particles are completely inflexible and incompressible, then the packed bed with a particle concentration of α_{max} must be supported from above or below, as indicated in the figure, depending on the value of j_{21}/v_∞. For very flexible particles, such as bubbles, the particle-particle forces can be so small that Eq. (4.14) gives a good representation up to values of α very close to unity.

The limit of operation in countercurrent flow is known as "flooding" and occurs when the line representing Eq. (4.2) is a tangent to the curve of j_{21} versus α. If the flow rate of either phase is increased beyond this point, no steady flow solution is possible and a change in behavior must occur. There can either be a change in flow regime or a rejection of excess material at the ends of the flow passage.

The j_{21}-versus-α curve can be assembled from its tangents if the corresponding flow rates at the flooding points are known. Figure 4.3 shows this procedure applied to the flooding data of Blanding and Elgin[5] for a liquid-liquid (water-naphtha) system in a vertical pipe.

Another way of representing the different modes of operation is to rewrite Eq. (4.2) in the form

$$j_2 = \frac{\alpha}{1-\alpha} j_1 + \frac{1}{1-\alpha} j_{21} \tag{4.15}$$

For a given system the last term in this equation is a function of α in view of Eq. (4.10). Therefore, the curves of constant α on a plot of j_2 versus j_1 will be straight lines of slope $\alpha/(1-\alpha)$. The intercepts on the axes will be $j_{21}/(1-\alpha)$ and $-j_{21}/\alpha$. The various regions of operation are shown on a graph of this kind in Fig. 4.4, which corresponds to the j_{21}-versus-α relationship shown in Fig. 4.1. These conclusions are illustrated by Deruaz's data[6] for the bubbly regime of gas-liquid flow in Fig. 4.5.

Using Eq. (4.14) and the geometrical definition of flooding shown in Fig. 4.1, it is readily shown that the corresponding flow rates at the flooding point are given parametrically as functions of α by the equations

$$j_2 = j_{21} - \alpha \frac{dj_{21}}{d\alpha} = \alpha^2 v_\infty n(1-\alpha)^{n-1} \tag{4.16}$$

$$j_1 = -j_{21} - (1-\alpha)\frac{dj_{21}}{d\alpha} = -(1-\alpha n)v_\infty(1-\alpha)^n \tag{4.17}$$

Example 4.2 For bubbly flow of a particular mixture it is found that $n = 2$ and $v_\infty = 1$ fps. What is the relationship between Q_f and Q_g for flooding in a vertical 6-in.-diam pipe?

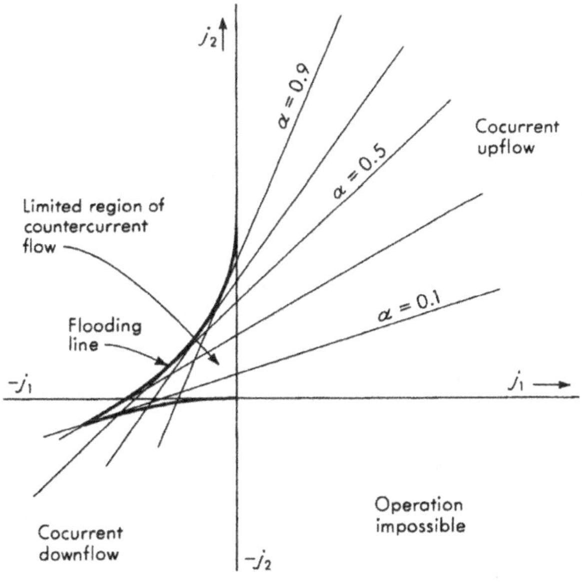

Fig. 4.4 The various regions of operation in one-dimensional vertical flow in terms of the component fluxes for a flow regime in which the drift flux is a function of α but independent of j_1 and j_2. The particular case $\rho_1 > \rho_2$ is shown.

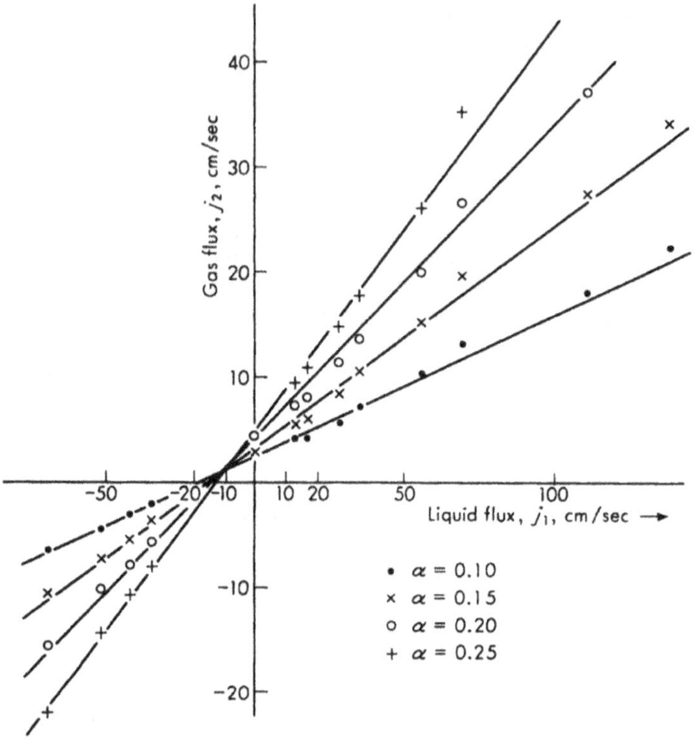

Fig. 4.5 Deruaz's data[6] for flow of bubbly mixtures in vertical pipes with 8- and 10-cm² cross section.

Solution From Eqs. (4.16) and (4.17), using the given values of n and v_∞, and identifying the gas as the discontinuous component 2, we have

$$j_g = 2\alpha^2(1 - \alpha) \qquad \text{fps}$$
$$j_f = -(1 - 2\alpha)(1 - \alpha)^2 \qquad \text{fps}$$

The area of cross section of the pipe is $(\pi/4)(\frac{1}{2})^2 = 0.196$ ft². Therefore $Q_f = 11.8j_f$ (in cubic feet per minute) and $Q_g = 11.8j_g$. The following table gives the predicted values at the flooding points:

α	0	0.1	0.2	0.3	0.4	0.5
Q_f, cfm	11.8	7.64	4.41	2.31	0.85	0
Q_g, cfm	0	0.21	0.76	1.49	2.26	2.95

It is assumed that no agglomeration of bubbles, leading to a change in flow regime, occurs.

4.4 CORRECTIONS TO THE SIMPLE THEORY

The simple one-dimensional gravity-dominated theory does not take account of variations in concentration and velocity across the cross section. The quantities α, v_1 and v_2, and j_1 and j_2 are merely averages across the direction of flow. Continuity and dynamic equations written in terms of these averages are not necessarily identical with the rigorous formulation of these equations as integrals over the flow field (e.g., the product of averages is not the same as the average of a product). This problem is already familiar to students of single-phase fluid mechanics who know, for example, that the momentum flux in a pipe with a velocity profile, but uniform density, is

$$\rho \int v^2\, dA = A\rho\langle v^2\rangle \tag{4.18}$$

and that this is not the same as

$$A\rho\left(\frac{\int v\, dA}{A}\right)^2 = A\rho\langle v\rangle^2 \tag{4.19}$$

where the $\langle\ \rangle$'s denote averages over the cross section defined by the equation

$$\langle X\rangle = \frac{\int X\, dA}{A} \tag{4.20}$$

However, just as in single-component flow the ratio between Eqs. (4.18) and (4.19) is usually expressed as a correction factor to the one-dimensional theory, so in two-component flow the ratio of suitable integrals is also used to define parameters which are equal to unity for truly one-dimensional flow and are not far from unity in the general case.

Of particular usefulness is the distribution parameter C_0, introduced by Zuber and Findlay[7] and defined as

$$C_0 = \frac{\langle \alpha j\rangle}{\langle\alpha\rangle\langle j\rangle} \tag{4.21}$$

C_0 represents the ratio of the average of the product of flux and concentration to the product of the averages.

A convenient definition of an average velocity of phase 2 is obtained from Eq. (1.23)

$$\bar{v}_2 = \frac{\langle j_2\rangle}{\langle\alpha\rangle} \tag{4.22}$$

Averaging Eq. (4.4) term by term across the duct, we have

$$\langle j_{21}\rangle = \langle j_2\rangle - \langle \alpha j\rangle \tag{4.23}$$

Combining Eqs. (4.22), (4.23), and (4.21) leads to

$$\bar{v}_2 = C_0\langle j\rangle + \frac{\langle j_{21}\rangle}{\langle\alpha\rangle} \tag{4.24}$$

Note that Eq. (4.23) represents a strict equation of averages across the channel. The quantity $\bar{v}_2$ in Eq. (4.22), however, is not equal to $\langle v_2\rangle$ in general since $\langle v_2\rangle$ is related to the flux and concentration distributions by the equation

$$\langle v_2\rangle = \left\langle\frac{j_2}{\alpha}\right\rangle \tag{4.25}$$

The average velocity represented by $\bar{v}_2$ is usually more convenient to use than $\langle v_2\rangle$ because it can be directly related to the overall flow rate Q_2 and the volumetric mean concentration $\langle\alpha\rangle$ by means of Eq. (4.22).

$\langle\alpha\rangle$ is a convenient average volumetric concentration because it is readily related to a simple experiment in which a section of the duct is suddenly isolated and the proportion of the volume occupied by component 2 is determined. $\langle\alpha\rangle$ is also the quantity which is usually measured by γ-ray scanning techniques.

In terms of the overall volumetric flow rates, Q_1 and Q_2, the following equations are valid:

$$\langle j\rangle = \frac{Q_1 + Q_2}{A} \tag{4.26}$$

$$\langle j_2\rangle = \frac{Q_2}{A} \tag{4.27}$$

$$\langle j_1\rangle = \frac{Q_1}{A} \tag{4.28}$$

In view of Eqs. (4.22), (4.26), and (4.25), Eq. (4.24) can be expressed as

$$\frac{Q_2}{\langle\alpha\rangle A} = C_0\frac{Q_1 + Q_2}{A} + \frac{\langle j_{21}\rangle}{\langle\alpha\rangle} \tag{4.29}$$

Therefore the volumetric mean value of α is

$$\langle\alpha\rangle = \frac{Q_2 - A\langle j_{21}\rangle}{C_0(Q_1 + Q_2)} \tag{4.30}$$

If $\langle j_{21}\rangle$ is small compared with Q_2/A, Eq. (4.30) reduces to

$$\langle\alpha\rangle = \frac{1}{C_0}\frac{Q_2}{Q_1 + Q_2} \tag{4.31}$$

The second factor in Eq. (4.31) is the same as we should obtain from homogeneous theory. Thus the effect of concentration variations, but not the effect of relative velocity, is to multiply the mean concentration calculated from homogeneous theory by a correction factor $1/C_0$. This correction factor is the same as the *flow parameter* K used by Bankoff[8] and the *Armand parameter*[9] which is used in the Russian literature.

The quantity $\langle j_{21} \rangle$ cannot be evaluated in general without a knowledge both of the dependence of j_{21} on α and also the variation of α across the channel. Simplification is, however, possible in two cases.

Case 1 j_{21} independent of α, i.e.,

$$j_{21} = \text{const} \tag{4.32}$$

The average value of j_{21} is then equal to this constant value.

Case 2 j_{21} varies linearly with α, i.e.,

$$j_{21} = b_0 + b_1\alpha \tag{4.33}$$

where b_0 and b_1 are constants.

In this case

$$\langle j_{21} \rangle = b_0 + b_1\langle \alpha \rangle \tag{4.34}$$

and Eq. (4.30) becomes

$$\langle \alpha \rangle = \frac{Q_2 - Ab_0}{C_0(Q_1 + Q_2) + b_1} \tag{4.35}$$

Equation (4.35) is applicable, for example, to the slug-flow and churn-turbulent bubble-flow regimes in gas-liquid systems.

Note that the local fluxes j in the above context denote the time average of the volumetric rate of flow across a given area. If time variations in the fluxes are important, more complicated equations will arise.

Example 4.3 An important use of simple one-dimensional drift-flux theory is for the interpretation of experimental results. In this example the technique is applied to the data of Smissaert for air-water flow in a vertical 2.75 in. diameter pipe reported in Ref. 10. The subscript 1 refers to the water and 2 to the air.

First the data are used to calculate j_{21} from Eq. (4.2) and the results are plotted up as a function of α for the lowest and highest liquid rates as in Fig. 4.6. The assumption that j_{21} is only a function of α is seen to be quite good, but there is a noticeable drift with j_1. Furthermore, the value of j_{21} appears to become very large at about $\alpha = 0.8$.

Explanation for these observations is sought in terms of the parameter C_0. Equation (4.22) is combined with Eq. (4.24) to give

$$j_{21} = (1 - C_0\alpha)j_2 - C_0\alpha j_1 \tag{4.36}$$

The "averaging" signs have been dropped for convenience. Now it is noticed that the high values of j_{21} in Fig. 4.6 correspond to large values of $j_2(\approx 30 \text{ fps})$. In order for j_{21} to stay small at these flow rates, it is necessary for $(1 - C_0\alpha)$ to

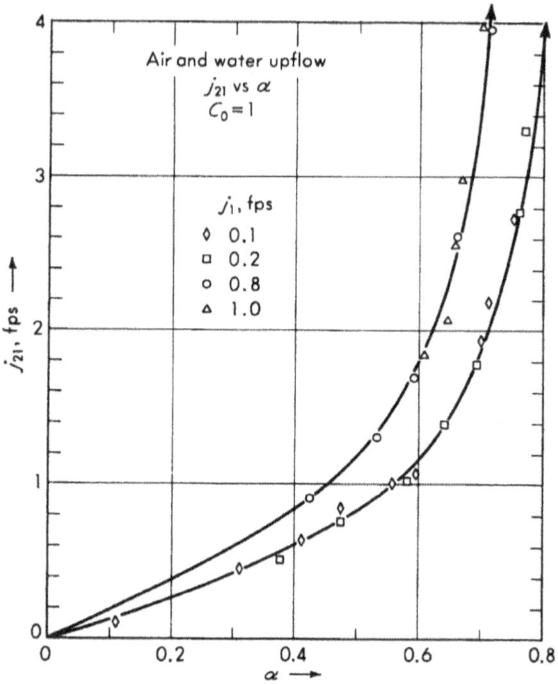

Fig. 4.6 Data of Smissaert[10] plotted as drift flux versus concentration, assuming $C_0 = 1$.

approach zero. Therefore α approaches the value $1/C_0$. Looking again at Fig. 4.6 we see that the limiting value for α is about 0.8, giving a value for C_0 of about 1.25.

Substituting $C_0 = 1.25$ in Eq. (4.36) and recalculating j_{21} for all of the data, we obtain the result shown in Fig. 4.7. The data scatter around a straight line, there is no systematic tendency with j_1, and the drift flux j_{21} does not take on excessively large values.

Having brought the data together by the use of the parameter C_0, we now test for the value of n in Eq. (4.14). Plotting j_{21}/α versus $(1 - \alpha)$ on log paper we expect a line of slope n and intercept v_∞ at $\alpha = 0$. Figure 4.8 shows that n is approximately zero and the value of v_∞ is about 1.2 fps. A change in flow regime is indicated as the value of α approaches 0.8. The scatter is of the same order of magnitude as the experimental repeatability.

In this example it was fortunate that the method of introducing the components led to a range of bubble sizes conducive to a constant value of v_∞. Thus it was possible to bring the data together by adjusting the parameter C_0 alone. In the case of the nitrogen-mercury results reported by Smissaert in the same report, a trend of v_∞ as a function of j_1 is evident, although an average value of $v_\infty = 2.5$ fps is within about 10 percent of all his data.

Since one has three parameters to play with (C_0, v_∞, and n), there are the usual pitfalls to be expected in correlating data in this way without any reference to the flow pattern or other important parameters such as bubble size. Of course the best technique is to make additional observations during the experiment which motivate an independent assessment of these parameters. This topic will be taken up again in later chapters when the individual flow regimes are discussed in detail.

The data have been plotted in Figs. 4.7 and 4.8 in a way which realistically represents the scatter and does not flatter the correlating scheme. Many published comparisons between data and correlations are plotted in such a way as to give the illusion of reduced scatter.

4.5 SIGN CONVENTIONS AND IDENTIFICATION OF COMPONENTS 1 AND 2

Since velocities and fluxes may be in either direction in one-dimensional flow, some sign convention is necessary. It does not matter which system is used as long as it is applied consistently.

Usually one is interested more in describing the motion and concentration of one component than the other. In fact, throughout this

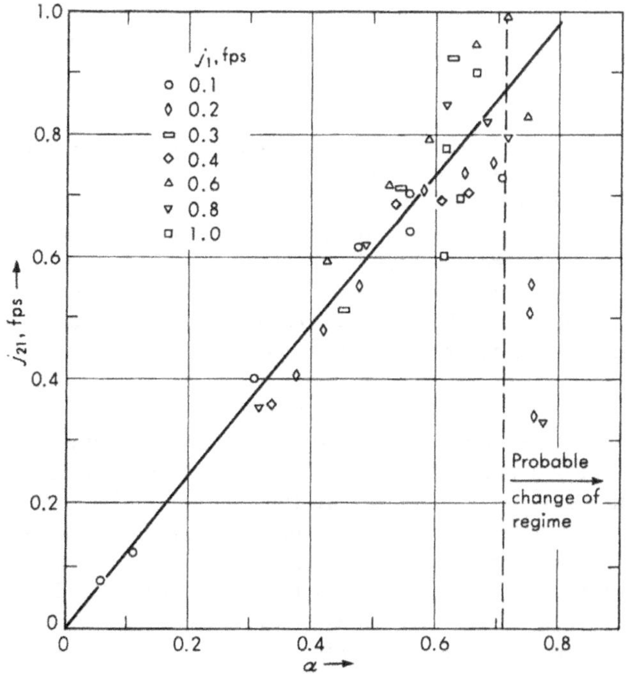

Fig. 4.7 Modified representation of Fig. 4.6 using $C_0 = 1.25$.

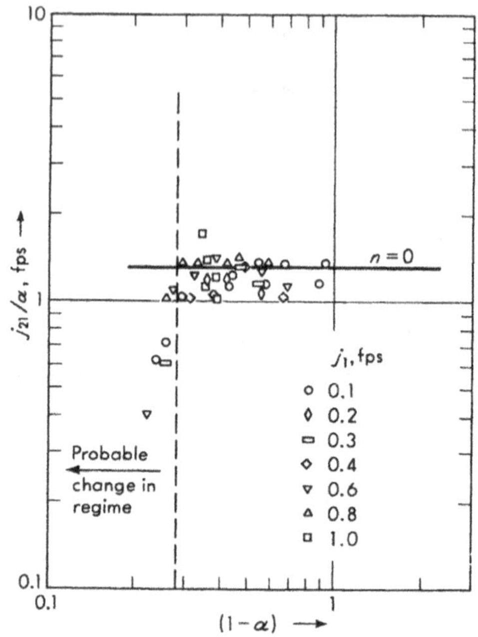

Fig. 4.8 Determination of n for data in Fig. 4.7.

chapter we have "favoritized" component 2 by calling its volumetric concentration α and leaving the concentration of 1 to be $1 - \alpha$. We have further selected j_{21} rather than j_{12} for analytical purposes although the defining equations are quite symmetrical. Moreover, we later stated that component 2 was to be chosen as the discontinuous component. The reason for this is that one is usually inclined to view the relative velocity v_{21} or the terminal velocity v_∞ as the velocity of a particle in a fluid medium rather than the velocity of the fluid which will bring particles to rest. Thus a convenient sign convention is often that in which velocities measured in the direction of drift of component 2 are positive, in which case j_{21} is always positive. Alternatively if a force balance is used to deduce j_{21}, as in Example 4.1, it may be useful to define the positive direction in terms of the direction of the gravitational field.

In later chapters it will be found convenient sometimes to use the inverse description and to focus attention on the motion of the continuous component, for example, when a fluid flows around approximately stationary particles in a fluidized bed. For this purpose we define ϵ as the volumetric concentration of the continuous component where

$$\epsilon = (1 - \alpha) \tag{4.37}$$

and work in terms of j_{12} rather than j_{21}. The analytical techniques remain unchanged.

4.6 UNSTEADY FLOW

The drift flux is a very useful parameter for analyzing unsteady flows by using the theories of wave motion which will be presented in Chap. 6.

PROBLEMS

4.1. Using the drift-flux model show that the average phase velocities are

$$v_1 = \frac{j}{1 + j_{21}/j_1} \qquad v_2 = \frac{j}{1 - j_{21}/j_2}$$

Deduce an expression for the momentum flux in a duct in one-dimensional flow.

4.2. Air and water flow in a vertical duct at 70°F and 36 psia. The drift flux is given as a function of void fraction by

$$j_{gf} = 0.8\,\alpha(1 - \alpha)^2 \qquad \text{fps}$$

What is the void fraction for the following values of j_f and j_g in feet per second?

j_f	0.5	1	0.5	2	5
j_g	0.5	0.5	1	2	5

For these conditions, compare the momentum flux and void fraction calculated from drift-flux theory with the values which would be predicted from homogeneous theory and the Martinelli correlation in a 2-in.-diam pipe.

4.3. Air is bubbled uniformly through stagnant water. What is the relationship between volumetric air flux and void fraction if $j_{gf} = v_\infty \alpha(1 - \alpha)^2$? What is the significance of the two possible values of void fraction? If the depth of the water before the bubbling is 100 cm and $v_\infty = 25$ cm/sec, what is the height of the two-phase mixture during bubbling as a function of the flow rate? When does flooding occur?

4.4. Sketch the curves of j_2 versus α as a function of j_1 and j_1 versus α as a function of j_2 for a drift flux given by Eq. (4.14). Identify the conditions of flooding, cocurrent and countercurrent flow.

4.5. Show graphically how the void fraction depends on the fluid flow rate in a fluidized bed for different particle fluxes, both upward and downward. What regions of the graph correspond to cocurrent and countercurrent, upward and downward flow? Over what regions of the graph is $\alpha = \alpha_{max}$ if the particles are unrestrained by the ends or the walls of the duct?

4.6. A flooding experiment yielded the following data for an air-water system.

$-j_f$, cm/sec	0	1	2	4	5	9	11	15	18
j_g, cm/sec	12	10	9	7	6	4	3	2	1

Derive the j_{gf}-versus-α relationship. What "best" values of v_∞ and n in Eq. (4.14) describe this relationship?

4.7. What does the graphical technique described in Fig. 4.1 become in the limiting case of homogeneous flow with $v_1 = v_2$?

4.8. One way which has been suggested for determining the values of C_0 and j_{21} is to plot j_2/α versus j. For what value of n in Eq. (4.14) will this relationship be linear? What characteristics of the line will then determine C_0 and v_∞? How does the drift velocity of component 2, v_{2j}, depend on α in this case?

4.9. For the churn-turbulent regime of bubbly flow and for the slug-flow regime, v_{gj} is approximately constant. If the flow is isothermal show that the choking condition is

$$\frac{j_g}{p}\left(G_g + \frac{G_f}{1 + v_{gj}/j_f}\right) = 1$$

Prove that, at the same mass flux of the components, the pressure at which choking occurs will be lower in this case than in homogeneous flow. Assume $C_0 = 1$.

4.10. Solve Prob. 4.9 if C_0 is constant but is not equal to 1.

4.11. When drift occurs the various expressions for the homogeneous flow frictional pressure drop are not identical; in particular we can choose between the relationships

$$\frac{2C_fG^2}{D\rho_m} \qquad \frac{2C_fGj}{D} \qquad \frac{2C_f\rho_m j^2}{D}$$

Compare these various predictions in terms of the relationship between j_{21} and the individual fluxes. Under what conditions will the predictions be most sensitive to the assumptions?

4.12. When drift occurs, the momentum flux differs from the homogeneous theory prediction. Express the momentum flux in terms of the ratios between j_{21} and the component fluxes, the mass flow rates and the overall flux. Show that the ratio between the momentum flux with drift and that without drift is

$$\frac{G_1/(1 + j_{21}/j_1) + G_2/(1 - j_{21}/j_2)}{G_1 + G_2}$$

4.13. Starting from Eqs. (3.39a) and (3.39b) and using the definition of Ω in Eq. (2.92), show that, in the presence of drift, the void propagation equation for constant fluid properties in vapor-liquid flow in a constant area duct is

$$\frac{\partial \alpha}{\partial t} + j\frac{\partial \alpha}{\partial z} + \frac{\partial}{\partial z} j_{gf} = \left[\frac{v_f}{v_{fg}} + (1 - \alpha)\right]\Omega$$

If j_{gf} is only a function of α and the system properties, use the relationship

$$\frac{\partial}{\partial z} j_{gf} = \frac{\partial \alpha}{\partial z}\frac{\partial}{\partial \alpha} j_{gf}$$

to prove that voids propagate with the continuity wave velocity

$$V_w = j + \frac{\partial}{\partial \alpha}(j_{gf})$$

with density decreasing as in Prob. 2.23.

4.14. Deduce the relationship between j_{21} and α from the data shown in Fig. 4.5. Comment on the form of the result.

4.15. Solve Prob. 2.30 if the drift flux is given by the expression $j_{gf} = 1.0\alpha$ fps.

4.16. Solve Prob. 2.31 if $j_{gf} = 1.0\alpha$ fps.

4.17. How do the lines shown in Fig. 4.4 compare with the predictions of (*a*) homogeneous flow, (*b*) the Martinelli correlation for viscous-viscous flow.

4.18. Under subcooled boiling conditions in vertical upflow, C_0 can be less than unity (Why?) while j_{gf} is positive. Investigate under what circumstances the void fraction in the duct can exceed the predictions of homogeneous theory.

4.19. Calculate the value of C_0 if α and j vary as power laws across a circular pipe. Discuss what ranges of values are reasonable under various conditions.

4.20. Suppose that bubbles supplied to a vertical pipe do not have a uniform size. If a probability distribution for bubble size can be estimated and j_{gf} can be expressed in terms of bubble size and the local value of α, how should one modify drift-flux theory in terms of appropriate averages?

REFERENCES

1. Wallis, G. B.: Paper No. 38, *Proc. Intern. Heat Transfer Conf.*, ASME, Boulder, Colo., vol. 2, pp. 319–340, 1961.
2. Wallis, G. B.: *Symp. Interaction Fluids Particles,* Inst. Chem. Engrs., London, pp. 9–16, 1962.
3. Wallis, G. B.: *Symp. Two-phase Flow,* Inst. Mech. Engrs., paper no. 3, pp. 11–20, 1962.
4. Wilhelm, R. H., and M. Kwauk: *Chem. Eng. Progr.*, vol. 44, pp. 201–217, 1948.
5. Blanding, F. H., and J. C. Elgin: *Trans. A. I. Ch. E.*, vol. 38, pp. 305–335, 1942.
6. Deruaz, R.: ANL transl. 61, 1964 (from the French, Centre D'Etudes Nucleaires de Grenoble, note TT#165, April, 1964).
7. Zuber, N., and J. Findlay: *Trans. ASME J. Heat Transfer,* ser. C, vol. 87, p. 453, 1965.
8. Bankoff, S. G.: *Trans. ASME J. Heat Transfer,* ser. C, vol. 82, p. 265, 1960.
9. Armand, A.: UKAEA, AERE transl. 828, 1959 (*Izvest. Vsesoyuz. Teplotekh. Inst.*, no. 1, pp. 16–23, 1946).
10. Smissaert, G. E.: ANL Rept. 6755, 1963.

5
Velocity and Concentration Profiles

5.1 INTRODUCTION

The next step in sophistication, beyond the simple "lumped" models of homogeneous and separated flow, is the consideration of velocity and concentration profiles across the duct. This is still a quasi-one-dimensional description of the flow because local velocities are allowed only in the principal direction of the motion. Any motion across the duct is either neglected or absorbed into parameters, such as "eddy diffusivity," which account for turbulent mixing. In turbulent flow the velocity and concentration profiles are averages over a long period of time.

The major use of velocity and concentration profiles in this book will be for motivating correction factors which can be applied to the simpler homogeneous or separated-flow models in order to increase their accuracy. In some cases it will be possible to derive analytical expressions for correlating parameters, such as friction factor and two-phase multipliers, rather than relying on the empiricism of previous chapters.

A further use of the theory will be the derivation of useful dimensionless groups.

5.2 QUALITATIVE ASPECTS

The concept of velocity profiles in single-phase flow is a familiar one. Certain simple cases, such as flows in circular pipes, have been investigated in great detail.

In two-phase flow the situation is far more complicated for the following reasons:

1. The concentration of a particular phase may be neither uniform nor symmetrical. Suspended particles tend to settle out in horizontal pipes. Liquid films stick to walls and are thicker at the bottom than at the top of the pipe. Bubbles rise preferentially near the middle of a vertical pipe. A series of large bubbles, or slugs, gives rise to large periodic variations in concentration.
2. The "concentration" is not usually an adequate means of describing the local two-phase properties. Large particles may sink to the bottom of a pipe whereas small ones remain suspended. Very small particles (below about 1 μ in diameter) follow the fluid streamlines, whereas large ones may not be significantly diverted within the length of the apparatus or may undergo many collisions with containing walls. Several flow configurations, such as drops, bubbles, and liquid films, may coexist at the same time.
3. Although there is no relative motion at the interface between the phases, there may be average relative motion on a scale which is larger than the distance at which a continuum description is meaningful. For example, the average droplet velocity in a suspension is not necessarily the same as the velocity of the suspending fluid. Thus there exist two velocity profiles, one for each phase, as shown in Fig. 5.1.

The general solution of these problems appears quite hopeless. A discussion of all the work which has been done in this field would itself require a complete book. All that will be done in this chapter is to derive a few simple results that are extensions of the homogeneous and separated-flow models and which provide material that will be useful in later chapters.

Velocity and concentration profiles can be analyzed using either a differential or an integral technique. In the former, the velocities and concentrations are derived as solutions to differential equations which are deduced from a study of small elements of the fluid field. In the latter,

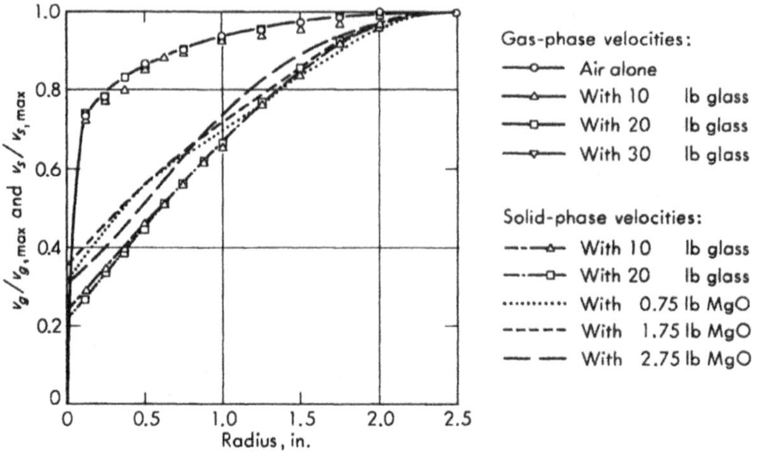

Fig. 5.1 Velocity distributions of gas and particles in a duct. (*Results of Soo, Trezek, Dimick, and Hohnstreiter.*[1])

the form of the profiles is assumed and the dynamic and geometrical conditions are satisfied in integral form.

5.3 DIFFERENTIAL ANALYSIS

VELOCITY PROFILES IN SINGLE-PHASE FLOW

In single-phase flows the "concentration" is usually uniform over the cross section. Fluid properties may vary normal to the flow due to temperature and composition changes, but it is usually possible to assign suitable local values of viscosity and density for use in the equations.

The well-known method for deriving a "universal" velocity profile for round pipes is first to calculate the shear stress distribution from a force balance, then relate the shear stress to the velocity gradient and finally integrate the velocity gradient to get the profile.[2] Some useful parameters are the *friction velocity* u^* and dimensionless forms of the velocity and distance from the wall, as follows:

$$u^* = \left(\frac{\tau_w}{\rho}\right)^{1/2} \tag{5.1}$$

$$u^+ = \frac{v}{u^*} \tag{5.2}$$

$$y^+ = \frac{\rho u^* y}{\mu} \tag{5.3}$$

Close to the wall the flow is assumed to be laminar and the shear stress is related to the velocity profile by the equation

$$\tau = \mu \frac{dv}{dr} \tag{5.4}$$

For flows with uniform density in round pipes, an approximate solution for small values of y is

$$u^+ = y^+ \tag{5.5}$$

and empirically this is found to be valid up to $y^+ \simeq 5$.

Numerous alternative expressions can be chosen to relate the shear stress to the velocity gradient in the region of turbulent flow. Perhaps the simplest is in terms of a *mixing length* l introduced by Prandtl,[3] as follows:

$$\tau = \rho l^2 \frac{dv}{dy} \left| \frac{dv}{dy} \right| \tag{5.6}$$

If it is further assumed that l varies linearly with y, we have

$$l = ky \tag{5.7}$$

In round pipes these equations lead to a logarithmic dimensionless velocity profile

$$u^+ = \frac{1}{k} \ln y^+ + C \tag{5.8}$$

Empirically it is found that $k \approx 0.4$, $C \simeq 5.5$, and that Eq. (5.8) is valid for $y^+ > 30$.

In between $y^+ = 5$ and $y^+ = 30$ there is a *buffer* layer in which the shear stress can be represented by a hybrid equation

$$\tau = (\mu + \epsilon\rho) \frac{dv}{dy} \tag{5.9}$$

and ϵ is given by an equation such as Deissler's[4]

$$\epsilon = n^2 vy(1 - e^{-\rho n^2 vy/\mu}) \tag{5.10}$$

n is an empirical dimensionless constant with a value of about 0.1.

If the flow is nonnewtonian, solutions can be derived in the laminar regime by replacing Eq. (5.4) by the appropriate relationship between shear stress and rate of strain.

VELOCITY PROFILES IN TWO-PHASE FLOW

The techniques which were described above can be applied to two-phase (or multiphase) flows as long as the flow is *locally homogeneous*. One velocity profile is then determined for both components together. In order to solve the equations the values of ρ, μ, and l are needed as a function of position. Usually the homogeneous density and equivalent viscosity are given as functions of void fraction or concentration α by the methods of Chap. 2. Thus, the definition of a flow regime in terms of a concentration profile and the properties of the components enables ρ and μ to be calculated everywhere. The mixing length presents more difficulty because comprehensive empirical expressions for it in two-phase flow have not yet been established. Two-phase mixing lengths are sometimes greater and sometimes smaller than in single-phase flows. For example, a study of the gas core in annular flow by Gill, Hewitt, and Lacey[5] revealed that Eq. (5.8) was valid but with values of k which depended on the flow rates, as shown in Fig. 5.2. The corresponding gas velocity profiles differed noticeably from the single-phase flow value (Fig. 5.3) in contrast to the results for a particle-fluid system shown in Fig. 5.1. General conclusions have yet to be drawn from studies of this type.

In any case a computation scheme can be set up which will readily accept information about property and mixing-length variations across the channel as they become available. For axially symmetric steady

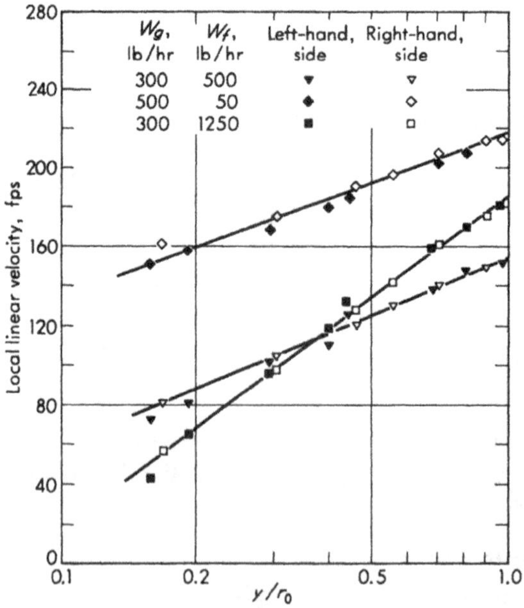

Fig. 5.2 Logarithmic plots of gas velocity profiles in annular-mist flow. (*Results of Gill, Hewitt, and Lacey.*[5])

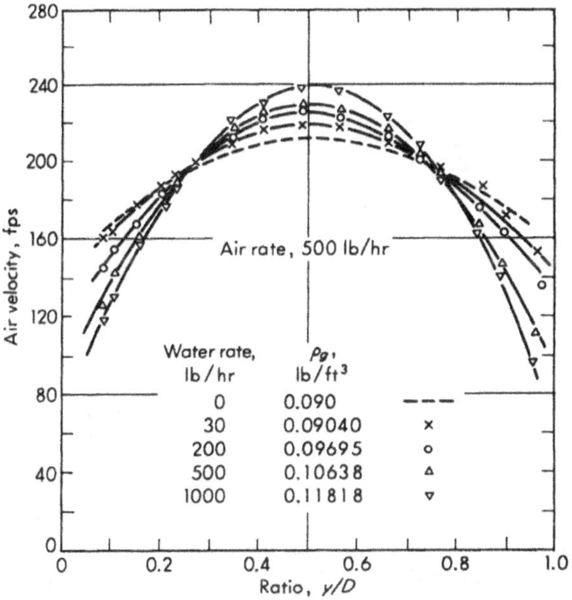

Fig. 5.3 Effect of injected water rate upon the air velocity profile in annular-mist flow. (*Gill, Hewitt, and Lacey.*[5])

flows without stratification, the shear stress distribution in a round pipe at an angle θ to the vertical is found, by balancing the forces on the cylinder of radius r shown in Fig. 5.4, to be

$$\tau = \frac{r}{2}\left(-\frac{dp}{dz} - \rho_2 g \cos\theta\right) - \frac{g(\rho_1 - \rho_2)\cos\theta}{2r}\int_0^r (1-\alpha)2r\,dr \tag{5.11}$$

We now introduce the dimensionless parameters

$$\tau^* = \frac{4\tau}{Dg(\rho_1 - \rho_2)} \tag{5.12}$$

$$\Delta p^* = \frac{-dp/dz - \rho_2 g \cos\theta}{g(\rho_1 - \rho_2)} \tag{5.13}$$

$$r^* = \frac{2r}{D} \tag{5.14}$$

and express Eq. (5.11) in the dimensionless form

$$\tau^* = r^*\,\Delta p^* - \frac{\cos\theta}{r^*}\int_0^{r^*} (1-\alpha)2r^*\,dr^* \tag{5.15}$$

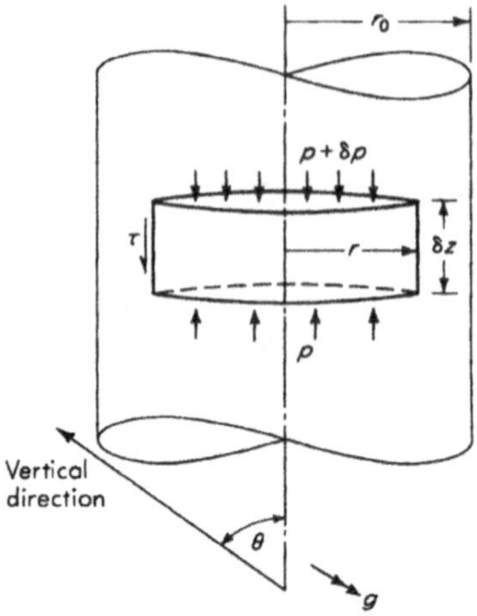

Fig. 5.4 Force balance on an elementary cylinder used to derive Eq. (5.11).

This shear stress distribution differs markedly from the familiar linear variation of single-phase flow. Depending upon the flow pattern, maxima and minima, changes of sign, and changes of curvature are all possible as the two terms in Eq. (5.15) interact.

Following Turner[6] we can also express Eqs. (5.4) and (5.6) in dimensionless form by defining the parameters

$$R_i = \left(\frac{\rho_i}{\rho}\right)^{1/2} \tag{5.16}$$

$$N_i = \frac{D\rho_i^{1/2}}{\mu_i}\,[Dg(\rho_1 - \rho_2)]^{1/2} \tag{5.17}$$

$$v_i^* = \frac{v\sqrt{\rho_i}}{[gD(\rho_1 - \rho_2)]^{1/2}} \tag{5.18}$$

$$l^* = \frac{2l}{D} \tag{5.19}$$

The subscript i refers to either phase and the phase subscripts are chosen so that $\rho_1 > \rho_2$. The properties or parameters with no subscripts apply to the local homogeneous average. For laminar flow, Eq. (5.4) becomes

$$\frac{dv_i^*}{dr^*} = -\frac{NR_i}{8}\tau^* \tag{5.20}$$

whereas for turbulent flow, from Eq. (5.6)

$$\frac{dv_i^*}{dr^*} = -\frac{HR_i}{l^*}|\tau^*|^{1/2} \tag{5.21}$$

H denotes the sign of the shear stress.

N in Eq. (5.20) is a function of the local homogeneous viscosity and density and can be written as

$$N = N_1\left(\frac{\rho}{\rho_1}\right)^{1/2}\frac{\mu_1}{\mu} = N_2\left(\frac{\rho}{\rho_2}\right)^{1/2}\frac{\mu_2}{\mu} \tag{5.22}$$

From Eq. (2.9) R_i can be expressed as a function of the local value of α and the density ratio ρ_1/ρ_2.

From any of Eqs. (2.56) to (2.61) the ratio μ/μ_i can be expressed in terms of α and the viscosity ratio μ_1/μ_2. Moreover

$$\frac{\mu_1}{\mu_2} = \frac{N_2}{N_1}\left(\frac{\rho_1}{\rho_2}\right)^{1/2} \tag{5.23}$$

Combining these results we eventually find that N is a function of N_1, N_2, ρ_1/ρ_2 and α.

The velocity profile is determined by integrating Eqs. (5.20) and (5.21), using Eq. (5.15) and information about the variation of α and l^* as functions of the radius r^*. The volumetric flow rate of each phase is obtained in dimensionless form by integrating the velocity profile. Thus

$$j_i^* = \int_0^1 \alpha_i v_i^* 2r^*\, dr^* \tag{5.24}$$

where α_i is the volumetric concentration of phase i and

$$j_i^* = j_i \rho_i^{1/2}[gD(\rho_1 - \rho_2)]^{-1/2} \tag{5.25}$$

In general, if the flow regime is specified so that the local value of α can be related to the average value $\langle\alpha\rangle$, we have

$$j_i^* = j_i^*\left(\Delta p^*, \langle\alpha\rangle, \frac{\rho_1}{\rho_2}, N_1, N_2, \theta\right) \tag{5.26}$$

Example 5.1 Determine expressions for j_f^* and j_g^* in laminar vertical annular gas-liquid flow.

Solution For vertical flow $\cos\theta = 1$. Since the flow is annular, the local value of α is unity in the gas core and zero in the liquid film. In terms of the overall void fraction therefore, Eq. (5.15) becomes

In the core $\quad \tau^* = r^*\,\Delta p^* \quad$ (5.27)

In the film $\quad \tau^* = r^*\,\Delta p^* - \dfrac{1}{r^*}\displaystyle\int_{\sqrt{\alpha}}^{r^*} 2r^*\, dr^* \quad$ (5.28)

Evaluating the integral in Eq. (5.28) we find

In the film $\quad \tau^* = r^*(\Delta p^* - 1) + \dfrac{\alpha}{r^*} \quad$ (5.29)

Substituting Eq. (5.29) into Eq. (5.20) with $i = f$ we obtain

$$\frac{dv_f^*}{dr^*} = -\frac{N_f}{8}\left[r^*(\Delta p^* - 1) + \frac{\alpha}{r^*}\right] \tag{5.30}$$

Integration from r^* to the wall, $r^* = 1$, $v_f^* = 0$, yields

$$v_f^* = \frac{N_f}{8}\left[\frac{(\Delta p^* - 1)(1 - r^{*2})}{2} - \alpha \ln r^*\right] \tag{5.31}$$

Substituting this value into Eq. (5.24) and performing the integration we have, since $\alpha_i = 1$ in the film and zero in the core,

$$j_f^* = \frac{N_f}{32}\{(\Delta p^* - 1)(1 - \alpha)^2 + 2\alpha[(1 - \alpha) + \alpha \ln \alpha]\} \tag{5.32}$$

The interface velocity is derived from Eq. (5.31) by putting $r^{*2} = \alpha$ and can be put in terms of the dimensionless gas velocity v_g^* by multiplying by the square root of the density ratio, thus

$$(v_g^*)_i = \frac{N_f}{16}\left(\frac{\rho_g}{\rho_f}\right)^{1/2}[(\Delta p^* - 1)(1 - \alpha) - \alpha \ln \alpha] \tag{5.33}$$

Now Eq. (5.27) is substituted into Eq. (5.20) to show that in the gas core

$$\frac{dv_g^*}{dr^*} = -\frac{N_g}{8} r^* \Delta p^* \tag{5.34}$$

Integrating from r^* to $\sqrt{\alpha}$ and using Eq. (5.33), we obtain the gas velocity profile

$$v_g^* = \frac{N_g}{16}\Delta p^* (\alpha - r^{*2}) + \frac{N_f}{16}\left(\frac{\rho_g}{\rho_f}\right)^{1/2}[(\Delta p^* - 1)(1 - \alpha) - \alpha \ln \alpha] \tag{5.35}$$

Using this in Eq. (5.24) with $\alpha_i = 1$ in the core and zero in the film, we find eventually that

$$j_g^* = \frac{N_g}{32}\left\{\alpha^2 \Delta p^* + 2\alpha \frac{N_f}{N_g}\left(\frac{\rho_g}{\rho_f}\right)^{1/2}[(\Delta p^* - 1)(1 - \alpha) - \alpha \ln \alpha]\right\} \tag{5.36}$$

Both Eqs. (5.32) and (5.36) are special cases of Eq. (5.26).

For a particular combination of components and flow regime in a given pipe, Eq. (5.26) expresses the two dimensionless flow rates j_1^* and j_2^* as functions of Δp^* and $\langle\alpha\rangle$. These two equations must be solved simultaneously if Δp^* and $\langle\alpha\rangle$ are to be deduced from known values of the flow rates.

Even when a theoretical solution to these equations cannot be obtained, the dimensionless parameters which have been derived can form the basis for a rational correlation scheme. The dimensionless parameters can also be combined in numerous ways to form new groupings when convenient.

Example 5.2 In the separated-flow model of annular vertical flow it has been suggested that the wall friction factor should be correlated versus the liquid film Reynolds number, $\text{Re}_f = j_f\rho_f D/\mu_f$. Show how this description is related to Eq. (5.26).

Solution The Reynolds number is readily expressed in terms of the previous dimensionless groups as

$$\mathrm{Re}_f = j_f^* N_f \tag{5.37}$$

The dimensionless wall shear stress, from Eq. (5.15), is

$$\tau_w^* = \Delta p^* - (1 - \alpha) \tag{5.38}$$

The wall friction factor can be expressed as follows:

$$C_{fw} = \frac{\tau_w(1-\alpha)^2}{\frac{1}{2}\rho_f j_f^2} = \frac{\tau_w^*(1-\alpha)^2}{2j_f^{*2}} \tag{5.39}$$

Therefore

$$C_{fw} = \frac{[\Delta p^* - (1-\alpha)](1-\alpha)^2}{2j_f^{*2}} \tag{5.40}$$

A correlation expressing C_{fw} as a function of Re_f thus implies the existence of a relationship between j_f^*, N_f, Δp^*, and α and is a special case of Eq. (5.26).

5.4 INTEGRAL ANALYSIS

To perform an integral analysis the form of the velocity and concentration profiles is assumed rather than calculated. Unknown parameters are either determined empirically or by making the profiles satisfy certain integral equations and boundary conditions. A familiar example from single-phase flow is the von Kármán integral technique for analyzing boundary layers. The method has not been widely used for two-phase flows, and only a few useful results are available.

The Bankoff *variable density model* for round tubes is based on the assumption of local homogeneity and no relative velocity. The velocity and concentration are assumed to vary according to power laws. Thus

$$\frac{v}{v_m} = \left(\frac{y}{r_0}\right)^{1/m} \tag{5.41}$$

$$\frac{\alpha}{\alpha_m} = \left(\frac{y}{r_0}\right)^{1/n} \tag{5.42}$$

v_m and α_m are the values at the center of the tube, y is the distance from the wall, and r_0 is the pipe radius. From Eq. (1.6) the average value of α is

$$\langle\alpha\rangle = \frac{1}{\pi r_0^2}\int_0^{r_0} \alpha_m\left(\frac{y}{r_0}\right)^{1/n} 2\pi(r_0 - y)\,dy = \frac{\alpha_m 2n^2}{(1+n)(1+2n)} \tag{5.43}$$

The mass flux of phase 2 is

$$G_2 = \frac{1}{\pi r_0^2}\int_0^{r_0} \alpha v \rho_2 2\pi(r_0 - y)\,dy$$

$$= \alpha_m v_m \rho_2 \frac{2n^2m^2}{(m+n+mn)(n+m+2nm)} \tag{5.44}$$

and the mass flux of phase 1 is

$$G_1 = \frac{1}{\pi r_0^2}\int_0^{r_0} (1 - \alpha) v \rho_1 2\pi (r_0 - y)\, dy$$
$$= \alpha_m v_m \rho_1 \left[\frac{2m^2}{(1 + m)(1 + 2m)\alpha_m} - \frac{2n^2m^2}{(m + n + mn)(n + m + 2nm)} \right] \tag{5.45}$$

Since

$$\frac{G_1}{G_2} = \frac{1 - x}{x} \tag{5.46}$$

Equations (5.44) and (5.45) can be combined to yield

$$1 + \frac{\rho_2}{\rho_1}\frac{1 - x}{x} = \frac{(m + n + nm)(n + m + 2nm)}{n^2(1 + m)(1 + 2m)\alpha_m} \tag{5.47}$$

Finally, elimination of α_m between Eqs. (5.43) and (5.47) shows that α is related to x by the equation

$$\alpha = \frac{K}{1 + (\rho_2/\rho_1)[(1 - x)/x]} \tag{5.48}$$

where

$$K = \frac{2(m + n + mn)(m + n + 2mn)}{(n + 1)(2n + 1)(m + 1)(2m + 1)} \tag{5.49}$$

Over a wide range of values of n between 0.1 and 5 and m between 2 and 7, K only varies between 0.5 and 1.

One interpretation of Eq. (5.48) is obtained by realizing that

$$\frac{\rho_2}{\rho_1}\frac{1 - x}{x} = \frac{j_1}{j_2} \tag{5.50}$$

Therefore, Eq. (5.48) is equivalent to

$$\alpha = K\frac{j_2}{j_1 + j_2} \tag{5.51}$$

and K is seen to be identical with the inverse of the parameter C_0 which was introduced in Chap. 4. The result of the analysis is a means for correcting homogeneous theory to allow for two-dimensional effects.

An advantage of having a model of this type is that it can be used to derive expressions for numerous quantities, such as momentum or kinetic-energy flux, in terms of the parameters m and n. Furthermore, since α is always zero at $y = 0$ (that is phase 1 must "wet" the wall) the wall shear stress can be related to the wall velocity gradient. The form of Eqs. (5.41) and (5.42), however, as well as the "locally homogeneous" assumption limits the number of flow regimes to which this technique can be applied.

The parameter K has been the basis for a variety of correlations relating void fraction to quality in gas-liquid flow.

Example 5.3 Show that, if K is constant, α tends to a limiting value as the quality is increased in an evaporator. How can greater values of α be achieved?

Solution As the quality is increased the ratio $(1 - x)/x$ tends to zero and Eq. (5.48) predicts that α tends to a limiting value equal to K. If this value is to be exceeded, the flow regime must change to one which has a higher value of K.

Equation (5.41) can be useful even when Eq. (5.42) is quite inappropriate, for example in annular gas-liquid flow. Assume that the liquid film can be regarded as behaving as part of an equivalent single-phase flow which fills the pipe. For turbulent flow $m \approx 7$ at low Reynolds numbers and the average velocity of the equivalent single-phase flow is found by integration to be

$$\bar{v} = {}^{49}\!/\!_{60} v_m \tag{5.52}$$

If the void fraction is α the average velocity in the liquid film is

$$v_f = \frac{1}{\pi r_0^2(1 - \alpha)} \int_0^{r_0(1-\sqrt{\alpha})} v_m \left(\frac{y}{r_0}\right)^{1/7} 2\pi(r_0 - y)\, dy \tag{5.53}$$

whence

$$v_f = {}^{49}\!/\!_{60} v_m \frac{(1 - \sqrt{\alpha})^{8/7}(1 + {}^8\!/\!_7\sqrt{\alpha})}{1 - \alpha} \tag{5.54}$$

Therefore, in view of Eq. (5.52)

$$\frac{v_f}{\bar{v}} = (1 - \sqrt{\alpha})^{8/7} \frac{(1 + {}^8\!/\!_7\sqrt{\alpha})}{(1 - \alpha)} \tag{5.55}$$

The wall friction factor is known as a function of $\bar{v}$ from single-phase flow theory and the wall shear stress is

$$\tau_w = C_{fw}\, {}^1\!/\!_2 \rho_f \bar{v}^2 \tag{5.56}$$

By substitution in Eq. (5.56) from Eq. (5.55), the wall shear stress can now be related to the average liquid velocity and calculated directly, if the void fraction is known. This approach will be discussed in more detail in Chap. 11.

5.5 MORE COMPLEX METHODS OF ANALYSIS

Velocity and concentration profiles in fully developed steady flow are perhaps the simplest examples of two-dimensional effects. A vast number of more complex problems have practical importance, including two-phase jets, developing profiles in inlet regions of pipe flow, boundary layer problems of film boiling, condensation, ablation and combustion, flows around bends and in centrifugal separators, interfacial waves, droplet break-up phenomena, and motion of large bubbles in ducts and nozzles.

Detailed development of equations which are suitably sophisticated to solve these problems is beyond the scope or intention of this text. However, the predictions of multidimensional analysis will often be quoted when the results are needed for incorporation into the one-dimensional theories. For instance, the analytical derivation of the equation governing the rise velocity of a large bubble in a vertical pipe will not be quoted but the result will be used extensively.

PROBLEMS

5.1. Derive Eq. (5.8) from Eqs. (5.6) and (5.7), using the single-phase shear stress distribution in a round pipe for a fluid of constant density.

5.2. Show how the shear stress distribution in vertical annular gas-liquid flow depends upon the values of Δp^* and α. Present the results graphically.

5.3. Deduce Eq. (5.51) from (5.41) and (5.42) and the definition of C_0 in Chap. 4 [Eq. (4.21)]. Show that $C_0 = K^{-1}$.

5.4. Evaluate the Bankoff parameter K for $n = 0.1$, 1, and 5 and $m = 2$, 4, and 7.

5.5. Show that the dimensionless pressure drop

$$\Delta p^* = \frac{-(dp/dz) - \rho_g g}{g(\rho_f - \rho_g)}$$

for a gas-liquid system in vertical flow represents the reading on a manometer connected to unit length of pipe, containing liquid of density ρ_f below gas with density ρ_g.

5.6. What dimensionless group would you expect to govern the rise velocity of a large bubble of gas in an inviscid fluid?

5.7. What dimensionless group would you expect to govern the rise velocity of a large bubble of gas in a very viscous fluid?

5.8. Combine the results of Probs. 5.6 and 5.7 to show that the rise velocity can be represented for liquids of any viscosity by plotting v_b^* versus N_f where v_b^* is a dimensionless bubble velocity

$$v_b^* = v_b(\rho_f)^{1/2}[gD_b(\rho_f - \rho_g)]^{-1/2}$$

What relationship between N_f and v_b^* represents the bubble in a viscous fluid?

5.9. In gas-liquid vertical upflow the definition of Δp^* is

$$\Delta p^* = \frac{(-dp/dz - \rho_g g)}{g(\rho_f - \rho_g)}$$

Consider laminar or turbulent flow of the gas or the liquid alone in a vertical pipe. For the usual laminar-flow relations, and for a given turbulent friction factor, obtain the following expressions relating Δp^* to other dimensionless groups. The coordinate z is measured upward.

(*a*) Laminar flow of liquid alone,

$$\Delta p^* = 1 + \frac{32j_f^*}{N_f}$$

(*b*) Laminar flow of gas alone,

$$\Delta p^* = \frac{32j_g^*}{N_g}$$

(*c*) Turbulent flow of liquid alone,

$$\Delta p^* = 1 + 2C_f j_f^{*2}$$

(*d*) Turbulent flow of gas alone,

$$\Delta p^* = 2C_f j_g^{*2}$$

5.10. Armand[10] suggests that $K = 0.833$ in horizontal flow. Predict the dependence of void fraction on quality for steam-water mixtures at 500, 1000, and 2000 psia and compare with the Martinelli and Thom correlations.

5.11. Perform an integral analysis of the liquid film in laminar annular flow, starting from the single-phase flow velocity distribution

$$v = 2\bar{v}\left[1 - \left(\frac{r}{r_0}\right)^2\right]$$

Show that $v_f = \bar{v}(1 - \alpha)$ and hence that the Martinelli parameter ϕ_f has the value

$$\phi_f = \frac{1}{1 - \alpha}$$

5.12. By assuming a constant value of mixing length, integrate Eq. (5.6) in single-phase flow to show that

$$\Delta p^* = (7l^*j^*)^2$$

and hence that

$$l^* = \frac{(2C_f)^{1/2}}{7}$$

Develop the same analysis for the separate-cylinders model, assuming that l^* is the same in each cylinder and equals the single-phase flow value. Show that in this case $n = 3.5$ in Eq. (3.30).

5.13. Equations (5.32) and (5.36) can be used to plot j_f^* versus j_g^* for constant values of α. Show that the lines of constant α are linear. What is the locus of points with constant Δp^*? Where is the flooding locus? Compare with Fig. 4.4.

For air and water at 70°F and 15 psia in a 1-in. pipe, evaluate N_f, N_g, and ρ_g/ρ_f. Compare the j_f^*-versus-j_g^* plot with the bubbly flow results shown in Fig. 4.5.

5.14. Determine an expression for momentum flux in a pipe from the Bankoff model. Compare with homogeneous flow theory.

5.15. If the single-phase friction factor C_f is related to the Reynolds number $\rho_f\bar{v}D/\mu_f$ by the Blasius equation,

$$C_f = A\left(\frac{\rho_f\bar{v}D}{\mu_f}\right)^{-1/4}$$

deduce the relationship between j_f^*, N_f, Δp^*, and α in horizontal flow, using the results of Example 5.2.

Show that the result is almost exactly the same as assuming that the wall friction factor defined by Eq. (5.39) is related to the film Reynolds number Re_f by a similar equation

$$C_{f_w} = A\left(\frac{\rho_f j_f D}{\mu_f}\right)^{-\frac{1}{4}}$$

5.16. Solve Example (5.1) for the case of horizontal flow. Show that

$$\frac{32j_g^*}{N_g\,\Delta p^*} = \alpha^2 + 2\frac{N_f}{N_g}\left(\frac{\rho_g}{\rho_f}\right)^{\frac{1}{2}}\alpha(1-\alpha)$$

$$\frac{32j_f^*}{N_f\,\Delta p^*} = (1-\alpha)^2$$

The left-hand sides of these equations are the same as the Martinelli parameters for viscous flow. For various values of (μ_f/μ_g), deduce the relationships between ϕ_f and ϕ_g, α and X, and compare with Figs. 3.6 and 3.8.

5.17. Use Eq. (5.42) and the values of local density and viscosity from Chap. 2 to analyze laminar gas-liquid flow in horizontal, vertical, and inclined pipes, using a differential technique.

5.18. Show that the axial velocity profile

$$v_z = -\frac{m\pi z}{2a\rho_g}\cos\frac{\pi y}{2a}$$

and the transverse velocity distribution

$$v_y = \frac{m}{\rho_g}\sin\frac{\pi y}{2a}$$

can represent frictionless flow with constant properties in between parallel plates distance $2a$ apart when uniform condensation is occurring on each plate at a mass rate m per unit area.[9] z is the axial coordinate and y the coordinate from the centerline normal to the plates. What is the pressure as a function of z and y? Show that the streamlines are normal to the walls and that the centerline dynamic head is twice the average. Deduce this latter conclusion by making an overall momentum balance and comparing with Bernoulli's equation for the center streamline.

5.19. Show that the solution to Problem 5.18 for small values of z is an approximation to the flow represented by the velocity potential

$$\phi = -\frac{2a}{\pi}\frac{m}{\rho_g}\cosh\frac{\pi z}{2a}\cos\frac{\pi y}{2a}$$

Show that this solution represents irrotational flow whereas the solution to Prob. (5.18) represents flow with constant rotation of each fluid element. Determine the pressure field in the duct.

5.20. Show that frictionless flow in a tube of radius r_0 with uniform condensation on the walls can be represented[9] by the velocity field

$$v_z = -\frac{\pi m z}{\rho_g r_0}\cos\frac{\pi r^2}{2r_0^2}$$

$$v_r = \frac{m r_0}{\rho_g r}\sin\frac{\pi r^2}{2r_0^2}$$

Sketch the velocity distributions and compare with the results of Olson and Eckert.[8] Show that the centerline dynamic head is twice the average. Determine the pressure field as a function of r and z.

5.21. Discuss how the solutions to Probs. 5.18, 5.19, and 5.20 can be used to represent reversible evaporation and condensation in a duct. These results may have commercial importance for the design of efficient two-phase separation equipment for distillation, desalination, and water purification and for increasing the efficiency of thermodynamic engines employing evaporation and condensation (e.g., refrigerators, power plants).

5.22. Show how it is possible, in the presence of condensation on the walls of a duct, to have either positive, negative, or zero wall shear stress depending on the shape of the velocity profile.

REFERENCES

1. Soo, S. L., G. J. Trezek, R. C. Dimick, and G. F. Hohnstreiter: *I & EC Fundamentals*, vol. 3, pp. 98–106, May, 1964.
2. Schlichting, H.: "Boundary Layer Theory," 4th ed., chap. 20, McGraw-Hill Book Company, New York, 1960.
3. Prandtl, L.: "Essentials of Fluid Dynamics," p. 118, Blackie and Son, Ltd., Glasgow, 1952 (transl. of "Führer durch die Strömungslehre," 3d ed., Vieweg u. sohn, Braunschweig, 1949).
4. Deissler, R. E.: NACA Tech. Notes, no. 2129, 1950; no. 2138, 1952; and no. 3145, 1959.
5. Gill, L. E., G. F. Hewitt, and P. M. C. Lacey: *Chem. Eng. Sci.*, vol. 19, p. 665, 1964.
6. Turner, J. M.: Ph.D. thesis, Dartmouth College, Hanover, N.H., 1966.
7. Bankoff, S. G.: *Trans. ASME J. Heat Transfer*, ser. C, vol. 82, p. 265, 1960.
8. Olson, R. M., and E. R. G. Eckert: *Trans. ASME J. Appl. Mech.*, ser. E, vol. 88, pp. 7–17, 1966.
9. Wallis, G. B.: *Trans. ASME J. Appl. Mech.*, ser. E, pp. 950–953, 1966.
10. Armand, A.: UKAEA, AERE transl. 828, 1959 (*Izvest. Vsesoyuz. Teplotekh. Inst.*, no. 1, pp. 16–23, 1946).

6
One-dimensional Waves

6.1 INTRODUCTION

This chapter will be devoted to the development of one-dimensional wave theory for both single-phase and two-phase flows. Single-phase flow is included because many of the results which are needed for later developments are, strangely enough, not to be found in any well-known textbook on the subject. As the concepts will be new to most readers, many examples will be used to illustrate the physical meaning of the mathematics.

Wave theory is a very powerful technique for analyzing unsteady flows and transient response. It also explains the processes such as choking and flooding which limit the performance of equipment. In some cases flow-regime changes can be attributed to instabilities which result from wave amplification.

Waves can either propagate continuous changes in some variables or can involve a step change or a finite discontinuity. The latter will be called shock waves, or "shocks" for short. Both waves and shocks can exist in many forms depending on the important physical processes which cause them. Two of the most important classes are *continuity*† and *dynamic* waves. Continuity waves are a quasi-steady-state phenomenon and occur whenever there is a relationship between flow rate and concentration. One steady-state value simply propagates into another one and there are no dynamic effects of inertia or momentum. Dynamic waves, on the other hand, depend for their existence on forces which will accelerate material through the wave as a result of concentration gradients. Since both the densities of the phases and their volumetric concentration are forms of concentration, many different kinds of waves can occur within the two major classes. In addition, the interaction between the various waves determines which, if any, dominates the motion and can also govern the stability of the flow.

Many two-phase wave phenomena, particularly interfacial waves, are two- and three-dimensional in character and will not be discussed in this chapter.

6.2 CONTINUITY WAVES IN SINGLE-PHASE FLOW

Continuity waves occur whenever the steady equilibrium flow rate of a substance depends on the amount of that substance which is present. For example, the flow rate of water in a river depends on the depth, the flow rate of cars on a highway depends on the traffic density, and the rate at which doctors interview patients depends on the number who are waiting outside their door. More specifically, the flow rate per unit width of a viscous fluid draining down a vertical wall is given as a function of thickness by the equation

$$q = \frac{g(\rho_f - \rho_g)\delta^3}{3\mu_f} \tag{6.1}$$

q is the flow rate and δ, the thickness, is the amount of fluid that is present.

In order to achieve some generality and to lead on to a variety of later developments, let us assume that the flow rate is expressed as a suitable flux j. Denote the amount of substance by a general variable α which can be called a concentration. α can be expressed in any units that are consistent with the definition of j such as pounds per cubic foot, cubic feet per cubic foot, molecules per cubic centimeter, cars per mile, liquid depth, patients per room, etc. All that is required is that j should

† I first heard of a "continuity wave" indirectly from someone who had heard Professor A. H. Shapiro use the term at a Massachusetts Institute of Technology seminar. This name is more descriptive than "kinematic" waves, which is Lighthill and Whitham's expression for the same thing, and is less easily confused with "dynamic" waves. Only very simple ideas of continuity are needed in order to understand what is going on and it is high time that the subject was given a place in elementary fluid mechanics texts.

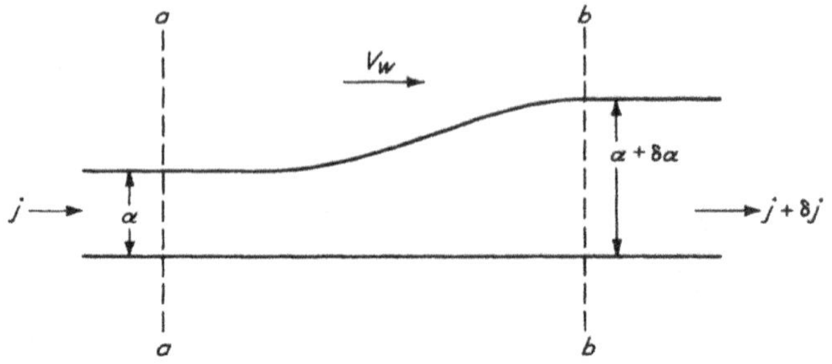

Fig. 6.1 Propagation of a continuity wave.

be a function of α, and should equal V times α, where V is the average velocity.

The continuity wave velocity is derived by considering continuity of matter across the control volume moving with the wave and bounded by the lines *aa* and *bb* in Fig. 6.1 which shows a wave propagating with velocity V_w from a region where the concentration is α to a region where the concentration is $\alpha + \delta\alpha$. Relative to the wave front, as much matter is approaching as is departing. Therefore

$$j - \alpha V_w = j + \delta j - V_w(\alpha + \delta\alpha) \tag{6.2}$$

whence

$$V_w = \left(\frac{\partial j}{\partial \alpha}\right)_f \tag{6.3}$$

If the change in concentration across the wave is finite, exactly the same reasoning leads to the result

$$V_s = \left(\frac{j_1 - j_2}{\alpha_1 - \alpha_2}\right)_f \tag{6.4}$$

Subscripts denote different sides of the wave front. Such a wave will be called a *continuity shock wave*.

The subscript f denotes that an equilibrium of forces is maintained on both sides of the wave and that there are no inertia effects. For example, continuity waves often occur in systems where gravity is balanced against drag forces. Removing the inertia terms and the pressure gradient from Eq. (3.45), written for a single-phase flow, we have

$$f + b = 0 \tag{6.5}$$

If b is a constant body force, then the drag force per unit volume will also be a constant as long as the flow is not accelerating. The fact that j is a function of α implies that this constant drag force is a function of both j and α. In fact, for a specific system, the drag law may be known in the form

$$f = f(j,\alpha) \tag{6.6}$$

Alternatively, since

$$j = \alpha V \tag{6.7}$$

Equation (6.6) is equivalent to

$$f = f(V,\alpha) \tag{6.8}$$

Using Eq. (6.7) the continuity wave velocity from Eq. (6.3) can be expressed in the alternative form

$$V_w = V + \alpha\left(\frac{\partial V}{\partial \alpha}\right)_f \tag{6.9}$$

which shows that the wave velocity exceeds the average velocity by the amount $\alpha(\partial V/\partial \alpha)_f$. One advantage of this formulation is that the result is independent of the units in which α is measured.

If only the relationship (6.8) is known explicitly, we may write

$$\left(\frac{\partial V}{\partial \alpha}\right)_f = -\frac{(\partial f/\partial \alpha)_V}{(\partial f/\partial V)_\alpha} \tag{6.10}$$

Using subscripts to denote partial differentiation, Eqs. (6.9) and (6.10) can be combined to give

$$V_w = V - \alpha\frac{f_\alpha}{f_v} \tag{6.11}$$

The following examples will illustrate the physical significance of these results. See Appendix B for graphical representation.

Example 6.1 Calculate the continuity wave velocity for a viscous fluid flowing down a vertical wall. Show that this velocity is three times the average velocity. Deduce the surface profile of a film which flows down the wall of a vessel which is drained from the bottom after being initially full.

Solution δ represents the amount of liquid per unit width and can be identified with α. q is the flow per unit width and can be identified with j. The average velocity of the liquid, from Eq. (6.7), is

$$V = \frac{q}{\delta} = \frac{g(\rho_f - \rho_g)\delta^2}{3\mu_f} \tag{6.12}$$

The continuity wave velocity is derived from Eq. (6.3). Thus

$$V_w = \frac{\partial q}{\partial \delta} = \frac{g(\rho_f - \rho_g)\delta^2}{\mu_f} \tag{6.13}$$

It follows from Eqs. (6.12) and (6.13) that

$$V_w = 3V \tag{6.14}$$

As the vessel drains, continuity waves will propagate values of film thickness, each with the appropriate velocity given by Eq. (6.12). If the surface is flat at $z = 0$ when the draining starts, waves corresponding to all values of δ start there at $t = 0$. During the subsequent motion the larger values of δ will propagate faster, in view of Eq. (6.13). After a time t a wave has gone a distance

$$z = V_w t \tag{6.15}$$

and the position of each wave can be represented in the zt plane by straight lines (Fig. 6.2*b*).

Substituting from Eq. (6.13) we have

$$z = \frac{g(\rho_f - \rho_g)t}{\mu_f}\delta^2 \tag{6.16}$$

which is the equation of the surface profile at time t (as deduced by Jeffreys[2] and shown in Fig. 6.2*a*).

Example 6.2 A viscous fluid is being poured down a wall at a steady rate corresponding to a thickness δ_1. The flow rate is suddenly reduced to a value corresponding to a new thickness δ_2, in steady flow. Describe what happens to the liquid surface profile.

Solution Waves will propagate from $z = 0$ exactly as in the previous example. Only waves corresponding to values of δ between δ_1 and δ_2 will exist. The initial step change will spread out as it moves down the wall (Fig. 6.3).

The two previous examples illustrate how the motion of continuity waves enables the "end conditions" to be propagated throughout the flow. This mechanism has a general significance. If continuity waves do not propagate in this way, the end conditions are unable to exert any influence.

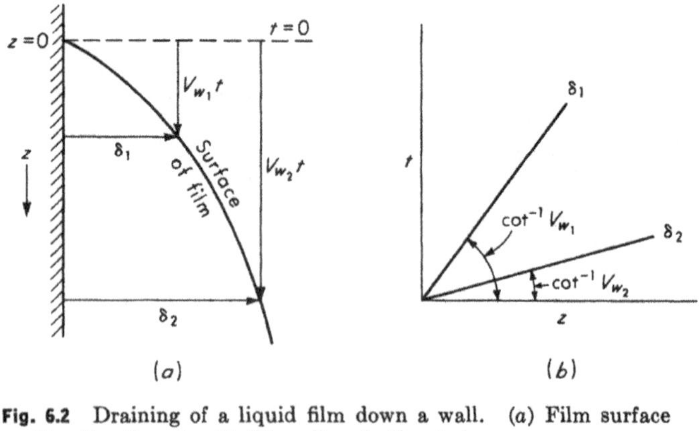

Fig. 6.2 Draining of a liquid film down a wall. (*a*) Film surface profile after time t. (*b*) Wave lines in the zt plane.

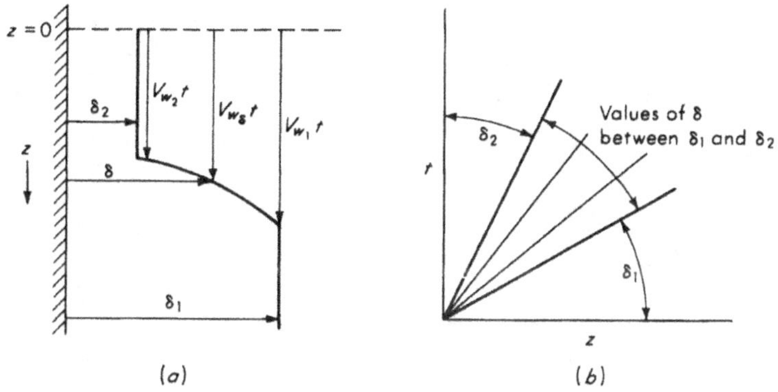

Fig. 6.3 Behavior of a falling liquid film after the flow rate is suddenly reduced. (*a*) Surface profile after time *t*. (*b*) Wave lines in the *zt* plane.

The limiting condition occurs when the continuity waves are brought to rest and is analogous to choking due to a Mach number of unity in the presence of compressibility waves.

Example 6.3 The friction factor in turbulent flow of a falling liquid film is approximately constant. What is the ratio between the continuity wave velocity and the average liquid velocity?

Solution The wall shear stress is

$$\tau_w = C_f \cdot \tfrac{1}{2}\rho V^2 \tag{6.17}$$

Identify δ, the thickness, as the concentration; then f, the force per unit volume is

$$f = \frac{\tau_w}{\delta} = \frac{C_f \rho V^2}{2\delta} \tag{6.18}$$

Using Eq. (6.11) we have

$$V_w = V - \delta \frac{-C_f \rho V^2/2\delta^2}{C_f \rho V/\delta}$$

whence

$$V_w = \frac{3V}{2} \tag{6.19}$$

Therefore the wave velocity is one and one-half times the average velocity.

THE FORMATION AND STABILITY OF CONTINUITY SHOCKS

If a fast continuity wave overtakes a slower one, they will agglomerate and form a shock. This process is represented on the *zt* plane by the intersection of the wave propagation lines to form a new line with intermediate slope.

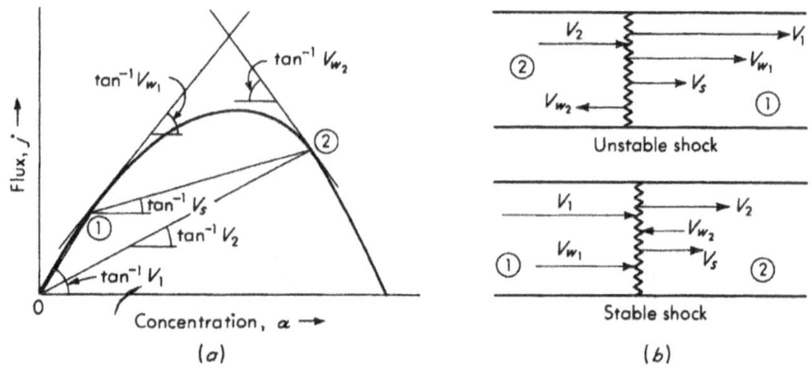

Fig. 6.4 The direction of stable shock propagation deduced from the curvature of the flux-concentration curve.

In order for a shock to maintain a steep front and not degenerate into a series of smaller shocks and waves, it is necessary that continuity waves run toward the shock on both sides. These effects determine whether a stable shock can occur and, if so, whether it propagates into fluid at state 1 or state 2. Without this consideration the direction of the shock would be indeterminate from Eq. (6.4).

For example, Fig. 6.4*a* shows a continuity shock represented on the curve of j versus α. In view of Eq. (6.4) the shock velocity is given by the slope of the chord joining points 1 and 2. The corresponding continuity wave velocities in the fluid adjacent to the shock are given by the slopes of tangents to the curve at 1 and 2. Figure 6.4*b* shows that state 2 must occur on the positive side of the shock if it is to be stable.

Example 6.4. On long uniform highways it is found experimentally that there is a relationship between the number of cars per mile (concentration) and the flow rate in cars per hour. There is no flow when there are either no cars at all ($\alpha = 0$) or when the cars are bumper to bumper ($\alpha = \alpha_{max}$). There is a maximum flow rate (j_{max}) at some intermediate value of α. Suppose that a steady flow of cars has existed for a long time on a highway with a value of $j_1 < j_{max}$ and that an accident occurs at z_0 which temporarily reduces the allowable flow rate in its vicinity to a value j_2 less than j_1. What happens and how can it be predicted from the j-versus-α relationship?

Solution Figure 6.5*a* shows j as a function of α. Since the steady flow j_1 has been established for a long time, it has probably propagated downstream and therefore has the lower of the two possible values of α (the one corresponding to the positive slope of the curve). When the accident limits the flow to j_2, a new value of α is propagated back up the highway in the negative direction. The curve must therefore have a negative slope at the point which represents this situation, and so the larger of the two values of α corresponding to j_2 occurs.

The conditions for a stable shock between α_1 and α_2 are satisfied and this wave moves back up the highway with the negative velocity V_s shown in the figure.

After the obstruction is removed at time t', the initial shock wave at the scene of the accident splits up into continuity waves corresponding to a continuous spectrum of values of α. Those to the left of the maximum in the curve propagate forward and those to the right propagate backward. The fastest forward-moving wave corresponds to $\alpha = \alpha_2'$ since this is the traffic density which was set up by the flow j_2 downstream of the accident. The fastest backward-moving wave propagates $\alpha_2 - \delta\alpha$ up the highway until it coalesces with the shock at time $t = t''$. As further waves join this shock it "weakens,"

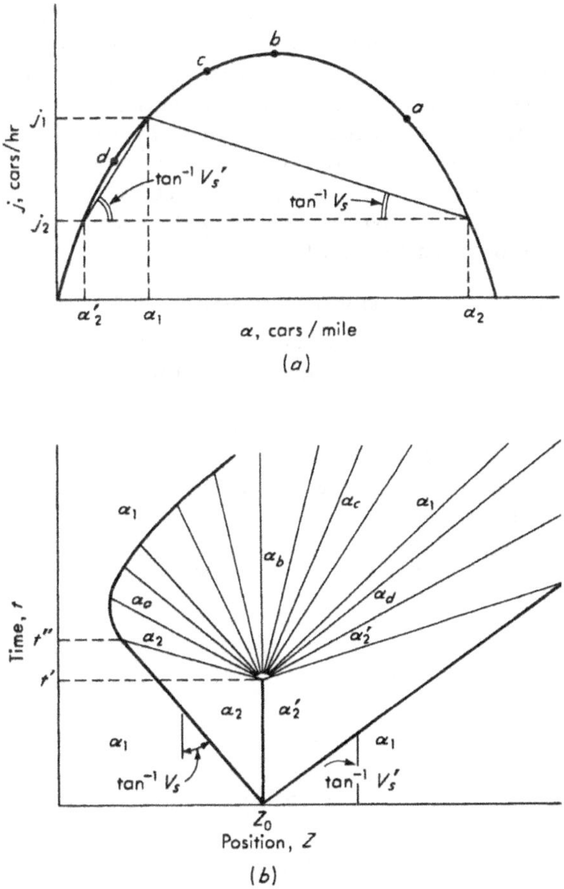

Fig. 6.5 Traffic density variations following a temporary restriction on a highway. In (*b*) shock waves are represented by heavy lines and continuity waves by thin lines.

swings around in the tz plane and asymptotically degenerates to an infinitesimal continuity wave corresponding to α_1. The fan of waves between α_2' and α_1 in Fig. 6.5*b* eventually merges with the shock from α_2' to α_1 which has propagated downstream from the accident.

The traffic density as a function of t and z is readily determined from Fig. 6.5*b*. In the large open areas of the graph the density is constant, whereas in the region covered by the fan of lines the density varies continuously.

STABILITY OF CONTINUITY WAVES

As long as the wavelength is sufficiently long, it is always possible to draw the sections *aa* and *bb* in Fig. 6.1 far enough away from the steeper parts of the wave front so that steady flow can be assumed. However, a quasi-steady-flow analysis, which assumes that the net forces on the flowing substance are everywhere in balance, is unable to account for the acceleration which must occur somewhere in the wave. Another mechanism must be postulated to account for this acceleration. In many cases this can be shown to be due to a force that is produced by concentration gradients (i.e., a *compressibility* effect). The effect of concentration gradient is conveniently analyzed by considering *dynamic waves* which will be discussed in a later section.

THE EFFECT OF A SOURCE OF MATTER

Continuity wave theory requires modification if there is an external source of matter being added to the flow (for example, tributaries discharging into a river, cars entering a highway from driveways, condensation on a liquid film). In this case we modify Fig. 6.1 to include a source of matter β, per unit length, and, for variety, consider the stationary control volume of length δz shown in Fig. 6.6. If the units of β are wisely chosen we have

$$\beta\delta z + j = (j + \delta j) + \frac{\partial\alpha}{\partial t}\,\delta z \tag{6.20}$$

whence

$$\frac{\partial\alpha}{\partial t} + \frac{\partial j}{\partial z} = \beta \tag{6.21}$$

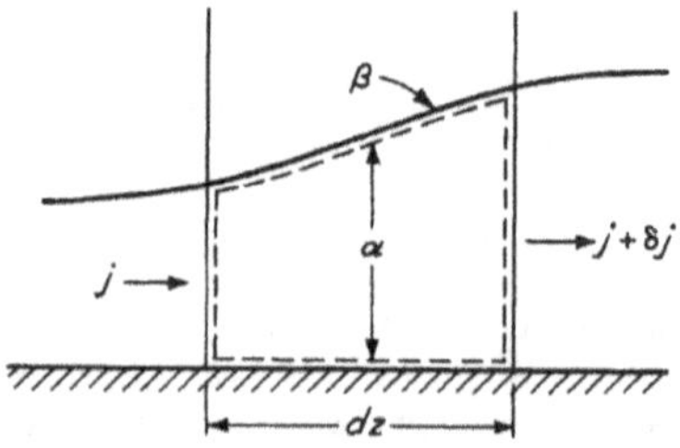

Fig. 6.6 Control volume for analyzing continuity waves when there is a source of matter β per unit length.

If j is a function of α, $\partial j/\partial z$ can be replaced by $(\partial j/\partial \alpha)(\partial \alpha/\partial z)$ and use of Eq. (6.3) gives

$$\frac{\partial \alpha}{\partial t} + V_w \frac{\partial \alpha}{\partial z} = \beta \tag{6.22}$$

The left-hand side of Eq. (6.22) is the total time derivative of α for a coordinate system moving with velocity V_w. Therefore

$$\left(\frac{d\alpha}{dt}\right)_{\text{moving with } V_w} = \beta \tag{6.23}$$

If β is zero, continuity waves propagate values of α unchanged, as before. Otherwise the waves will grow or decay, depending on the sign of β.

Example 6.5 Solve Example 6.1 if there is uniform condensation on the film surface at a constant rate β (volume of liquid per unit length per unit width per unit time).

Solution Identify α with δ, the film thickness. Then Eq. (6.23) becomes for a given wave

$$\frac{d\delta}{dt} = \beta \tag{6.24}$$

Since β is constant, Eq. (6.24) integrates to predict that

$$\delta - \delta_0 = \beta(t - t_0) \tag{6.25}$$

where the subscript 0 represents initial conditions which serve to identify the wave.

Since the wave velocity is V_w, we have, from Eq. (6.13), again for a given wave

$$\frac{dz}{dt} = \frac{g(\rho_f - \rho_g)\delta^2}{\mu} \tag{6.26}$$

Combining Eqs. (6.24) and (6.26) gives

$$\frac{d\delta}{dz} = \frac{d\delta/dt}{dz/dt} = \frac{\beta\mu}{g(\rho_f - \rho_g)\,\delta^2} \tag{6.27}$$

and, on integration,

$$\frac{(\delta^3 - \delta_0^3)g(\rho_f - \rho_g)}{3\mu} = \beta(z - z_0) \tag{6.28}$$

Equation (6.28) gives the wave paths in the δz plane. Elimination of δ between Eqs. (6.25) and (6.28) gives the wave line in the zt plane in terms of the initial parameters. Thus

$$[\delta_0 + \beta(t - t_0)]^3 = \delta_0^3 + \frac{3\mu\beta(z - z_0)}{g(\rho_f - \rho_g)} \tag{6.29}$$

On the other hand, elimination of δ_0 between Eqs. (6.25) and (6.28) gives the value of δ as a function of z at a given time and is the surface profile,

$$\delta^3 = [\delta - \beta(t - t_0)]^3 + \frac{3\mu\beta(z - z_0)}{g(\rho_f - \rho_g)} \tag{6.30}$$

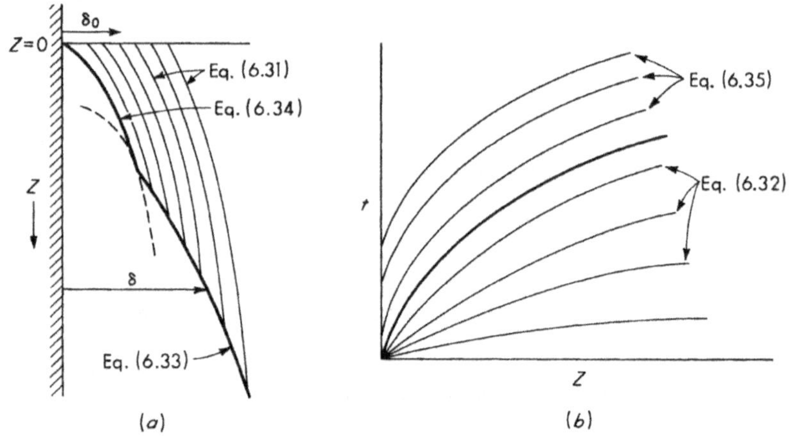

Fig. 6.7 Transient draining of a viscous-liquid film in the presence of uniform condensation. (*a*) Surface profile and wave paths. (*b*) Wave lines.

For the present problem there are two families of waves, those which start at $t_0 = 0$ from $z_0 = 0$ with all values of δ and those which start at later times from $z_0 = 0$ with the value $\delta_0 = 0$. For the first family we have, from Eqs. (6.28) to (6.30):

Wave path in δz plane:

$$\delta^3 = \delta_0{}^3 + \frac{3\mu\beta z}{g(\rho_f - \rho_g)} \tag{6.31}$$

Wave line in zt plane:

$$(\delta_0 + \beta t)^3 = \delta_0{}^3 + \frac{3\mu\beta z}{g(\rho_f - \rho_g)} \tag{6.32}$$

Surface profile:

$$\delta^3 = (\delta - \beta t)^3 + \frac{3\mu\beta z}{g(\rho_f - \rho_g)} \tag{6.33}$$

For the second family the wave paths and the surface profile are coincident and are represented by

$$\delta^3 = \frac{3\mu\beta z}{g(\rho_f - \rho_g)} \tag{6.34}$$

which is the equation of the film profile in the final steady state. In the zt plane the wave lines are the parallel curves, deduced from Eq. (6.29),

$$t = t_0 + \left[\frac{3\mu z}{g(\rho_f - \rho_g)\beta^2}\right]^{1/3} \tag{6.35}$$

These results are illustrated in Fig. 6.7.

6.3 CONTINUITY WAVES IN INCOMPRESSIBLE TWO-COMPONENT FLOW

Consider two incompressible components, 1 and 2, flowing in a duct of constant cross section. For continuity reasons the overall flow rate and hence the mean volumetric flux j is constant throughout the duct (although it may vary with time due to changes in the end condition). Denoting the volumetric fluxes of the components by j_1 and j_2, we have

$$j = j_1 + j_2 \tag{6.36}$$

Let the volumetric concentration of component 2 be α. The equilibrium condition can then be expressed as before:

$$f(j_1, j_2, \alpha) = \text{const} \tag{6.37}$$

The technique used to derive Eqs. (6.3) and (6.4) may then be applied in exactly the same way to show that the continuity wave and shock-wave velocities are

$$V_w = \left(\frac{\partial j_2}{\partial \alpha}\right)_{j,f} = \left[\frac{\partial j_1}{\partial(1-\alpha)}\right]_{j,f} \tag{6.38}$$

$$V_s = \left[\frac{(j_2)_1 - (j_2)_2}{\alpha_1 - \alpha_2}\right]_{j,f} = \left[\frac{(j_1)_1 - (j_1)_2}{\alpha_2 - \alpha_1}\right]_{j,f} \tag{6.39}$$

To visualize the meaning of Eqs. (6.38) and (6.39), consider that Eq. (6.37) is plotted to give α as a function of j_2, for various values of j_1 (Fig. 6.8). In gas-liquid flow, for instance, this would be a plot of void fraction versus gas rate for various liquid rates. In going from point 1 to point 2 on the graph across a wave one must follow a line $j_1 + j_2 = \text{const}$

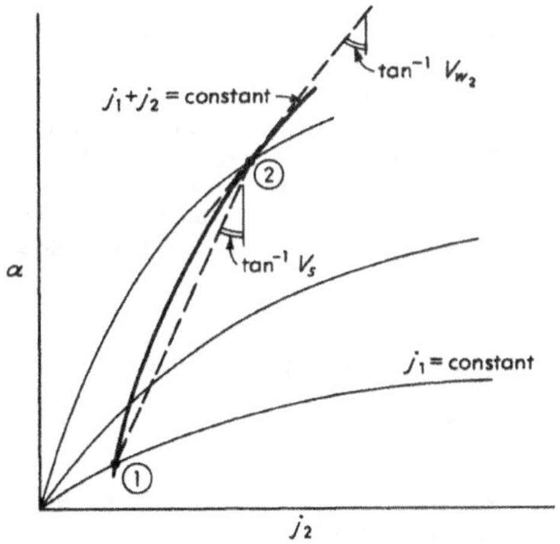

Fig. 6.8 Continuity waves in incompressible two-component flow. The continuity wave velocity is given by the slope of the tangent to the curve $j_1 + j_2 = \text{const}$. The shock velocity is the slope of the chord joining points on the same curve.

from Eq. (6.36). Equations (6.38) and (6.39) then give the required wave velocities directly. Graphically, the continuity wave and shock velocities are represented by the slopes of tangents and chords to the curves $j_1 + j_2 =$ const.

An important modification can be made by focusing attention on the drift flux j_{21} which was introduced in Chap. 4.

Substituting Eq. (4.4) into Eqs. (6.38) and (6.39), we obtain

$$V_w = j + \left(\frac{\partial j_{21}}{\partial \alpha}\right)_{j,f} \tag{6.40}$$

$$V_s = j + \left[\frac{(j_{21})_1 - (j_{21})_2}{\alpha_1 - \alpha_2}\right]_{j,f} \tag{6.41}$$

Equations (6.40) and (6.41) are always valid. They are particularly useful in cases where the drift flux is only a function of concentration and does not depend on the overall flux. In this case the wave velocities differ from the volumetric average velocity by an amount which depends only on the value of α and the system properties. The relationship between j_{21} and α then fulfills the same function for two-phase flow as the j-versus-α relationship does for single-phase flow. Wave and shock velocities can be represented by the slopes of tangents and chords, as before.

In later chapters continuity wave theory will be applied in detail to explain unsteady one-dimensional behavior in bubble columns, foams, sedimentation, and fluidization.

Example 6.6 For a bubble column the expression for j_{gf} is determined empirically to be

$$j_{gf} = v_\infty \alpha (1 - \alpha)^2 \tag{6.42}$$

It is also known that foams with $\alpha > \frac{1}{2}$ are very unstable and lead to rapid bubble bursting and agglomeration.

A depth H of liquid is put into a vertical vessel of constant diameter. Gas is then bubbled through with a flux $j_g = v_\infty/8$. What is the new height of the bubbly mixture? If the flow rate of gas is changed slightly, how long does it take before the whole column has adjusted to the new conditions?

Solve the same problem for a gas flux of $v_\infty/4$.

Solution Since j_f is zero we have, from Eq. (4.2) with component 2 identified as the gas,

$$j_{gf} = (1 - \alpha) j_g \tag{6.43}$$

Combining this with Eq. (6.42) gives

$$j_g = v_\infty \alpha (1 - \alpha) \tag{6.44}$$

With $j_g = v_\infty/8$, the solutions to Eq. (6.44) are $\alpha = \frac{1}{2} \pm \sqrt{2}/4$. Normally the lower value of α will occur at the bottom of the column and the higher value at the top (because continuity waves propagate in from the ends). However, we are told that the foam is unstable so only $\alpha = \frac{1}{2} - \sqrt{2}/4$ occurs. Since liquid is conserved, the new height H' is given by the expression

$$\frac{H' - H}{H'} = \alpha$$

Therefore

$$H' = \frac{H}{1 - \alpha} = \frac{H}{\frac{1}{2} + \sqrt{2}/4} = 1.17H \tag{6.45}$$

When the flow rate is changed slightly, the new conditions will propagate with the continuity wave velocity given by Eq. (6.40). Since $j = j_g$ we have, using Eq. (6.42)

$$V_w = j_g + v_\infty(1 - \alpha)(1 - 3\alpha) \tag{6.46}$$

With the given values of j_g and the calculated values of α, therefore

$$V_w = 0.605 v_\infty \tag{6.47}$$

The time taken by the wave to traverse the whole column and propagate the new value of α is

$$\frac{H'}{V_w} = 1.935 \frac{H}{v_\infty} \tag{6.48}$$

If the value of j_g had been $v_\infty/4$, the solution to Eq. (6.43) would have been $\alpha = \frac{1}{2}$ which is a double root corresponding to flooding. Equation (6.46) then predicts $V_w = 0$, showing that waves cannot propagate at all, thus giving another explanation for the mechanism of flooding.

6.4 DYNAMIC WAVES

DYNAMIC WAVES IN SINGLE-COMPONENT FLOW

Dynamic waves occur whenever a net force on a flowing substance is produced by a concentration gradient. Let this force per unit volume be denoted by f. For a concentration gradient $\partial\alpha/\partial z$ the net force produced, as long as we stay in the range over which a linear relationship applies, is

$$f = \left[\frac{\partial f}{\partial(\partial\alpha/\partial z)}\right]\frac{\partial\alpha}{\partial z} \tag{6.49}$$

Using a subscript notation for derivatives and replacing $\partial\alpha/\partial z$ by the more convenient $\nabla\alpha$, Eq. (6.49) can be rewritten

$$f = f_{\nabla\alpha}\frac{\partial\alpha}{\partial z} \tag{6.50}$$

Consider the case shown in Fig. 6.9 in which a dynamic wave is traveling with velocity c relative to stationary fluid or particles. The

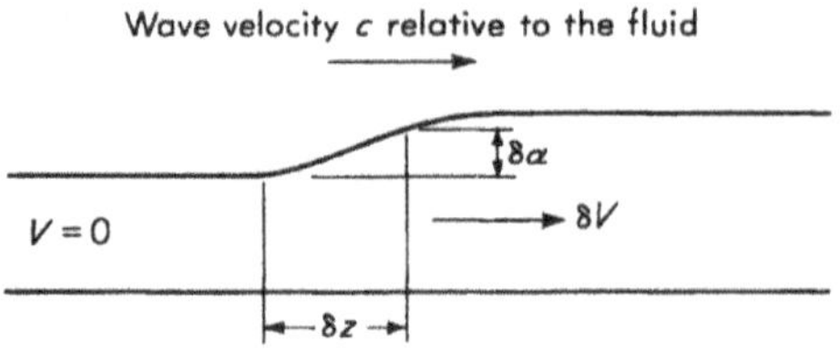

Fig. 6.9 Propagation of a dynamic wave.

wave may be brought to rest by imposing a velocity $-c$ on the whole system. Then the continuity equation through the wave is

$$c \frac{d\alpha}{dz} + \alpha \frac{dV}{dz} = 0 \tag{6.51}$$

and the equation of motion is

$$\rho c \frac{dV}{dz} = f_{\nabla\alpha} \frac{d\alpha}{dz} \tag{6.52}$$

Eliminating dV/dz and $d\alpha/dz$ from Eqs. (6.51) and (6.52) we obtain

$$c^2 = -\frac{\alpha f_{\nabla\alpha}}{\rho} \tag{6.53}$$

Therefore the wave velocity is given by the expression

$$c = \pm \left(\frac{-\alpha f_{\nabla\alpha}}{\rho} \right)^{1/2} \tag{6.54}$$

Relative to the fluid, dynamic waves run in both directions with speed c, whereas continuity waves run in only one direction. The condition that waves exist at all is that $f_{\nabla\alpha}$ is negative. Physically this means that the system under consideration resists compression or expansion (otherwise it would either explode or collapse catastrophically if left to itself in an environment at constant pressure).

EXAMPLES OF DYNAMIC WAVES IN SINGLE-COMPONENT FLOW

Long waves in a canal of constant width The concentration can be measured in terms of volume per unit width per unit length, i.e., depth. Therefore α can be identified with the water depth y in the canal. Thus

$$\alpha = y \tag{6.55}$$

The force on a vertical slice of fluid, width b, caused by a small change in depth δy is easily found to be

$$F = -\rho g y b\,(\delta y) \tag{6.56}$$

Therefore the force per unit volume averaged over the section is

$$f = \frac{F}{ybdz} = -\rho g \frac{dy}{dz} \tag{6.57}$$

Hence, using Eq. (6.50),

$$f_{\nabla\alpha} = -\rho g \tag{6.58}$$

Substituting Eqs. (6.58) and (6.55) into Eq. (6.54) gives at once

$$c = \pm (gy)^{1/2} \tag{6.59}$$

which is a well-known result.

Waves in a homogeneous compressible fluid The concentration α is conveniently measured in terms of mass per unit volume, i.e.,

$$\alpha = \rho \tag{6.60}$$

The force on a fluid element per unit volume is

$$f = -\frac{dp}{dz} = -\frac{\partial p}{\partial \rho}\frac{d\rho}{dz} \tag{6.61}$$

Hence, from Eq. (6.50),

$$f_{\nabla\alpha} = -\frac{\partial p}{\partial \rho} \tag{6.62}$$

Substituting Eqs. (6.62) and (6.60) into Eq. (6.54) then gives the familiar expression

$$c = \pm \left(\frac{\partial p}{\partial \rho}\right)^{1/2} \tag{6.63}$$

The derivative $\partial p/\partial \rho$ must be evaluated along a specified path. For gases the isentropic path is usually appropriate; however, in some special cases, other paths (e.g., isothermal) may be applicable.

DYNAMIC WAVES IN INCOMPRESSIBLE TWO-COMPONENT FLOW IN A CONSTANT AREA DUCT

Consider two components flowing steadily in a straight duct of constant area with velocities v_1 and v_2. Apply a velocity $-U$ to the whole system to bring any dynamic waves to rest. The velocities in the new frame of reference are

$$v_1' = v_1 - U \tag{6.64}$$
$$v_2' = v_2 - U \tag{6.65}$$

The continuity equations are

$$\frac{d}{dz}[v_1'(1-\alpha)] = 0 \tag{6.66}$$

$$\frac{d}{dz}(v_2'\alpha) = 0 \tag{6.67}$$

and the equations of motion are, from Eqs. (3.45) and (3.46),

$$\rho_1 v_1' \frac{dv_1'}{dz} = -\frac{dp}{dz} + f_1 + b_1 \tag{6.68}$$

$$\rho_2 v_2' \frac{dv_2'}{dz} = -\frac{dp}{dz} + f_2 + b_2 \tag{6.69}$$

Subtracting Eq. (6.69) from Eq. (6.68) we get

$$\rho_1 v_1' \frac{dv_1'}{dz} - \rho_2 v_2' \frac{dv_2'}{dz} = f_1 - f_2 + b_1 - b_2 \tag{6.70}$$

Dynamic waves occur when the right-hand side of Eq. (6.70) depends linearly on the concentration gradient, thus

$$f_1 - f_2 + b_1 - b_2 = -f_{\nabla\alpha}\frac{d\alpha}{dz} \tag{6.71}$$

Substituting Eq. (6.71) into Eq. (6.70) and using Eqs. (6.66) and (6.67) to eliminate dv_1'/dz and dv_2'/dz gives

$$\frac{\rho_1 v_1'^2}{1-\alpha} + \frac{\rho_2 v_2'^2}{\alpha} + f_{\nabla\alpha} = 0 \tag{6.72}$$

Expressing v_1' and v_2' in terms of the original velocities from Eqs. (6.64) and (6.65) leads to a quadratic in U,

$$U^2\left(\frac{\rho_1}{1-\alpha} + \frac{\rho_2}{\alpha}\right) - 2U\left(\frac{\rho_1 v_1}{1-\alpha} + \frac{\rho_2 v_2}{\alpha}\right) + \frac{\rho_1 v_1^2}{1-\alpha} + \frac{\rho_2 v_2^2}{\alpha} + f_{\nabla\alpha} = 0 \tag{6.73}$$

whence

$$U = \frac{\dfrac{v_1\rho_1}{1-\alpha} + \dfrac{v_2\rho_2}{\alpha} \pm \left[\dfrac{-\rho_1\rho_2(v_1-v_2)^2}{\alpha(1-\alpha)} - \left(\dfrac{\rho_1}{1-\alpha} + \dfrac{\rho_2}{\alpha}\right) f_{\nabla\alpha}\right]^{1/2}}{\rho_1/(1-\alpha) + \rho_2/\alpha} \tag{6.74}$$

Defining a weighted mean velocity (which is not the same as the velocity of the center of gravity) by

$$V_0 = \frac{v_1\rho_1/(1-\alpha) + v_2\rho_2/\alpha}{\rho_1/(1-\alpha) + \rho_2/\alpha} \tag{6.75}$$

and a dynamic wave velocity by

$$c = \pm \left(\frac{-(v_1 - v_2)^2}{\alpha/\rho_2 + (1 - \alpha)/\rho_1} - f_{\nabla\alpha} \right)^{1/2} \left(\frac{\rho_1}{1 - \alpha} + \frac{\rho_2}{\alpha} \right)^{-1/2} \tag{6.76}$$

Equation (6.74) becomes

$$U = V_0 \pm c \tag{6.77}$$

Dynamic waves therefore move relative to the weighted mean velocity V_0 with a velocity $\pm c$ given by Eq. (6.76).

From Eq. (6.76) it can be seen that the quantity $f_{\nabla\alpha}$ must not only be negative but must also be sufficiently large to overcome the destabilizing effect of the relative motion.

AN EXAMPLE OF DYNAMIC WAVES IN INCOMPRESSIBLE TWO-COMPONENT FLOW; WAVES IN A RECTANGULAR HORIZONTAL DUCT

Consider two fluids flowing with velocities v_1 and v_2 in a horizontal duct of depth H and uniform width. There are no body forces acting in the direction of flow; however, under the influence of a gradient in concentration, the variation of hydrostatic pressure across the duct will give rise to forces f in the direction of flow which are not contained in the mean pressure gradient.

Let the pressure at the top of the duct be p. Then the net force per unit width on the element of lighter fluid shown shaded in Fig. 6.10 is

$$F_1 = (1 - \alpha)H\,\delta p \tag{6.78}$$

However, for the heavier fluid the net force is

$$F_2 = \alpha H[\delta p - g(\rho_2 - \rho_1)H\,\delta\alpha] \tag{6.79}$$

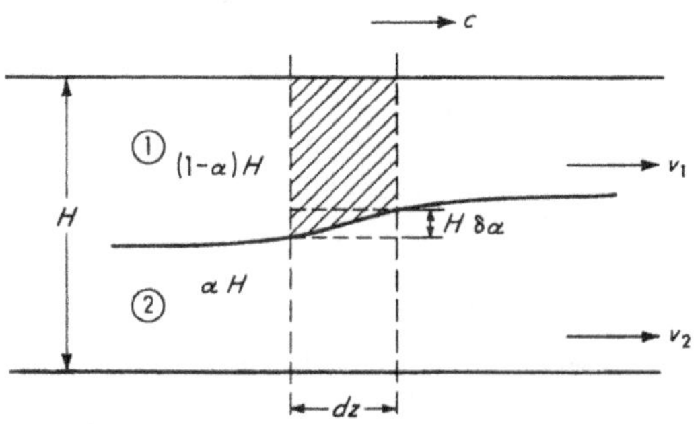

Fig. 6.10 Dynamic wave propagation in two incompressible stratified components in a horizontal duct.

Subtracting the total force per unit volume of fluid 2 from the force per unit volume of fluid 1, we obtain

$$f_1 - f_2 = \frac{\delta p\ (1-\alpha)H}{\delta z\ (1-\alpha)H} - \frac{\alpha H}{\alpha H\ \delta z}[\delta p - g(\rho_2 - \rho_1)H\ \delta\alpha] \tag{6.80}$$

Therefore

$$f_1 - f_2 = (\rho_2 - \rho_1)gH\frac{\partial\alpha}{\partial z} \tag{6.81}$$

The quantity $f_{\nabla\alpha}$ is therefore, from Eq. (6.71),

$$f_{\nabla\alpha} = -(\rho_2 - \rho_1)gH \tag{6.82}$$

From Eq. (6.76) the velocity of dynamic waves relative to V_0 is then

$$c = \pm\left[\frac{-(v_1 - v_2)^2}{\alpha/\rho_2 + (1-\alpha)/\rho_1} + (\rho_2 - \rho_1)gH\right]^{1/2}\left(\frac{\rho_1}{1-\alpha} + \frac{\rho_2}{\alpha}\right)^{-1/2} \tag{6.83}$$

The relative velocity is seen to have a destabilizing effect because it decreases the dynamic wave velocity. In fact, for sufficiently high relative velocity c^2 becomes negative and the flow is unstable. This occurs when

$$(v_1 - v_2)^2 > (\rho_2 - \rho_1)gH\left(\frac{\alpha}{\rho_2} + \frac{1-\alpha}{\rho_1}\right) \tag{6.84}$$

The above equations become invalid for wavelengths smaller than the duct dimensions because the one-dimensional approximation breaks down. However, they can be shown to be the limit for long wavelengths of the two-dimensional flow solution given by Lamb.[3]

Further analysis of the flow of stratified fluids is developed by Long.[4–6] Of particular interest is his discussion of solitary waves with a hyperbolic secant profile. For waves with an amplitude δ we define the following nomenclature.

$$\delta^* = \frac{\delta}{H} \tag{6.85}$$

$$j_1^* = j_1\rho_1^{1/2}[gH(\rho_2 - \rho_1)]^{-1/2} \tag{6.86}$$

$$j_2^* = j_2\rho_2^{1/2}[gH(\rho_2 - \rho_1)]^{-1/2} \tag{6.87}$$

An equation given by Long[6] for the condition at the wave crest is

$$\frac{j_1^{*2}}{(1-\alpha)^3}\left(1 - \frac{\delta^*}{1-\alpha}\right) + \frac{j_2^{*2}}{\alpha^3}\left(1 + \frac{\delta^*}{\alpha}\right) = 1 \tag{6.88}$$

The condition under which Eq. (6.88) is satisfied by all values of δ^*, i.e., waves of any amplitude, is

$$\frac{j_1^{*2}}{(1-\alpha)^4} = \frac{j_2^{*2}}{\alpha^4} \tag{6.89}$$

In this case Eq. (6.88) reduces to

$$\frac{j_1^{*2}}{(1-\alpha)^3} + \frac{j_2^{*2}}{\alpha^3} = 1 \tag{6.90}$$

Now, if we let α vary in Eq. (6.90) the resulting relationships between j_1 and j_2 will be ellipses with an envelope which bounds the range of allowable flow rates. The envelope is determined by differentiating Eq. (6.90) with respect to α to get

$$\frac{j_1^{*2}}{(1-\alpha)^4} - \frac{j_2^{*2}}{\alpha^4} = 0 \tag{6.91}$$

which is identical to Eq. (6.89) (if it were not, there would be a lack of consistency).

Eliminating α from Eqs. (6.90) and (6.91) leads to the result

$$j_1^{*1/2} + j_2^{*1/2} = 1 \tag{6.92}$$

This equation represents the maximum possible flow rates of the components that are possible without the formation of stationary waves of indeterminate amplitude. Thus Eq. (6.92) defines the locus of the flow rates at "flooding" of the channel.

The above derivation is of interest because of the remarkable similarity between Eq. (6.92) and empirical flooding correlations for vertical annular flow (Chap. 11), although the dynamics are quite different in the latter case and gravity does not act directly as a restoring force on the interface.

THE EFFECT OF COMPRESSIBILITY ON DYNAMIC WAVES IN TWO-COMPONENT FLOW

When compressibility effects are important, the propagation of dynamic waves is governed by the gradients of three "concentrations," the densities of the two components as well as the volumetric concentration of one of them.

The continuity Eqs. (6.66) and (6.67) now become

$$\frac{d}{dz}[v_1'\rho_1(1-\alpha)] = 0 \tag{6.93}$$

$$\frac{d}{dz}(v_2'\rho_2\alpha) = 0 \tag{6.94}$$

Changes in density are related to changes in pressure by the equations

$$\frac{d\rho_1}{dz} = \frac{dp/dz}{\partial p/\partial \rho_1} \tag{6.95}$$

$$\frac{d\rho_2}{dz} = \frac{dp/dz}{\partial p/\partial \rho_2} \tag{6.96}$$

From Eqs. (6.93) and (6.95), therefore,

$$\frac{1}{v_1'}\frac{dv_1'}{dz} - \frac{1}{1-\alpha}\frac{d\alpha}{dz} + \frac{1}{\rho_1(\partial p/\partial \rho_1)}\frac{dp}{dz} = 0 \tag{6.97}$$

and from Eqs. (6.94) and (6.96), similarly,

$$\frac{1}{v_2'}\frac{dv_2'}{dz} + \frac{1}{\alpha}\frac{d\alpha}{dz} + \frac{1}{\rho_2(\partial p/\partial \rho_2)}\frac{dp}{dz} = 0 \tag{6.98}$$

Making the previous assumption that the f's in Eqs. (6.68) and (6.69) depend on concentration gradient, and ignoring body forces, we have

$$\rho_1 v_1'\frac{dv_1'}{dz} - f_{1\nabla\alpha}\frac{d\alpha}{dz} + \frac{dp}{dz} = 0 \tag{6.99}$$

$$\rho_2 v_2'\frac{dv_2'}{dz} - f_{2\nabla\alpha}\frac{d\alpha}{dz} + \frac{dp}{dz} = 0 \tag{6.100}$$

Equations (6.97) to (6.100) are four equations for four unknowns and are compatible only if

$$\begin{vmatrix} \frac{1}{v_1'} & 0 & \frac{-1}{1-\alpha} & \frac{1}{\rho_1(\partial p/\partial \rho_1)} \\ 0 & \frac{1}{v_2'} & \frac{1}{\alpha} & \frac{1}{\rho_2(\partial p/\partial \rho_2)} \\ \rho_1 v_1' & 0 & -f_{1\nabla\alpha} & 1 \\ 0 & \rho_2 v_2' & -f_{2\nabla\alpha} & 1 \end{vmatrix} = 0 \tag{6.101}$$

Evaluating the determinant and multiplying throughout by v_1' and v_2' we obtain

$$\left(\frac{\rho_2 v_2'^2}{\alpha} + f_{2\nabla\alpha}\right)\left(1 - \frac{v_1'^2}{\partial p/\partial \rho_1}\right) = \left(f_{1\nabla\alpha} - \frac{\rho_1 v_1'^2}{1-\alpha}\right)\left(1 - \frac{v_2'^2}{\partial p/\partial \rho_2}\right) \tag{6.102}$$

If the compressibilities are zero, the previous result, Eq. (6.72), is obtained. If, on the other hand, the f's can be neglected (as, for example, in stratified flow at low relative velocity), the result is

$$\frac{\alpha}{\rho_2 v_2'^2}\left(1 - \frac{v_2'^2}{\partial p/\partial \rho_2}\right) + \frac{1-\alpha}{\rho_1 v_1'^2}\left(1 - \frac{v_1'^2}{\partial p/\partial \rho_1}\right) = 0 \tag{6.103}$$

in agreement with Eq. (3.61).

If the relative velocity is zero and both components have the velocity v', Eq. (6.103) reduces to

$$v'^2 = \frac{\dfrac{\alpha}{\rho_2} + \dfrac{1-\alpha}{\rho_1}}{\dfrac{\alpha}{\rho_2(\partial p/\partial \rho_2)} + \dfrac{1-\alpha}{\rho_1(\partial p/\partial \rho_1)}} \tag{6.104}$$

Therefore compressibility waves move with velocities

$$U = V \pm c_{cs} \tag{6.105}$$

where c_{cs} denotes the velocity of a compressibility wave in stratified flow, with no relative velocity which is, from Eqs. (6.104) and (6.105),

$$c_{cs} = \pm \left(\frac{\dfrac{\alpha}{\rho_2} + \dfrac{1-\alpha}{\rho_1}}{\dfrac{\alpha}{\rho_2 c_2{}^2} + \dfrac{1-\alpha}{\rho_1 c_1{}^2}} \right)^{1/2} \tag{6.106}$$

where $c_1{}^2$ and $c_2{}^2$ are the velocities of compressibility waves in each component separately. Equation (6.106) shows that the wave velocity lies between the values of c_1 and c_2.

It should be stressed that the velocity c_{cs} is not the velocity of a compressibility wave in a homogeneous mixture in which all relative velocity is suppressed by means of the forces between the components. If the f's are zero in Eqs. (6.68) and (6.69) it is impossible for both the conditions $v_1' = v_2'$ and $dv_1'/dz = dv_2'/dz$ to be satisfied. For truly homogeneous flow, therefore, there must be sufficient friction between the components. If there are no external forces apart from this interaction, then

$$f_1(1 - \alpha) + f_2\alpha = 0 \tag{6.107}$$

since the action and reaction between the components are equal and opposite. Treating the flow as homogeneous and adding $1 - \alpha$ times Eq. (6.68) to α times Eq. (6.69) and neglecting body forces, we obtain

$$[(1 - \alpha)\rho_1 + \alpha\rho_2]v' \frac{dv'}{dz} = -\frac{dp}{dz} \tag{6.108}$$

which is simply the equation of motion of a fluid with mean density ρ_m given by Eq. (2.9)

Eliminating $d\alpha/dz$ from Eqs. (6.97) and (6.98) and assuming homogeneous flow, we get

$$\frac{1}{v'}\frac{dv'}{dz} + \frac{dp}{dz}\left[\frac{1-\alpha}{\rho_1(\partial p/\partial \rho_1)} + \frac{\alpha}{\rho_2(\partial p/\partial \rho_2)}\right] = 0 \tag{6.109}$$

From Eqs. (6.108) and (6.109) it follows at once, in agreement with Eq. (2.50), that

$$v_1'^2 = c_{ch}{}^2 = \left\{[\alpha\rho_2 + (1 - \alpha)\rho_1]\left(\frac{1-\alpha}{\rho_1 c_1{}^2} + \frac{\alpha}{\rho_2 c_2{}^2}\right)\right\}^{-1} \tag{6.110}$$

where c_{ch} is the compressibility wave velocity in a strictly homogeneous mixture. Unlike c_{cs}, c_{ch} does not necessarily lie between c_1 and c_2 and in some circumstances may be far less than either. If ρ_2 is much less than ρ_1 and c_2^2 is less than c_1^2, Eq. (6.110) becomes

$$c_{ch}^2 = \frac{c_2^2 \rho_2}{\rho_1 \alpha(1-\alpha)} \tag{6.111}$$

The lowest value of c_{ch} is therefore at $\alpha = \frac{1}{2}$ and is

$$(c_{ch})_{\min} = 2c_2 \left(\frac{\rho_2}{\rho_1}\right)^{1/2} \tag{6.112}$$

For air and water at atmospheric pressure, for example $c_2 \approx 1100$ fps, $\rho_2/\rho_1 = 0.0012$, and $(c_{ch})_{\min}$ is then equal to 75 fps.

If only the void fraction of a two-component mixture is specified, there is obviously a problem in deciding whether to use Eq. (6.106) or Eq. (6.110) or, indeed, some compromise between the two. Qualitatively, one would expect Eq. (6.110) to be true for a fine dispersion of bubbles in a liquid, whereas Eq. (6.106) should apply when two fluid streams flow side by side with no drag forces between them. Transient drag forces would probably be both amplitude and frequency dependent, and therefore the wave velocity would be a function of these variables as well as the properties of the components. At present, knowledge of these phenomena is very limited.

Example 6.7 In Ref. 7 an equation is derived for the speed of sound in a bubbly mixture as follows:

$$c = \frac{1+\delta}{\delta}(\gamma_g R T)^{1/2} \tag{a}$$

where

$$\delta = \frac{Q_g}{Q_f} \tag{b}$$

$$R = \frac{\beta}{1+\beta} R_g \tag{c}$$

$$\beta = \frac{W_g}{W_f} \tag{d}$$

and R_g is the gas constant for the gas. Deduce this result from Eq. (6.110).

Solution Neglecting the liquid compressibility, so that $\rho_1 c_1^2$ can be regarded as large compared with $\rho_2 c_2^2$, and rewriting Eq. (6.110) in gas-liquid nomenclature we find that

$$c^2 = \frac{\rho_g c_g^2}{\alpha \rho_m} \tag{e}$$

where

$$\rho_m = \alpha \rho_g + (1-\alpha)\rho_f \tag{f}$$

The void fraction in homogeneous flow is related to the volumetric flow-rate ratio as follows:

$$\alpha = \frac{Q_g}{Q_f + Q_g} = \frac{\delta}{1 + \delta} \tag{g}$$

The mass flow-rate ratio is given in terms of the void fraction by the equation

$$\beta = \frac{W_g}{W_f} = \frac{\alpha \rho_g}{(1 - \alpha)\rho_f} \tag{h}$$

Using Eqs. (*h*) and (*f*) we obtain

$$\rho_m = \alpha \rho_g \frac{1 + \beta}{\beta} \tag{i}$$

Substituting Eqs. (*i*) and (*g*) into Eq. (*e*) gives

$$c^2 = c_g^2 \frac{\beta}{1 + \beta} \left(\frac{1 + \delta}{\delta} \right)^2 \tag{j}$$

Now, for the adiabatic compressibility wave in the gas alone we have

$$c_g^2 = \gamma_g R_g T \tag{k}$$

Substituting Eq. (*k*) into Eq. (*j*), taking the square root and using Eq. (*c*) eventually gives the desired result

$$c = \frac{1 + \delta}{\delta} (\gamma_g R T)^{1/2} \tag{l}$$

It is also readily shown, as in Ref. 7, that the specific heats for the mixture in thermal equilibrium are

$$c_p = \frac{c_f + \beta c_{pg}}{1 + \beta} \tag{m}$$

$$c_v = \frac{c_f + \beta c_{vg}}{1 + \beta} \tag{n}$$

The isentropic exponent for expansion with the two phases in continual thermal equilibrium is therefore

$$\gamma' = \frac{c_p}{c_v} = \frac{c_f + \beta c_{pg}}{c_f + \beta c_{vg}} \tag{o}$$

This equation is equivalent to Eq. (2.25) of Example 2.2. Equation (*l*) will therefore be true only if there is no mutual heat transfer; if equilibrium is maintained, γ' from Eq. (*o*) should replace γ_g.

THE EFFECT OF PHASE CHANGE

If phase change can occur during the passage of a wave, then the wave speed depends on the degree to which equilibrium is achieved. In a very dispersed steam-water flow, for example, the value of $\partial p/\partial \rho$ for the homogeneous mixture can be determined from steam tables or a property chart and the sonic velocities plotted as shown in Fig. 6.11. Many nonequilib-

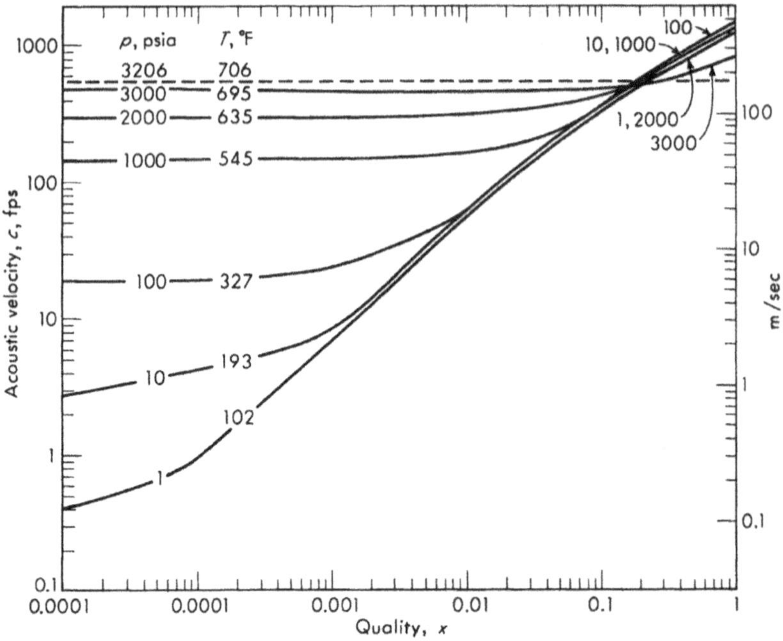

Fig. 6.11 Theoretical values for the velocity of sound in equilibrium, homogeneous steam-water mixtures. (*Gouse and Brown.*[8])

rium effects complicate this situation and bedevil devices such as low-pressure flash evaporators. Surface-tension and nucleation effects can also be important.

6.5 THE INTERACTION BETWEEN DYNAMIC AND CONTINUITY WAVES

SINGLE-PHASE FLOW

Dynamic and continuity waves occur simultaneously in many practical systems. The condition that both kinds of wave shall exist is simply that the force per unit volume on the fluid or particles shall be a function of all three variables—the velocity, the concentration, and the concentration gradient—i.e.,

$$f = f\left(V, \alpha, \frac{\partial \alpha}{\partial z}\right) \tag{6.113}$$

In uniform steady flow without acceleration the equilibrium condition is

$$f + b = 0 \qquad \frac{\partial \alpha}{\partial z} = 0 \tag{6.114}$$

For unsteady flow the continuity and momentum equations are

$$\frac{\partial \alpha}{\partial t} + V \frac{\partial \alpha}{\partial z} + \alpha \frac{\partial V}{\partial z} = 0 \tag{6.115}$$

$$\rho \left(\frac{\partial V}{\partial t} + V \frac{\partial V}{\partial z} \right) = f + b \tag{6.116}$$

Now, consider an initially steady flow satisfying Eq. (6.114) with concentration α everywhere and with a coordinate system chosen so that $V = 0$. For a small perturbation α' and v from equilibrium, Eqs. (6.115) and (6.116) become

$$\frac{\partial \alpha'}{\partial t} + \alpha \frac{\partial v}{\partial z} = 0 \tag{6.117}$$

$$\rho \frac{\partial v}{\partial t} = f_v v + f_\alpha \alpha' + f_{\nabla\alpha} \frac{\partial \alpha'}{\partial z} \tag{6.118}$$

where the subscripts denote partial differentiation as before.

Taking the z derivative of Eq. (6.118) and using Eq. (6.117) to eliminate v from the equations yields

$$\rho \frac{\partial^2 \alpha'}{\partial t^2} + \alpha f_{\nabla\alpha} \frac{\partial^2 \alpha'}{\partial z^2} - f_v \frac{\partial \alpha'}{\partial t} + \alpha f_\alpha \frac{\partial \alpha'}{\partial z} = 0 \tag{6.119}$$

Using the results derived previously we can define the following quantities which are identical with the continuity and dynamic wave velocities relative to the average velocity:

$$v_w = V_w - V = -\frac{f_\alpha}{f_v} \alpha \tag{6.120}$$

$$c = \left(-\frac{\alpha}{\rho} f_{\nabla\alpha} \right)^{1/2} \tag{6.121}$$

Since in almost any conceivable system f_v is negative (otherwise particles would tend to accelerate without limit), we can also define

$$B = -\frac{f_v}{\rho} \tag{6.122}$$

where B is a positive quantity which represents "damping."

Equation (6.119) can be rewritten with the help of Eqs. (6.120) to (6.122). Thus

$$\frac{\partial^2 \alpha'}{\partial t^2} - c^2 \frac{\partial^2 \alpha'}{\partial z^2} + B \left(\frac{\partial \alpha'}{\partial t} + v_w \frac{\partial \alpha'}{\partial z} \right) = 0 \tag{6.123}$$

The physical significance of Eq. (6.123) is most easily shown by substituting the expression

$$\alpha' = \alpha_0 e^{at + i\omega(t - z/U)} \tag{6.124}$$

which represents a wave with frequency $\omega/2\pi$ and velocity U and which grows or decays with time, depending on the sign of a. The quantities a and ω are, by definition, real. Performing the substitution we obtain, after separating real and imaginary parts and solving for a and ω^2,

$$a = \frac{B}{2}\left(\frac{v_w}{U} - 1\right) \tag{6.125}$$

$$\omega^2 = \frac{B^2}{4}\frac{v_w{}^2 - U^2}{U^2 - c^2} \tag{6.126}$$

From Eq. (6.126) it can be shown that the requirement that ω^2 should be positive limits the value of U^2 to values between $v_w{}^2$ and c^2. Therefore, if $v_w{}^2$ is greater than c^2, v_w/U will be greater than unity for waves moving in the direction of v_w, and these waves will grow. On the other hand, if $v_w{}^2$ is less than c^2, v_w/U is always less than unity and all waves will attenuate. The stability of the flow is governed entirely by the relative magnitude of the dynamic and continuity wave velocities, c and v_w. If continuity waves overtake dynamic waves, the flow is unstable. Since the coordinate system has been chosen with $V = 0$, v_w must be expressed relative to the average velocity.

A familiar illustration of the above conclusion is given by the analogous behavior of traffic streams. Lighthill and Whitham[1] have shown how continuity wave theory may be used to describe unsteady flow in a stable stream of traffic, and their concluding remarks suggest that instabilities similar to "roll waves" may occur in some cases. We can see qualitatively what occurs by considering the dynamic waves that are the result of the response of a driver (and his brakes and accelerator) to changes in local gradients of concentration. If brakes are bad, the road surface is slippery, or if vision is obscured by dense fog then the dynamic wave velocity is decreased. If it is decreased below the continuity wave velocity the result is a "pile up" involving large numbers of cars. Similar effects occur under normal conditions if the drivers are in a hurry and drive too closely together so that the continuity wave velocity is increased beyond the dynamic wave velocity. An experienced driver will adjust his speed and separation from other cars intuitively so that the dynamic wave velocity is always a "factor of safety" greater than the local continuity wave velocity.

It can also be checked that Eqs. (6.125) and (6.126) lead to the usual results in the case where one or other wave motion only is present. For example, if $B = 0$, all frequencies are possible if and only if $c^2 = U^2$ and in this case $a = 0$, and there is no amplification or attenuation (dynamic waves). On the other hand when ω becomes very small (long waves), $v_w{}^2 = U^2$ and waves propagate unchanged with the velocity v_w since a is then zero from Eq. (6.125) (continuity waves).

Example 6.8 Examine the stability of long one-dimensional waves on a viscous film falling down an inclined plane.

Solution Let the plane make an angle θ with the vertical. The components of gravity along and perpendicular to the plane are $g \cos \theta$ and $g \sin \theta$. Substituting these values into Eqs. (6.12), (6.13), and (6.59) for a film of thickness δ, we find

$$V = \frac{g \cos \theta \,(\rho_f - \rho_g)\delta^2}{3\mu_f} \tag{6.127}$$

$$V_w = \frac{g \cos \theta \,(\rho_f - \rho_g)\delta^2}{\mu_f} \tag{6.128}$$

$$c = \pm(\delta g \sin \theta)^{1/2} \tag{6.129}$$

Instability will occur if the continuity waves overtake the downward-moving dynamic waves, that is, if

$$V_w > V + c \tag{6.130}$$

Substituting from Eqs. (6.127) to (6.129) the condition for instability is found to be

$$\frac{2g \cos \theta \,(\rho_f - \rho_g)\delta^2}{3\mu_f} > (\delta g \sin \theta)^{1/2} \tag{6.131}$$

Alternatively, substituting from Eq. (6.127) and rearranging,

$$\frac{4V\delta(\rho_f - \rho_g)}{\mu_f} > 3 \tan \theta \tag{6.132}$$

The left-hand side of Eq. (6.132) is the usual expression for the Reynolds number of a falling film if $\rho_f \gg \rho_g$.

Benjamin,[9] by a more exact method, obtained the factor $\frac{10}{3}$ rather than 3 on the right-hand side of Eq. (6.132).

The instability is manifested by the appearance of *roll waves* such as those which are observed on windows and inclined streets after rainfall and on walls which are painted too thickly.

INCOMPRESSIBLE TWO-COMPONENT FLOW

Consider the flow of two incompressible components without change of phase in a constant-area duct. The continuity equations for the components are

$$\frac{\partial \alpha}{\partial t} + v_2 \frac{\partial \alpha}{\partial z} + \alpha \frac{\partial v_2}{\partial z} = 0 \tag{6.133}$$

$$-\frac{\partial \alpha}{\partial t} - v_1 \frac{\partial \alpha}{\partial z} + (1 - \alpha) \frac{\partial v_1}{\partial z} = 0 \tag{6.134}$$

The equations of motion are, as usual,

$$\rho_1 \left(\frac{\partial v_1}{\partial t} + v_1 \frac{\partial v_1}{\partial z} \right) = -\frac{\partial p}{\partial z} + b_1 + f_1 \tag{6.135}$$

$$\rho_2 \left(\frac{\partial v_2}{\partial t} + v_2 \frac{\partial v_2}{\partial z} \right) = -\frac{\partial p}{\partial z} + b_2 + f_2 \tag{6.136}$$

Subtracting Eq. (6.136) from Eq. (6.135) we get

$$\rho_1\left(\frac{\partial v_1}{\partial t} + v_1\frac{\partial v_1}{\partial z}\right) - \rho_2\left(\frac{\partial v_2}{\partial t} + v_2\frac{\partial v_2}{\partial z}\right) = b_1 - b_2 + f_1 - f_2 \tag{6.137}$$

The right-hand side of Eq. (6.137) may be replaced by a new variable f. Thus

$$f = -[(b_1 - b_2) + (f_1 - f_2)] \tag{6.138}$$

f is a function of the velocity of both components, the concentration and the concentration gradient (and possibly other derivatives of velocity or concentration), i.e.,

$$f = f\left(v_1, v_2, \alpha, \frac{\partial \alpha}{\partial z}, \text{etc.}\right) \tag{6.139}$$

Considering a small perturbation v_1, v_2, α', δf from a uniform steady flow V_1, V_2, α and retaining only the first-order terms in Eqs. (6.133), (6.134), and (6.137), we get

$$\frac{\partial \alpha'}{\partial t} + V_2\frac{\partial \alpha'}{\partial z} + \alpha\frac{\partial v_2}{\partial z} = 0 \tag{6.140}$$

$$-\frac{\partial \alpha'}{\partial t} - V_1\frac{\partial \alpha'}{\partial z} + (1-\alpha)\frac{\partial v_1}{\partial z} = 0 \tag{6.141}$$

$$\rho_1\left(\frac{\partial v_1}{\partial t} + V_1\frac{\partial v_1}{\partial z}\right) - \rho_2\left(\frac{\partial v_2}{\partial t} + V_2\frac{\partial v_2}{\partial z}\right) = -\delta f \tag{6.142}$$

The quantities v_1 and v_2 can be eliminated from the above equations by differentiating Eq. (6.142) with respect to z and substituting from Eqs. (6.140) and (6.141). The final result is

$$\frac{\partial^2 \alpha'}{\partial t^2}\left(\frac{\rho_1}{1-\alpha} + \frac{\rho_2}{\alpha}\right) + 2\frac{\partial^2 \alpha'}{\partial z\,\partial t}\left(\frac{V_1\rho_1}{1-\alpha} + \frac{V_2\rho_2}{\alpha}\right) + \frac{\partial^2 \alpha'}{\partial z^2}\left(\frac{\rho_1 V_1^2}{1-\alpha} + \frac{\rho_2 V_2^2}{\alpha}\right) = \frac{\partial(\delta f)}{\partial z} \tag{6.143}$$

The quantity δf is made up of contributions due to each perturbation as follows:

$$\delta f = f_\alpha \alpha' + f_{v_1} v_1 + f_{v_2} v_2 + f_{\nabla\alpha}\frac{\partial \alpha'}{\partial z} \tag{6.144}$$

Differentiating Eq. (6.144) with respect to z and substituting for $\partial v_1/\partial z$ and $\partial v_2/\partial z$ from Eqs. (6.140) and (6.141) gives

$$\frac{\partial(\delta f)}{\partial z} = \frac{\partial \alpha'}{\partial t}\left(\frac{f_{v_1}}{1-\alpha} - \frac{f_{v_2}}{\alpha}\right) + \frac{\partial \alpha'}{\partial z}\left(f_\alpha + \frac{V_1 f_{v_1}}{1-\alpha} - \frac{V_2 f_{v_2}}{\alpha}\right) + f_{\nabla\alpha}\frac{\partial^2 \alpha'}{\partial z^2} \tag{6.145}$$

Equation (6.145) may be reconciled with general continuity wave theory by writing f as a function of the velocity of one component and the total volumetric flux (or volumetric mean velocity) j. Thus

$$f = f'(V_2, j, \alpha) \tag{6.146}$$

where

$$j = V_1(1 - \alpha) + V_2\alpha \tag{6.147}$$

The derivatives in Eq. (6.145) can be evaluated in terms of Eq. (6.146) as follows:

$$f_\alpha = (V_2 - V_1)\frac{\partial f'}{\partial j} + \frac{\partial f'}{\partial \alpha} \tag{6.148}$$

$$f_{v_2} = \frac{\partial f'}{\partial v_2} + \alpha \frac{\partial f'}{\partial j} \tag{6.149}$$

$$f_{v_1} = (1 - \alpha)\frac{\partial f'}{\partial j} \tag{6.150}$$

Substituting Eqs. (6.148) to (6.150) into Eq. (6.145), we have

$$\frac{\partial(\delta f)}{\partial z} = -\frac{\partial \alpha'}{\partial t}\left(\frac{1}{\alpha}\frac{\partial f'}{\partial v_2}\right) + \frac{\partial \alpha'}{\partial z}\left(\frac{\partial f'}{\partial \alpha} - \frac{V_2}{\alpha}\frac{\partial f'}{\partial v_2}\right) + f_{\nabla\alpha}\frac{\partial^2 \alpha'}{\partial z^2} \tag{6.151}$$

$$\frac{\partial(\delta f)}{\partial z} = -\frac{1}{\alpha}\frac{\partial f'}{\partial v_2}\left[\frac{\partial \alpha'}{\partial t} + \frac{\partial \alpha'}{\partial z}\left(V_2 - \alpha\frac{f'_\alpha}{f'_{v_2}}\right)\right] + f_{\nabla\alpha}\frac{\partial^2 \alpha'}{\partial z^2} \tag{6.152}$$

The factor multiplying $\partial\alpha'/\partial z$ in Eq. (6.152) is seen with the help of Eq. (6.146) to be

$$V_2 + \alpha\left(\frac{\partial V_2}{\partial \alpha}\right)_{f',j} = \frac{\partial}{\partial \alpha}(V_2\alpha)_{f',j} = \left(\frac{\partial j_2}{\partial \alpha}\right)_{f',j} \tag{6.153}$$

and therefore represents the continuity wave velocity V_w that was derived previously, Eq. (6.38).

Equations (6.152) and (6.153) can be substituted into Eq. (6.143) to give an equation of the form

$$\frac{\partial^2 \alpha'}{\partial t^2} + 2V_0\frac{\partial^2 \alpha'}{\partial z\,\partial t} + A\frac{\partial^2 \alpha'}{\partial z^2} + B\left(\frac{\partial \alpha'}{\partial t} + V_w\frac{\partial \alpha'}{\partial z}\right) = 0 \tag{6.154}$$

The quantity V_0 is a weighted mean velocity given [see Eq. (6.75)] by

$$V_0 = \frac{V_1\rho_1/(1 - \alpha) + V_2\rho_2/\alpha}{\rho_1/(1 - \alpha) + \rho_2/\alpha} \tag{6.155}$$

and the specific values of A and B are

$$A = \frac{\rho_1 V_1^2/(1 - \alpha) + \rho_2 V_2^2/\alpha + f_{\nabla\alpha}}{\rho_1/(1 - \alpha) + \rho_2/\alpha} \tag{6.156}$$

$$B = \frac{-f_{v_2}/\alpha + f_{v_1}/(1 - \alpha)}{\rho_1/(1 - \alpha) + \rho_2/\alpha} \tag{6.157}$$

Equation (6.154) is interpreted by making the substitution

$$\alpha' = \alpha_0 e^{at+i\omega(t-z/U)} \tag{6.158}$$

then defining all wave velocities in terms of their differences from V_0. Thus

$$U = V_0 + u \tag{6.159}$$
$$V_w = V_0 + v_w \tag{6.160}$$

Then a new quantity is defined which is the same as the square of the dynamic wave velocity given in Eq. (6.76).

$$c^2 = V_0^2 - A \tag{6.161}$$

The eventual result is

$$a = \frac{B}{2}\left(\frac{v_w}{u} - 1\right) \tag{6.162}$$

$$\omega^2 = \frac{B^2}{4}\frac{U^2}{u^2}\frac{v_w^2 - u^2}{u^2 - c^2} \tag{6.163}$$

The qualitative conclusions that can be drawn from Eqs. (6.162) and (6.163) for two-component flow are exactly the same as those which were drawn from Eqs. (6.125) and (6.126) for single-component flow. Dynamic waves move with velocity $\pm c$ relative to the weighted average velocity V_0 defined by Eq. (6.155). Instability results when $v_w^2 > c^2$ in which case waves grow in the direction of v_w at a rate governed by Eq. (6.162). If c^2 is negative, the flow is always unstable This approach has successfully predicted the instability of stratified flow.[13]

6.6 DYNAMIC SHOCK WAVES

The theory of dynamic shock waves can be derived from the continuity, momentum, and energy laws across finite discontinuities. The *normal shock* of gas dynamics and the *hydraulic jump* in hydraulics are both special cases of dynamic shock waves.

NORMAL COMPRESSIBILITY SHOCKS

Consider a stationary normal shock in a two-phase gas-liquid flow as shown in Fig. 6.12. The continuity equation across the shock is simply

$$G_1 = G_2 \tag{6.164}$$

The momentum equation is

$$p_1 + (G_f v_f + G_g v_g)_1 = p_2 + (G_f v_f + G_g v_g)_2 \tag{6.165}$$

The energy equation is

$$\left[G_f\left(h_f + \frac{v_f^2}{2}\right) + G_g\left(h_g + \frac{v_g^2}{2}\right)\right]_1 = \left[G_f\left(h_f + \frac{v_f^2}{2}\right) + G_g\left(h_g + \frac{v_g^2}{2}\right)\right]_2 \quad (6.166)$$

Thermodynamic relationships will enable further manipulations to be performed between pressure, enthalpy, and density.

The solution of these equations in detail is left for later chapters. However, as an example, we shall develop the solution for the simple case of the isothermal homogeneous shock wave with only one of the components compressible and obeying the perfect gas laws. Denote the compressible component by subscript g and the other by subscript f (although it could be a solid).

In this case Eq. (6.165) reduces to

$$p_1 + Gj_1 = p_2 + Gj_2 \quad (6.167)$$

Furthermore, since only the gas is compressible,

$$j_2 - j_1 = (j_g)_2 - (j_g)_1 \quad (6.168)$$

Combining Eq. (6.168) with Eq. (6.167) gives

$$p_1 - p_2 = G[(j_g)_2 - (j_g)_1] \quad (6.169)$$

Moreover, for isothermal flow of the gas

$$p_1(j_g)_1 = p_2(j_g)_2 \quad (6.170)$$

Combining Eqs. (6.169) and (6.170) we find

$$p_1 = G(j_g)_2 \quad (6.171)$$
$$p_2 = G(j_g)_1 \quad (6.172)$$

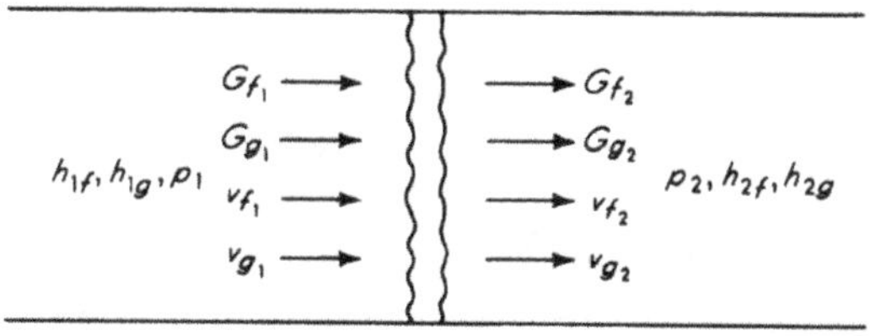

Fig. 6.12 A normal dynamic shock wave in two-phase gas-liquid flow.

In view of Eq. (2.81) the isothermal homogeneous Mach numbers before and after the shock are

$$M_1^2 = \frac{G(j_g)_1}{p_1} \tag{6.173}$$

$$M_2^2 = \frac{G(j_g)_2}{p_2} \tag{6.174}$$

Combining Eqs. (6.171) to (6.174) leads to the pleasantly simple result

$$M_1^2 = \frac{1}{M_2^2} = \frac{(j_g)_1}{(j_g)_2} = \frac{p_2}{p_1} \tag{6.175}$$

Eddington[10] has confirmed these relationships for gas-liquid flow at about 50 percent void fraction.

OBLIQUE SHOCK WAVES

Oblique shocks can be treated in the usual way by resolving the motion along and normal to the wave line.

Consider, for example, the isothermal homogeneous oblique shock, as shown in Fig. 6.13.

Along the wave the velocity is unchanged. Therefore

$$j_1 \cos \beta = j_2 \cos (\beta - \theta) \tag{6.176}$$

Normal to the wave we have, from the velocity triangles,

$$j_1 \sin \beta = j_{1N} \tag{6.177}$$

$$j_2 \sin (\beta - \theta) = j_{2N} \tag{6.178}$$

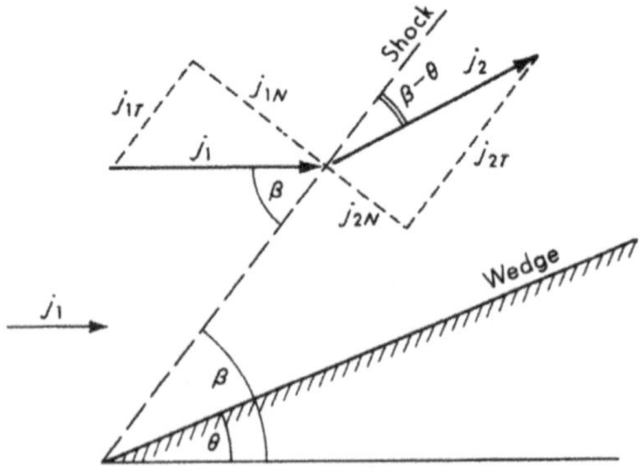

Fig. 6.13 An oblique two-phase, dynamic shock wave.

From Eq. (6.175) for the normal components we find

$$M_1^2 \sin^2 \beta = M_{1N}^2 = \frac{(j_g)_{1N}}{(j_g)_{2N}} = \frac{(j_g)_{1N}}{j_{2N} - j_{fN}} \tag{6.179}$$

The elimination of j_{2N} between Eqs. (6.176) to (6.179) yields

$$\frac{(j_g)_{1N}}{M_1^2 \sin^2 \beta} = j_{1N} \cot \beta \tan (\beta - \theta) - j_{fN} \tag{6.180}$$

Now, since all the normal incident fluxes are equal to sin β times the fluxes in the original flow direction, rearrangement of Eq. (6.180) leads to an implicit expression for the shock angle in terms of the known incident parameters and the wedge angle θ:

$$\frac{(j_g)_1}{M_1^2 j_f} = \frac{j_1}{j_f} \cos \beta \sin \beta \tan (\beta - \theta) - \sin^2 \beta \tag{6.181}$$

The derivation of further interesting quantities, such as the pressure ratio, follows at once by using the value of β to resolve the incident flow normal to the wave.

These results obtain good confirmation from the work of Eddington.[10]

RELAXATION PHENOMENA

Since velocities and thermodynamic properties change very rapidly across a shock wave, the two phases may take a significant time to come to equilibrium again. The mechanism of return to equilibrium is known as *relaxation*. If relaxation is slow, the overall shock thickness can be considerable.

For example, consider a suspension of small spherical particles in a fluid. Let the dilution be such that the fluid properties, velocities, and temperature remain constant while the particles come to equilibrium. Denoting the particles by the subscript s we have, for acceleration in laminar flow, following the path of each particle (Lagrangian viewpoint),

$$\frac{\rho_s \pi d^3}{6} \frac{Dv_s}{Dt} = 3\pi\mu_g d(v_g - v_s) \tag{6.182}$$

whence

$$\frac{d^2\rho_s}{18\mu_g} \frac{Dv_s}{Dt} = (v_g - v_s) \tag{6.183}$$

The inverse time constant for velocity relaxation is therefore

$$t_v = \frac{d^2\rho_s}{18\mu_g} \tag{6.184}$$

Similarly, for thermal relaxation we have

$$\frac{c_s \rho_s \pi d^3}{6} \frac{DT_s}{Dt} = 2\pi k_g d(T_g - T_s) \tag{6.185}$$

and the inverse time constant for thermal relaxation is

$$t_T = \frac{d^2 \rho_s c_s}{12 k_g} \tag{6.186}$$

For example, for aluminum particles [$\rho_s = 169$ lb/ft^3, $c_s = 0.21$ Btu/(lb)(°F)] of diameter 0.001 in. in air at 68°F, $t_v = 5.45$ msec and $t_T = 4.93$ msec. In this time the particles will go several feet in a gas stream with velocity about 1000 fps.

PROBLEMS

6.1. On a particular highway the relationship between j (cars per hour) and α (cars per mile) is

$$j = \frac{\alpha}{2}(120 - \alpha)$$

What is the speed limit? What is the maximum capacity of the highway? If a flow of 1000 cars per hour is stopped at a traffic light for 2 min and then released, what happens?

6.2. What is the maximum capacity of the highway in Prob. 6.1 if the traffic light operates continuously, being alternately red for 1 min and green for 1 min indefinitely?

6.3. The equation in Prob. 6.1 represents each lane of a three-lane highway. If the flow rate is 4000 cars/hr and one lane is closed by an accident for ½ hr, how far back up the highway is the influence of the accident felt?

6.4. One function of the dean of a college faculty is to circulate interesting documents. These are sent sequentially to faculty members in alphabetical order. If the rate at which these are read by each professor is proportional to the number in his in-tray raised to some power n, what happens

(*a*) If the dean only issues documents on Mondays.

(*b*) If one faculty member is away for a month.

Is any special value of n particularly desirable?

6.5. If the flow rate in Example 6.2 is increased suddenly back to the original value, show that a shock wave is formed. How should the flow rate be varied in order that δ may vary as a function of time as a symmetrical triangular wave at the top of the wall? How will this wave change its shape as it propagates? When will the first shock wave form?

6.6. Solve Example 6.3, using Eq. (6.3) rather than Eq. (6.11). What is the answer if the friction factor is proportional to the Reynolds number to the power $-n$?

6.7. Prove that a stable continuity shock cannot occur between two points on the j-versus-α curve if the line joining these points cuts the curve at some intermediate point.

6.8. Develop continuity wave theory for single-phase flow in a duct with variable area. Show that increase in area causes the waves to attenuate, and vice versa.

6.9. Show that, if the friction factor is constant, turbulent falling films are unstable if

$$\cot \theta > 2C_f$$

6.10. From the result of Prob. 6.9 find the critical Froude number for the formation of white water in rivers.

6.11. Solve Prob. 6.9 if the friction factor varies as the Reynolds number to the power $-n$.

6.12. Derive the relationship between film thickness and flow rate relative to the ground for a laminar liquid film on a belt of unit width moving vertically with speed V. What is the maximum possible flow rate? For flow rates less than the maximum there are two possible values of thickness.[11] Which values of thickness occur before and after an obstacle such as a sharp edge which is inserted at right angles to the flow? Why?

6.13. Consider a liquid film on the inner wall of a rotating drum. If the drum speed is V and viscous and gravity forces alone are important, derive the relationship between circumferential liquid flow rate per unit width and film thickness for various positions around the drum. Assume laminar flow. Show that there are two possible configurations of the liquid film, as shown in Fig. 6.14, depending on the amount of liquid which is present. What is the reason for the change in configuration? At what point in regime 2 is the continuity wave velocity zero? In terms of the drum dimensions and fluid properties, what is the critical amount of liquid which is needed to bring about regime 2?

6.14. Solve Prob. 6.13 for turbulent flow with a constant value of C_f. Under what conditions will centrifugal and inertia forces be significant?

6.15. What is the continuity wave velocity in homogeneous incompressible two-phase flow in a constant-area duct?

6.16. Steam condenses at a constant rate on the inside of a rotating drum and is removed at circumferential locations (rotating with the drum) distance L apart, where L is much less than the radius of the drum. As the drum rotates the condensate film sloshes to and fro between the removal points. If the film is laminar and inertia and centrifugal forces

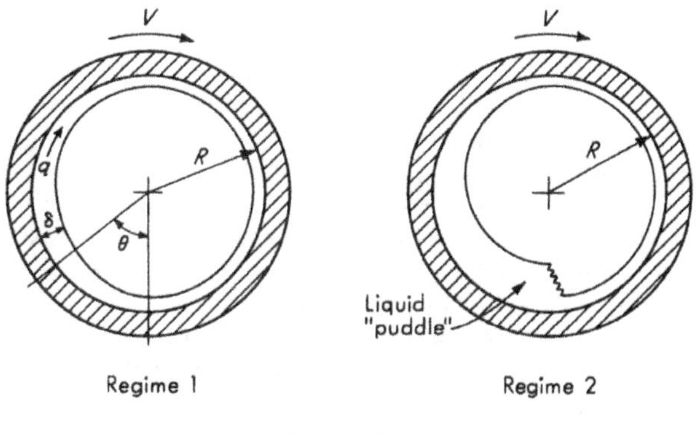

Fig. 6.14 Two regimes of liquid film behavior inside a rotating drum. (*Prob.* 6.13.)

are small, show by idealizing the sinusoidal gravitational field (resolved along the film) by a square wave that:

(*a*) There are either three or four distinct groups of waves on the film, depending on whether T, the time for one-half revolution, is greater than or less than

$$T_c = \left[\frac{3L\pi\mu_f}{2g\beta^2(\rho_f - \rho_g)}\right]^{1/3}$$

(*b*) The dimensionless mean film thickness $\bar{\delta}/L$ can be represented as a function of T/T_c and $\beta T/L$.

(*c*) No shock wave is formed.

Represent the various regimes of waves on both the $z\delta$ and zt planes and explain the behavior in detail.

6.17. Plot the homogeneous compressibility wave velocity as a function of α for air-water mixtures at 70°F and 15, 100, and 1000 psia.

6.18. Equation (6.92) predicts flooding due to the growth of stationary waves. Moving waves can also be unstable in view of Eq. (6.84). Show how Eq. (6.84) leads to lower limits on the flow rates if flooding is to be avoided in a long horizontal duct. In short ducts, only those waves which are almost stationary will grow enough to cause flooding.

6.19. Substitute

$$\alpha' = \alpha_0 e^{a'z + i\omega'(t - z/U)}$$

into Eq. (6.123) and solve for a' and ω'. Interpret the result physically and compare with Eqs. (6.125) and (6.126). Is the stability condition unchanged?

6.20. For a liquid film on a vertical surface, $c = 0$ and the motion is always unstable. Consider a turbulent falling film on which there are disturbances of wavelength $\lambda = 2\pi n\delta$. Show that, if n is large enough for the one-dimensional idealizations to apply, the value of U given by Eq. (6.126) is

$$U = \frac{v}{[2(1 + \sqrt{1 + 4/n^2C_f^2})]^{1/2}}$$

If $n = 10$ and $C_f = 0.01$, show that waves will grow by a factor of e in a distance of about 92 times the film thickness.

6.21. Solve Prob. 6.20 for laminar flow and show how the wave amplification depends on the Reynolds number.

6.22. A normal shock wave is moving at 200 fps and the conditions in the stationary air-water mixture ahead of it are $p = 15$ psia, $T = 75$°F, and $\alpha = 0.3$ (α is the gas void fraction). What are the conditions behind the shock? If the shock "bounces" normally off a stationary wall, what values of p and α are set up near the wall?

6.23. A flow of helium and water with $\alpha = 1/2$, $T = 100$°F, and $p = 20$ psia impinges squarely on a wedge with an included angle 2θ of 20°. Predict the shock angle and the pressure behind the shock. The approach velocity is 400 fps.

6.24. Using the definition of flooding from Chap. 4, show that the continuity wave velocity is always zero at the flooding point.

6.25. In constant property two-phase flow with phase change in a duct of constant area show that continuity waves propagate so that

$$\left(\frac{d\alpha}{dt}\right)_{\text{moving with } V_w} = \left[\frac{v_f}{v_{fg}} + (1 - \alpha)\right]\Omega$$

where Ω is defined by Eq. (2.92). Compare with Eq. (6.23).

6.26. The following four Mach numbers can be defined by forming the ratio of the phase velocities and the velocities of compressibility and dynamic waves:

$$M_{d1} = \frac{v_1}{\sqrt{[(1-\alpha)/\rho_1]f_{1\nabla\alpha}}} \qquad M_{d2} = \frac{v_2}{\sqrt{(-\alpha/\rho_2)\,f_{2\nabla\alpha}}}$$

$$M_1 = \frac{v_1}{\sqrt{\partial p/\partial \rho_1}} \qquad M_2 = \frac{v_2}{\sqrt{\partial p/\partial \rho_2}}$$

Show that Eq. (6.102) can be rearranged to the form

$$\frac{\alpha}{\rho_2 v_2'^2}\,\frac{1 - M_2'^2}{1 - 1/M_{d2}'^2} + \frac{1-\alpha}{\rho_1 v_1'^2}\,\frac{1 - M_1'^2}{1 - 1/M_{d1}'^2} = 0$$

and that if Eq. (6.107) is satisfied this reduces to

$$\alpha\,\frac{1 - M_2'^2}{1 - M_{d2}'^2} + (1-\alpha)\,\frac{1 - M_1'^2}{1 - M_{d1}'^2} = 0$$

6.27. If the relative velocity is small compared with c_1 and c_2, show that the wave velocity in stratified flow is

$$U = \frac{V_2\alpha/\rho_2 + V_1(1-\alpha)/\rho_1}{\alpha/\rho_2 + (1-\alpha)/\rho_1} \pm c_{cs}$$

6.28. Derive Eq. (6.90) starting from Bernoulli's equation and the condition of pressure equality at the interface.

6.29. Show that the two-component flow results in this chapter reduce to the single-component flow results when only one component is present.

6.30. The system described in Example 6.6 initially contains a depth H of pure liquid. The steady gas flux $j_g = v_\infty/8$ is then suddenly turned on and kept constant thereafter. If there is no bubble bursting or agglomeration, describe in detail what happens. The system extends upward to a height several times greater than H, at which point any foam is allowed to overflow. There are at least three stages in the process.

6.31. Under what conditions can continuity waves result from the flow or flux being a function of thermodynamic density rather than concentration? Develop continuity wave theory for the case where both densities and volumetric concentration influence the flow rate.

6.32. In slug flow, large cylindrical gas bubbles and liquid plugs alternate in series. What is the mean sonic velocity[12] for this flow pattern? How does it compare with the predictions of homogeneous and separated flow [Eqs. (6.110) and (6.106)]? Is the slug-flow acoustic velocity always greater than in the other flow patterns?

6.33. (*a*) Use Eqs. (6.164), (6.165), and (6.166) to develop a theory of dynamic shock waves in equilibrium single-component vapor-liquid flow.

(*b*) Steam at 10 psia and 90% quality is flowing in a duct and enters a stationary dynamic shock wave. Determine the velocities on each side of the shock as a function of the pressure on the downstream side.

6.34. (*a*) Use Eqs. (6.164), (6.165), and (6.166) to show how to analyze "condensation shocks" which occur when droplets nucleate in supercooled vapor traveling at high speeds.

(*b*) Solve Prob. 6.33*b* if the steam is dry and supercooled but has enthalpy corresponding to 90% quality. How does this problem differ from Prob. 6.33*b*? Assume thermodynamic equilibrium behind the shock.

REFERENCES

1. Lighthill, M. J., and G. B. Whitham: *Proc. Roy. Soc.* (London), vol. 229A, p. 281, 1955.
2. Jeffreys, H.: *Proc. Cambridge Phil. Soc.*, vol. 26, pp. 204–205, 1930.
3. Lamb, Sir Horace: "Hydrodynamics," 6th ed., p. 371, Dover Publications, Inc., New York, 1945.
4. Long, R. R.: *Tellus*, vol. 5, no. 7, pp. 42–57, 1953.
5. Long, R. R.: *Tellus*, vol. 6, no. 2, pp. 97–115, 1954.
6. Long, R. R.: *Tellus*, vol. 8, no. 4, pp. 460–471, 1956.
7. Huey, C. T., and R. A. A. Bryant: ASME paper 65-WA/FE-S, 1965.
8. Gouse, S. W. Jr. and G. A. Brown: M.I.T., E.P.L. Report DSR 8040-1, April 1963.
9. Benjamin, T. B.: *J. Fluid Mech.*, vol. 2, p. 554, 1957.
10. Eddington, R. B.: *AIAA J.*, Investigation of Shock Phenomena in a Supersonic Two-phase Tunnel, AIAA paper 66–87, 1966.
11. Van Rossum, J. J.: *Appl. Sci. Res.*, sec. A, vol. 7, pp. 121–144, 1958.
12. Henry, R. E., and H. K. Fauske: *Trans. Am. Nucl. Soc.*, vol. 11, no. 1, p. 364, June, 1968.
13. Crowley, C. J., G. B. Wallis, and J. J. Barry, *Int. J. Multiphase Flow*, Vol. 18, No. 2, pp. 249–271, 1992.

7
Interfacial Phenomena

7.1 INTRODUCTION

The behavior of surfaces and interfaces is quite fascinating, involving numerous physical and chemical effects. For instance the presence of minute quantities of impurities can dramatically alter the appearance of condensation and boiling, the stability of foams, the waviness of lakes, and the tendency of mirrors and spectacles to "mist up."

For the purpose of analyzing two-phase flows, the major importance of interfacial phenomena is the way in which they affect the boundary conditions that the various equations must satisfy. In single-phase flow the usual requirements are that the stress and velocity fields should be continuous. In two-phase flows finite discontinuities in certain components of both velocity and stress are possible at interfaces.

7.2 VELOCITY BOUNDARY CONDITIONS

If there is no phase change or mass transfer, the velocity compatibility condition is the same in one-component and two-component flows. With

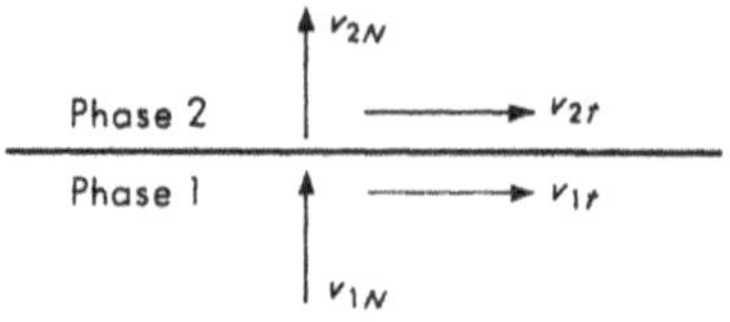

Fig. 7.1 Velocity boundary conditions at an interface.

phase change, however, there is a possibility of a finite velocity across the interface. Flow across the interface must satisfy continuity; therefore if there is a density change, there will also be a velocity change. Relative to the interface shown in Fig. 7.1, therefore, the following compatibility conditions must be satisfied in the directions along and perpendicular to the interface.

1. Continuous tangential velocities:

$$v_{1t} = v_{2t} \tag{7.1}$$

2. Continuity across the interface:

$$\rho_1 v_{1N} = \rho_2 v_{2N} = m \tag{7.2}$$

where m represents a flux of mass crossing the interface normally. It is usual to assume that the average continuum velocity during phase change is perpendicular to the interface.

If the interface is moving, its velocity is simply superimposed on the above velocities.

7.3 STRESS BOUNDARY CONDITIONS

SURFACE-TENSION EFFECTS

Continuity of the stress field across an interface is modified by the effect of surface tension. If the interface is curved, the pressure on one side differs from the pressure on the other by an amount

$$p_2 - p_1 = \sigma\left(\frac{1}{R_a} + \frac{1}{R_b}\right) \tag{7.3}$$

where R_a and R_b are radii of curvature of the interface in a pair of perpendicular directions (Fig. 7.2).

If the surface tension is uniform, the shear stress is continuous across an interface. However, gradients of surface tension can be set up by the presence of impurities, dust, or surface-active agents, or by tempera-

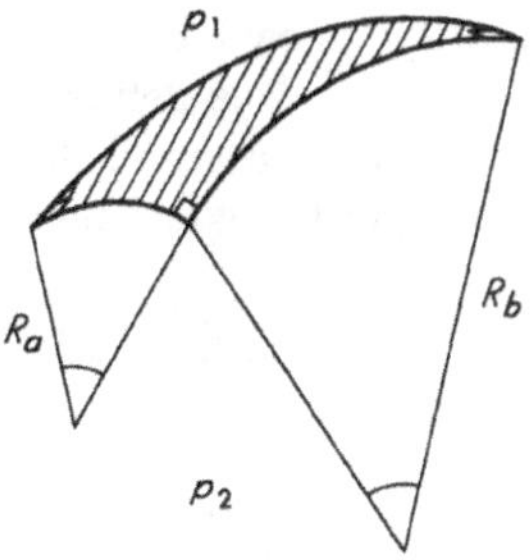

Fig. 7.2 Pressure boundary condition at an interface. R_a and R_b are radii of curvature in two perpendicular directions.

ture gradients along the interface. In this case the shear stress jumps by an amount equal to the surface-tension gradient, thus

$$\boldsymbol{\tau}_1 + \nabla\sigma = \boldsymbol{\tau}_2 \tag{7.4}$$

Equation (7.4), which is a vector expression, can be readily proved by referring to Fig. 7.3 which shows an element of surface in which the x direction is chosen in the direction of $\nabla\sigma$. The force from fluid 1 is $\tau_1\,dx\,dy$, the force from fluid 2 is $-\tau_2\,dx\,dy$, and the surface tension force is $[-\sigma\,dy + (\sigma + \nabla\sigma\,dx)\,dy]$.

In the x direction, therefore, a force balance yields

$$(\tau_1)_x + \nabla\sigma - (\tau_2)_x = 0 \tag{7.5}$$

whereas, in the y direction at right angles to x,

$$(\tau_1)_y - (\tau_2)_y = 0 \tag{7.6}$$

Equations (7.5) and (7.6) are the scalar components of Eq. (7.4).

When an interface meets a surface or another fluid, the three resulting *interfacial tensions* must be in equilibrium at the point where they

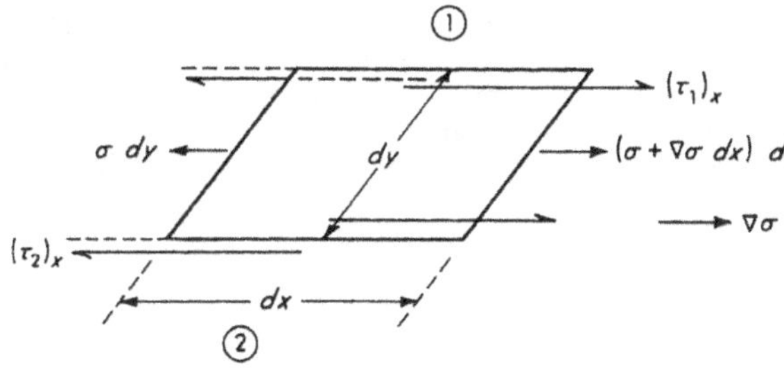

Fig. 7.3 Shear stress boundary conditions at an interface.

meet. For three soap bubbles, for instance, the three surfaces must meet at 120° to each other. For a gas and a liquid at a plane solid surface the *contact angle* is defined as the angle between the gas-liquid interface and the solid, measured through the liquid. For equilibrium of the interface (or from the equivalent condition of minimum energy),

$$\cos \beta = \frac{\sigma_{sg} - \sigma_{sf}}{\sigma_{fg}} \tag{7.7}$$

The contact angle is an elusive parameter to measure because the interface responds not only to the surface tensions but to small variations in surface cleanliness and microscopic surface roughness. Local oxidation or adsorbed gases on the surface also can produce significant effects. For advancing and receding interfaces the contact angle may be quite different, especially if motion is rapid.

Surface tension is of great importance to both the hydrodynamic and heat-transfer characteristics of gas-liquid systems. The waviness and stability of an interface and the formation, entrainment, and atomization of bubbles and drops are all governed by surface tension. Bubble nucleation in boiling depends on the way in which minute bubbles grow from defects or cavities on the heated surface. Dropwise condensation is also governed by contact angle and wetting phenomena.

Example 7.1 What is the difference in pressure between the inside and outside of a vapor bubble of radius 10^{-4} in. if $\sigma = 58.9$ dynes/cm?

Solution Since the curvature is the same in all directions, Eq. (7.3) becomes

$$p_2 - p_1 = \frac{2\sigma}{R}$$

Therefore

$$p_2 - p_1 = \frac{(2)(58.9)(2.54)}{10^{-4}} (2.248 \times 10^{-6}) = 6.73 \text{ psi}$$

Example 7.2 Consider the stability of a plane horizontal interface between fluids of density ρ_1 and ρ_2. Let ρ_1 be the density of the upper fluid and let it be greater than ρ_2. Will a small sinusoidal perturbation of the interface represented by $\eta = \eta_0 \sin 2\pi z/L$ tend to grow or collapse?

Solution Since η is small the curvature of the interface is given by

$$\frac{1}{R} = \frac{d^2\eta}{dz^2} = -\frac{4\pi^2}{L^2} \eta_0 \sin \frac{2\pi z}{L}$$

The pressure change across the interface is therefore, from Eq. (7.3),

$$p_1 - p_2 = \frac{4\sigma\pi^2}{L^2} \eta_0 \sin \frac{2\pi z}{L}$$

The excess hydrostatic pressure due to the perturbation is

$$g(\rho_1 - \rho_2)\eta_0 \sin \frac{2\pi z}{L}$$

The perturbation will tend to collapse if $p_1 - p_2$ exceeds the excess hydrostatic pressure, that is, if

$$\frac{4\sigma\pi^2}{L^2} > g(\rho_1 - \rho_2)$$

or, alternatively,

$$L < L_c = 2\pi \sqrt{\frac{\sigma}{g(\rho_1 - \rho_2)}} \tag{7.8}$$

There is therefore a critical wavelength L_c given by Eq. (7.8) that must be exceeded if the perturbation is to grow. Analysis of the rate of growth of this perturbation, including the effects of inertia, reveals that the fastest growing wavelength is $L = \sqrt{3}\, L_c$. This phenomenon is known as Taylor instability.[1]

7.4 THE EFFECT OF PHASE CHANGE ON INTERFACIAL STRESSES

When phase change occurs across an interface, there is a transfer of mass across the interface, and, if the two components are not moving at the same velocity throughout, this mass transfer has an effect on whatever momentum transfer may be occurring simultaneously.

The evaporating or condensing vapor crossing the interface undergoes a change in normal velocity, according to Eq. (7.2); therefore there is a pressure exerted from the vapor on the liquid in either evaporation or condensation of an amount

$$(\Delta p)_N = m(v_{gN} - v_{fN}) = m^2 v_{fg} \tag{7.9}$$

This effect is usually small.

The effect of phase change on the shear stresses is more difficult to determine. For laminar flow on a flat plate with condensation, the situation is parallel to *boundary-layer suction* as discussed by Schlichting.[2] In this case there is an asymptotic solution which has the effect of removing all viscous drag on the surface and retaining only the momentum transfer term; i.e., for a difference in velocity between the streams of v_{12}, the interfacial stress is simply

$$\tau_{12} = m v_{12} \tag{7.10}$$

This shear stress apparently acts on the liquid during condensation and on the gas during vaporization so it is not possible at this stage to regard condensation as merely negative evaporation. Qualitatively it appears that condensation has the effect of reducing viscous shear on the gas stream, whereas evaporation reduces the shear on the liquid stream.

A simple model for predicting the interfacial shear in turbulent stratified gas-liquid flow with phase change has been developed by Silver and Wallis[3] using the Reynolds flux concept which was introduced in Chap. 3. According to this model the shear stress on an interface without phase change is due to fluid from the main stream striking the wall and bouncing back again after sharing its momentum with the wall. If a mass ϵ_0 strikes per unit area per unit time, the shear stress is

$$\tau_0 = \epsilon_0(v_g - v_f) \tag{7.11}$$

In a very elementary model it can be assumed that the mass flux is made up of two streams, one moving toward the wall and one moving away from it, each occupying one-half of the flow area and traveling with average speed u_0 (Fig. 7.4). Focusing attention on the vapor stream, we have

$$\epsilon_0 = \tfrac{1}{2}u_0\rho_g \tag{7.12}$$

Let there be a mass flux of m due to phase change in the direction from liquid to vapor. This flux can be brought about by superimposing the velocity in Eq. (7.2) on the existing turbulence pattern. The stream with velocity u_0 toward the interface now acquires a velocity $(u_0 - v_{gn})$, whereas the returning stream acquires a velocity $(u_0 + v_{gn})$. The amount of vapor which transfers momentum to the gas stream per unit area is therefore $\tfrac{1}{2}\rho_g(u_0 + v_{gn})$, whereas the net mass flux which transfers momentum to the liquid is only $\tfrac{1}{2}\rho_g(u_0 - v_{gn})$.

Assuming that the material which is transferred is initially in equilibrium with the stream from which it came, the "drag" forces on the vapor and liquid streams per unit volume of the duct are

$$F_g = -\frac{P_i}{A}(v_g - v_f)\tfrac{1}{2}\rho_g(u_0 + v_{gN}) \tag{7.13}$$

$$F_f = \frac{P_i}{A}(v_g - v_f)\tfrac{1}{2}\rho_g(u_0 - v_{gN}) \tag{7.14}$$

where P_i is the interfacial perimeter.

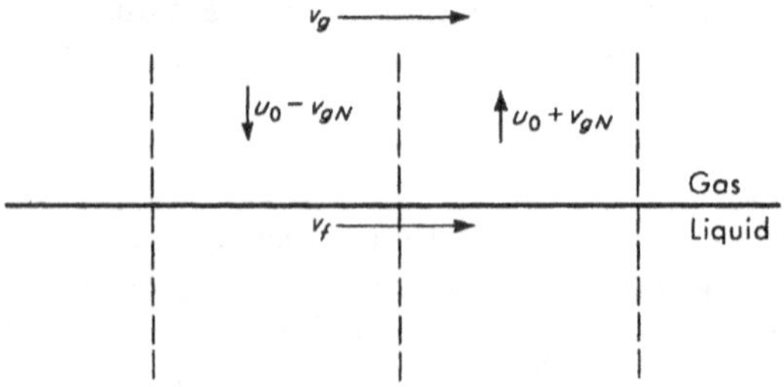

Fig. 7.4 Reynolds flux with phase change, according to Silver and Wallis.[3]

Substituting in terms of ϵ_0 and m, we have

$$F_g = -\frac{P_i}{A}(v_g - v_f)\left(\epsilon_0 + \frac{m}{2}\right) \tag{7.15}$$

$$F_f = \frac{P_i}{A}(v_g - v_f)\left(\epsilon_0 - \frac{m}{2}\right) \tag{7.16}$$

If condensation instead of boiling were occurring, the sign of m would simply be reversed in these equations. Therefore we may replace m by m_{fg} which denotes a vaporization flux and can be either positive or negative.

Identifying the vapor as phase 2, making use of Eqs. (3.42) and (3.43) in Eqs. (3.45) and (3.46), and adding forces F_{wf} and F_{wg} to account for wall shear stresses, we obtain, eventually, for steady flow the following equations of motion:

$$\rho_f v_f \frac{dv_f}{dz} = -\frac{dp}{dz} - \rho_f g \cos\theta - \frac{F_{wf}}{1-\alpha} + \frac{P_i(v_g - v_f)}{A(1-\alpha)}\left(\epsilon_0 - \frac{m_{fg}}{2}\right) \tag{7.17}$$

$$\rho_g v_g \frac{dv_g}{dz} = -\frac{dp}{dz} - \rho_g g \cos\theta - \frac{F_{wg}}{\alpha} - \frac{P_i(v_g - v_f)}{A\alpha}\left(\epsilon_0 + \frac{m_{fg}}{2}\right) \tag{7.18}$$

These results are identical with Eqs. (3.106) and (3.107).

It is sometimes convenient to separate the mass transfer into components of "flow" and "recirculation," since the flux m may not necessarily be in equilibrium with the wall (if it is injected through slots, for instance). The recirculating or mixing flux for phase change in either direction is

$$\epsilon_m = \epsilon_0 - \frac{|m|}{2} \tag{7.19}$$

and is decreased by either boiling or condensation. Equations (7.15) and (7.16) then result by adding the flow m to the appropriate stream.

A similar modification of the wall shear stress as a result of mass transfer or gas injection occurs in developing turbulent boundary layers. In their book, Kutateladze and Leontev[4] report an equation which is valid for injection of gas of the same molecular weight as the mainstream and, with the present notation, amounts to

$$\frac{\epsilon_m}{\epsilon_0} = \left(1 - \frac{m}{4\epsilon_0}\right)^2 \left(1 + \frac{m}{4\epsilon_0}\right)^{-1/2} \tag{7.20}$$

This result is shown in Fig. 7.5 compared with the results of several authors who studied flow over a permeable flat plate. Equation (7.19)

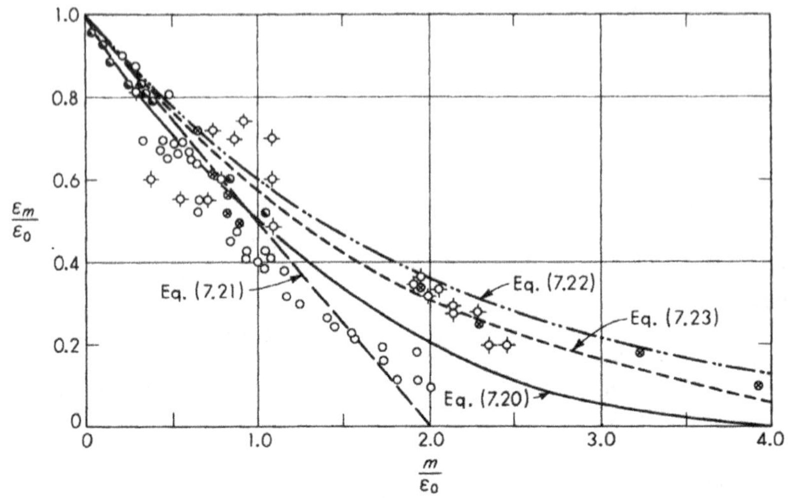

Fig. 7.5 Gas injection into a turbulent boundary layer. Data reproduced from Kutateladze and Leontev[4] compared with various theories.

in the following form is shown for comparison:

$$\frac{\epsilon_m}{\epsilon_0} = 1 - \frac{1}{2}\frac{m}{\epsilon_0} \tag{7.21}$$

Also shown is an equation which was deduced from a more sophisticated analysis by Silver and Wallis,[3]

$$\frac{\epsilon_m}{\epsilon_0} = e^{-m/2\epsilon_0} \tag{7.22}$$

and an equation from Spalding,[5]

$$\frac{\epsilon_m}{\epsilon_0} = \frac{m}{\epsilon_0}\,(e^{m/\epsilon_0} - 1)^{-1} \tag{7.23}$$

Rather surprisingly, the Reynolds flux model also gives reasonable predictions in the case of laminar boundary layers.[7] It is also useful for predicting the effect of phase change on interfacial heat- and mass-transfer coefficients. For example, evaporating drops tend to shield themselves from heat transfer from the surrounding gas. For Prandtl and Schmidt numbers close to unity, the ratio ϵ_m/ϵ_0 is the same thing as the ratio of the heat- and mass-transfer coefficients to the value with no evaporation. Many applications of Reynolds flux theory are discussed by Spalding[5] in his book.

The value of ϵ_0 is usually determined from a knowledge of single-phase flow under the same conditions. For example, if the shear stress is given by a friction factor correlation for flow in a pipe,

$$\tau_0 = C_f \tfrac{1}{2}\rho(v_g - v_f)^2 \qquad (7.24)$$

Equations (7.11) and (7.12) then show that

$$\epsilon_0 = \tfrac{1}{2}C_f\rho(v_g - v_f) \qquad (7.25)$$
$$u_0 = C_f(v_g - v_f) \qquad (7.26)$$

Therefore C_f could be interpreted as being the ratio between the mixing velocity and the relative velocity.

An alternative representation of turbulent mixing in single-phase flows is based on the *friction velocity*,

$$u^* = \left(\frac{\tau_0}{\rho}\right)^{1/2} \qquad (7.27)$$

Evidently u^* and u_0 are related by the expression

$$\frac{u_0}{u^*} = \sqrt{2C_f} \qquad (7.28)$$

In general, u^* is representative of velocity fluctuations in the main stream, whereas u_0 is a suitable average which characterizes the mixing process across the entire boundary layer.

If interfacial waves occur on the liquid film, the value of ϵ_0 is increased in proportion to the interface friction factor.

7.5 FURTHER EFFECTS

The numerous effects of interfaces on two-phase thermodynamics and transport phenomena are beyond the scope of this text. The interested reader is referred to the book by Davies and Rideal.[6] Surface electrical and adhesive effects can be included, if necessary, in the general one-dimensional flow theory by making suitable additions to the forces f_1 and f_2 in the momentum equations (3.40). Some additional phenomena will be discussed in later chapters.

PROBLEMS

7.1. Calculate m, v_{fn}, and v_{gn} in Eq. (7.2) for steam evaporating from a surface at 1, 10, 100, and 1000 psia corresponding to surface heat fluxes of 10^4, 10^5, and 10^6 Btu/(hr)(ft^2).

7.2. Calculate the pressure difference in Eq. (7.9) for the conditions of Prob. 7.1.

7.3. Liquid is sucked by the influence of surface tension into a long straight horizontal capillary tube which is exposed to the same pressure at both ends. Show that the distance which the interface has penetrated after time t is given by

$$z = \left(\frac{D\sigma t \cos \beta}{4\mu}\right)^{1/2}$$

7.4. Dynamic waves in annular vertical gas-liquid flow are influenced by surface tension. Using Eq. (6.158), assuming a thin film, and accounting for the pressure differences due to surface tension, show that Eq. (6.162) is unchanged but Eq. (6.163) becomes

$$\frac{\omega^2}{U^2}\left\{u^2 - c^2 + \frac{D\sigma(4/D^2 - \omega^2/U^2)}{4[\rho_g/\alpha + \rho_f/(1-\alpha)]}\right\} = \frac{B^2}{4}\frac{v_w^2 - u^2}{u^2}$$

Show that a spectrum of wavelengths is unstable. What is the shortest wavelength which can grow? Which wavelength grows the fastest?

7.5. When a droplet evaporates in a highly superheated vapor, its surface is close to the saturation temperature and the evaporation rate is governed by heat transfer. Show from an energy balance that

$$\frac{\epsilon_m}{\epsilon_0} = \frac{h_{fg}}{c_p\,\Delta T}\frac{m}{\epsilon_0}$$

where ΔT is the vapor superheat and c_p the vapor specific heat. By solving this equation simultaneously with Eq. (7.21), show that the evaporation rate is

$$m = \frac{c_p\,\Delta T\,\epsilon_0}{h_{fg}} \cdot \frac{1}{1 + 0.5\,c_p\,\Delta T/h_{fg}}$$

and that the second factor represents the effect of mass transfer in reducing the effective heat transfer. How big is this effect for

(*a*) water at 1000 psi in steam superheated by 200°F?

(*b*) gasoline at 14.7 psi in a 2000°F flame?

7.6. Use the Reynolds flux model to relate the molal concentration of noncondensable gases, $(c_a)_w$, near a surface at which condensation occurs at a rate m to the concentration in the free stream, $(c_a)_s$. Show that any of Eqs. (7.21), (7.22), and (7.23) predict[7]

$$\frac{(c_a)_w}{(c_a)_s} = 1 + \frac{m}{\epsilon_0} + \frac{1}{2}\left(\frac{m}{\epsilon_0}\right)^2 + \text{higher-order terms}$$

7.7. Air flows over a flat plate which is oriented parallel to the original flow direction. If the air is at 70°F, 14.7 psia, and has a velocity of 2 fps, plot the shear stress versus distance from the leading edge if air is blown uniformly from the surface of the plate with a velocity of 0.01 fps.

7.8. Paper passing over a heated solid cylinder is dried at a rate of 20 lb water/(ft²)(hr). It is suggested that blowing at the paper with air jets will increase the rate of drying. What minimum mass-transfer coefficient must exist under the jets if they are to be effective? (*Hint:* First relate the mass-transfer coefficient to u_0.)

7.9. In the derivation of Eqs. (7.15) and (7.16) the interface velocity was assumed equal to the liquid velocity. In general, however, the situation shown in Fig. 7.4 will apply to both sides of the interface, with ϵ_{0g} and ϵ_{0f} being the Reynolds flux in the

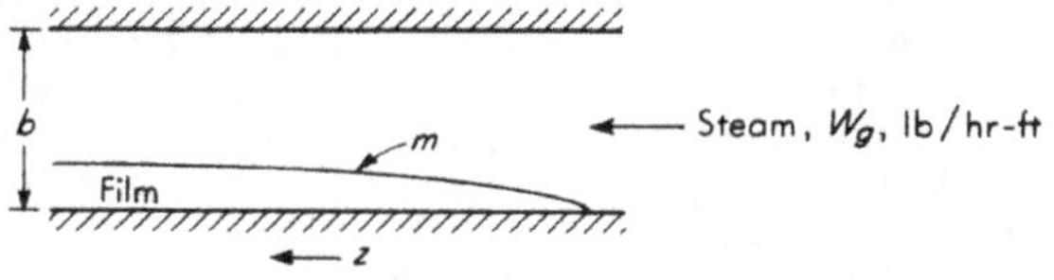

Fig. 7.6 Sketch for Prob. 7.13.

vapor and liquid streams, respectively. Show that in this case

$$F_g = -P_i \frac{(v_g - v_f)(\epsilon_{0g} + m/2)(\epsilon_{0f} + m/2)}{\epsilon_{0g} + \epsilon_{0f}}$$

$$F_f = P_i \frac{(v_g - v_f)(\epsilon_{0g} - m/2)(\epsilon_{0f} - m/2)}{\epsilon_{0g} + \epsilon_{0f}}$$

and that Eqs. (7.15) and (7.16) will be correct to first order in m/ϵ_0 if ϵ_0 is chosen so that

$$\frac{1}{\epsilon_0} = \frac{1}{\epsilon_{0f}} + \frac{1}{\epsilon_{0g}}$$

What is the value of the interface velocity?

7.10. Estimate the flow rate of air required for the elimination of friction drag on a flat plate mounted parallel to an air stream with velocity 200 fps if $p = 15$ psia, $T = 80°F$.

7.11. At what rate of vaporization will the interfacial shear vanish on a smooth water film on the inside wall of a 1-in.-diam pipe through which saturated steam is flowing at 500 psi and 100 fps.

7.12. What is the heat-transfer coefficient across a horizontal laminar film of condensate carrying a liquid flow of Γ (pounds per hour per foot width) if $v_g \gg v_f$ and if the rate of condensation is such that $m \gg \epsilon_0$?

7.13. Use the result of Prob. 7.12 to deduce the variation of film thickness and heat-transfer coefficient as a function of position for the situation shown in Fig. 7.6. Assume uniform properties and a constant value of m. How does the plate separation b influence the heat-transfer coefficient for a given steam flow rate?

7.14. Explain why infants have more difficulty breathing when they catch colds than do adults.

7.15. Solve Prob. 7.3 if the tube is vertical.

7.16. Use the Reynolds flux model to show that an approximate expression for the minimum transport velocity which will prevent settling of particles flowing in suspension in a horizontal pipe is

$$v = \frac{v_\infty}{C_f}$$

If $C_f = 0.005$, show that at the condition of minimum transport

$$\frac{v_\infty}{u^*} \approx 0.1$$

7.17. (*a*) An air bubble is stationary in water in which there is a temperature gradient. This causes surface-tension variations around the bubble which set up convective currents in the liquid. In which direction is the resulting force on the bubble?

(*b*) Does a bubble swim up or down the temperature gradient as a result of the force experienced in part *a*? Is the direction of motion compatible with "minimum energy" requirements for equilibrium?

(*c*) Explain how a gas bubble can be held down against an upward facing surface as a result of temperature gradients.

7.18. Example 7.1 refers to a steam bubble in water at atmospheric pressure. Use the steam tables to determine the liquid superheat which is needed for the bubble to be in equilibrium.

REFERENCES

1. Taylor, G. I.: *Proc. Roy. Soc.* (London), vol. A201, p. 192, 1950.
2. Schlichting, H.: "Boundary Layer Theory," 4th ed., pp. 487, 502–533, McGraw-Hill Book Company, New York, 1960.
3. Silver, R. S., and G. B. Wallis: *Proc. Inst. Mech. Engrs.*, vol. 180, part I, pp. 36–40, 1965–1966.
4. Kutateladze, S. S., and A. I. Leontev: "Turbulent Boundary Layers in Compressible Gases," pp. 70–71, transl. by D. B. Spalding, Academic Press Inc., New York, 1964.
5. Spalding, D. B.: "Convective Mass Transfer," McGraw-Hill Book Company, New York, 1963.
6. Davies, J. T., and E. K. Rideal: "Interfacial Phenomena," 2d ed., Academic Press Inc., New York, 1963.
7. Wallis, G. B.: *Intern. J. Heat Mass Transfer*, vol. 11, pp. 445–472, 1968.

part two

Practical Applications

8
Suspensions of Particles in Fluids

8.1 INTRODUCTION

One example of a two-phase system is a suspension of particles in a fluid. Because of the weaker influence of surface tension, subdivision, and distortion, this system is usually more simple than the equivalent gas-liquid or liquid-liquid dispersion. Nevertheless, many complications arise in practice due to factors such as the wide variety of sizes and shapes of the particles, nonuniform flow patterns, agglomeration, and interparticle forces.

Engineering applications include: fluidized beds for reduction of uranium ore, building shell molds, and plastic coating; settling tanks, filtration beds, ground-water flows to wells, soil compaction; pneumatic and hydraulic conveying; plasma spraying, sand blasting, rocket exhausts containing ash or unburnt metal powders; flow of paints, slurries, printing inks, soups, and paper fibers in suspension.

In this chapter we shall examine a series of *regimes* of fluid-particle flows each of which is governed by a balance between a limited number of forces. For example, the dynamics of sedimentation are determined by the interaction between buoyancy and drag forces, while the important forces in high-speed nozzle flows are pressure, inertia, and interphase drag. General equations containing numerous terms but not leading to any practical solutions will be avoided. The reader who is interested in more complicated systems which are not discussed in this chapter will need to modify the general techniques of Part One to suit his particular problem.

8.2 ONE-DIMENSIONAL VERTICAL FLOW OF A UNIFORM INCOMPRESSIBLE DISPERSION WITH NO WALL FRICTION

GENERAL THEORY OF UNIFORM STEADY FLOW

Consider the steady flow of a system of particles and fluid in a vertical duct of uniform cross section. If velocities are low enough and the duct is large enough, wall friction can be neglected and overall motion relative to the container can have no influence on the relative motion between the particles and the fluid. For a given system, therefore, the relative motion is only a function of the local concentration, properties, and the gravitational field and does not depend on the net flow rates of the components. Adequate confirmation of this assumption is provided in the detailed experiments of Lapidus *et al.*[1,2]

The arguments which were presented in Chap. 4 lead to the definition of a drift flux, j_{fs}, which depends only on concentration. Denoting the volumetric concentration of the fluid (the void fraction) by ϵ, we obtain as before

$$j_{fs} = (1 - \epsilon)j_f - \epsilon j_s \tag{8.1}$$

where the subscript f refers to the fluid and s to the particles, and the positive direction is chosen to be upward. It is more usual to use ϵ to describe fluid-solid systems, although an equally satisfactory description can be obtained in terms of $\alpha = 1 - \epsilon$.

Remembering that ϵ refers to the continuous component, the empirical correlation (4.14) is seen to correspond to

$$j_{fs} = v_\infty \epsilon^n (1 - \epsilon) \tag{8.2}$$

where v_∞ is the terminal speed of a single particle in an infinite stationary liquid. All the analytical techniques derived in Chap. 4 can then be applied if one can predict the quantities v_∞ and n from first principles.

TERMINAL VELOCITY OF A SINGLE PARTICLE

The terminal velocity of a single particle is determined by balancing the gravitational and drag forces. For spherical particles of diameter d, for example,

$$\frac{\pi}{6} d^3 g(\rho_s - \rho_f) = C_{D\infty} \frac{\pi}{4} d^2 \frac{1}{2} \rho_f v_\infty^2 \tag{8.3}$$

where $C_{D\infty}$ is the drag coefficient and is a function of the Reynolds number, Re_∞, defined by

$$\mathrm{Re}_\infty = \frac{v_\infty d \rho_f}{\mu_f} \tag{8.4}$$

Simplifying Eq. (8.3) and using Eq. (8.4) to eliminate v_∞, we obtain

$$C_{D\infty}\mathrm{Re}_\infty^2 = \frac{4}{3} \frac{d^3 \rho_f g(\rho_s - \rho_f)}{\mu_f^2} \tag{8.5}$$

The quantity $C_{D\infty}\mathrm{Re}_\infty^2$ is therefore independent of v_∞ and can be evaluated solely from known quantities.

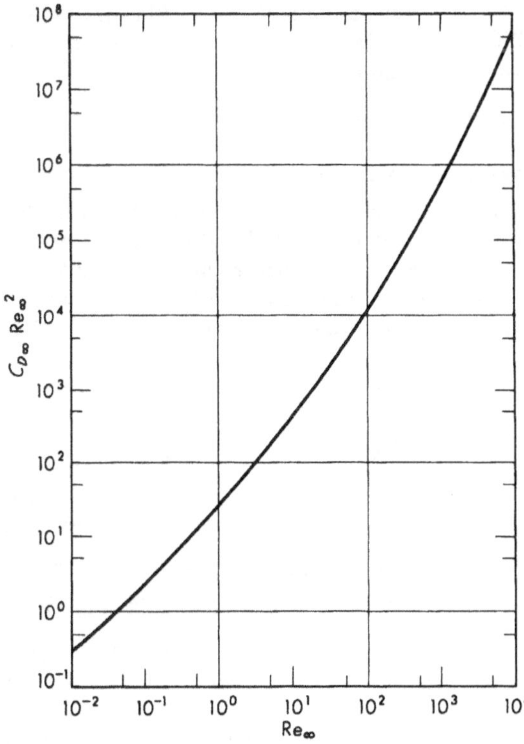

Fig. 8.1 Variation of $C_{D\infty}\mathrm{Re}_\infty^2$ with Re_∞ for spheres. (*Rowe.*[3])

Since $C_{D\infty}$ is a function of Re_∞, $C_{D\infty}\text{Re}_\infty^2$ can be expressed analytically or graphically[3] (Fig. 8.1) as a function of Re_∞. Using this relationship and Eq. (8.5) the quantity Re_∞ and hence v_∞ can be derived.

Most particle-fluid systems of practical interest have Reynolds numbers less than 1000 for which the drag coefficient may be related to the Reynolds number by the equation[3,4]

$$C_{D\infty} = \frac{24}{\text{Re}_\infty}(1 + 0.15\text{Re}_\infty^{0.687}) \tag{8.6}$$

For Reynolds numbers greater than about 1000, the drag coefficient is constant and is approximately

$$C_{D\infty} = 0.44 \tag{8.7}$$

Modifications to these equations for particles which are not spherical are described by Heywood.[5]

The right-hand side of Eq. (8.5) is readily expressed in terms of the dimensionless inverse viscosity N_f (sometimes called the "Grashof number"), and the result is

$$C_{D\infty}\text{Re}_\infty^2 = \tfrac{4}{3}N_f^2 \tag{8.8}$$

EVALUATION OF THE INDEX *n*

The index n in Eq. (8.2) has been shown by Richardson and Zaki[6] to be primarily a function of the Reynolds number which was defined in Eq. (8.4). A correction factor can also be introduced in terms of the ratio of the particle diameter d to the tube diameter D. The correlation of Richardson and Zaki over the whole range of Reynolds numbers is

$$\text{Re}_\infty < 0.2 \qquad n = 4.65 + 19.5\frac{d}{D} \tag{8.9a}$$

$$0.2 < \text{Re}_\infty < 1 \qquad n = \left(4.35 + 17.5\frac{d}{D}\right)\text{Re}_\infty^{-0.03} \tag{8.9b}$$

$$1 < \text{Re}_\infty < 200 \qquad n = \left(4.45 + 18\frac{d}{D}\right)\text{Re}_\infty^{-0.1} \tag{8.9c}$$

$$200 < \text{Re}_\infty < 500 \qquad n = 4.45\,\text{Re}_\infty^{-0.1} \tag{8.9d}$$

$$500 < \text{Re}_\infty \qquad n = 2.39 \tag{8.9e}$$

Much larger values of n can be obtained with particles which *flocculate*.[7] This usually occurs with particles with dimensions in the micron range and particularly those particles of odd shapes which can group themselves into spongelike flocs which remain intact. In these cases it is advisable to deduce the dependence of drift flux on concentration experimentally.

FORCES ON THE PARTICLES AND THE FLUID

Consider an experiment in which an array of particles is maintained stationary in a horizontal fluid stream and the force on a typical particle is measured in terms of the fluid velocity, properties, and concentration (Fig. 8.2). The total force necessary to restrain a particle with volume $\mathcal{V}_p$ will be

$$F' = \mathcal{V}_p\left(f_s - \frac{dp}{dz}\right) \tag{8.10}$$

where f_s is defined as in Chap. 3. F' can be correlated in terms of the fluid flux and properties by defining a drag coefficient, for a given ϵ, as follows:

$$(C_D)_\epsilon = \frac{F'}{\frac{1}{2}\rho_f j_{f0}^2 A_p} \tag{8.11}$$

where A_p is a characteristic cross section for the particle. The drag coefficient contains both the effect of the net pressure gradient and the effect of forces contained in f_s. j_{f0} is the fluid flux relative to stationary particles.

The entire volume between lines *aa* and *bb* in Fig. 8.2 must be in equilibrium. There is no fluid momentum change and "end effects" can be made vanishingly small by choosing a sufficiently large volume. Therefore, since the external force on the particles balances the pressure gradient,

$$F'(1 - \epsilon) = -\left(\frac{dp}{dz}\right)_F \mathcal{V}_p \tag{8.12}$$

The subscript F denotes that the pressure gradient is entirely "frictional." The drag force on the particles must be equal and opposite to the drag

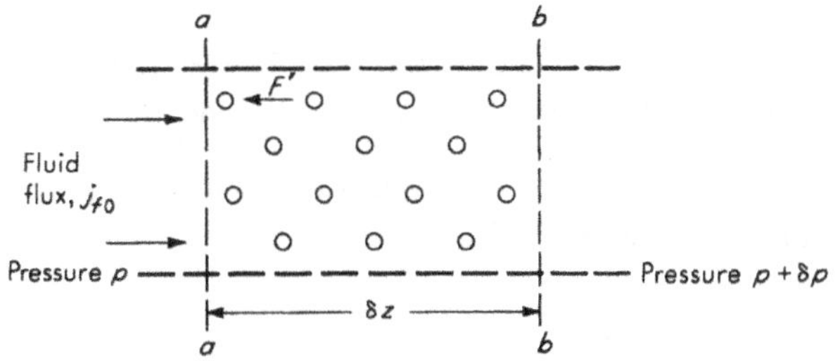

Fig. 8.2 Forces on particles in an array. The particles are held against the fluid drag by a force F' per particle.

force on the fluid. Therefore (as in Example 3.5),

$$F_{fs} = \epsilon f_f = -(1 - \epsilon)f_s \tag{8.13}$$

A further useful parameter is

$$f_{fs} = f_f - f_s \tag{8.14}$$

Combining Eqs. (8.10) to (8.14) all of the various forces can be expressed in terms of each other as follows:

$$\frac{F'}{\mathcal{V}_p} = (C_D)_\epsilon \tfrac{1}{2}\rho_f j_{f0}{}^2 \frac{A_p}{\mathcal{V}_p} = -\frac{1}{1-\epsilon}\left(\frac{dp}{dz}\right)_F = -\frac{f_f}{1-\epsilon}$$
$$= \frac{f_s}{\epsilon} = -f_{fs} = -\frac{F_{fs}}{\epsilon(1-\epsilon)} \tag{8.15}$$

If the drag coefficient $(C_D)_\epsilon$ can be determined as a function of void fraction and fluid properties, all of the forces can be evaluated.

For example, for a stationary fluidized bed in which particles are supported against gravity by the upward flow of fluid around them, Eqs. (3.45) and (3.46) reduce to

$$\frac{dp}{dz} = -g\rho_f + f_f = -g\rho_s + f_s \tag{8.16}$$

Combining this with Eq. (8.15) we find that

$$(C_D)_\epsilon = \frac{g(\rho_s - \rho_f)\dfrac{\mathcal{V}_p}{A_p}}{\tfrac{1}{2}\rho_f j_{f0}{}^2} \tag{8.17}$$

Now, Rowe's experiments[8] suggest that $(C_D)_\epsilon$ can be expressed as the product of a function of Reynolds number and a function of void fraction. Thus

$$(C_D)_\epsilon = C_{Ds}\psi(\epsilon) \tag{8.18}$$

The quantity $\psi(\epsilon)$ is the ratio between the drag force on a particle in an assembly and the same particle alone in a fluid stream with the same volumetric flux relative to the particles. C_{Ds} is given as a function of the "superficial" Reynolds number in the usual way [i.e., Eqs. (8.6) and (8.7)], where the Reynolds number to be used is defined as

$$\text{Re}_s = \frac{\rho_f j_{f0} d}{\mu_f} \tag{8.19}$$

For spherical particles supported against gravity the equilibrium condition equivalent to Eq. (8.8) is then

$$(C_D)_\epsilon \text{Re}_s{}^2 = \tfrac{4}{3}N_f{}^2 \tag{8.20}$$

Equations (8.20) and (8.5) therefore give, for particles in equilibrium,

$$C_{D_\infty}\mathrm{Re}_\infty^2 = (C_D)_\epsilon \mathrm{Re}_s^2 \tag{8.21}$$

Using Eqs. (8.6) and (8.7) to express the drag coefficient and substituting Eq. (8.18), Eq. (8.21) becomes

Re < 1000:

$$\mathrm{Re}_\infty(1 + 0.15\mathrm{Re}_\infty^{0.687}) = \psi(\epsilon)\mathrm{Re}_s(1 + 0.15\mathrm{Re}_s^{0.687}) \tag{8.22}$$

Re > 1000:

$$\mathrm{Re}_\infty^2 = \psi(\epsilon)\mathrm{Re}_s^2 \tag{8.23}$$

For given fluid properties, Eqs. (8.22) and (8.23) are relationships between v_∞, ϵ, and j_{f0}, and may be used to derive the value of j_{fs} in the usual way.

Equations (8.22) and (8.23) can be compared with Richardson and Zaki's correlations [Eq. (8.9)] by using Eqs. (8.1) and (8.2) to deduce that

$$j_{f0} = v_\infty \epsilon^n \tag{8.24}$$

and multiplying Eq. (8.24) by $\rho_f d/\mu_f$ to get

$$\mathrm{Re}_s = \mathrm{Re}_\infty \epsilon^n \tag{8.25}$$

For Reynolds numbers greater than 1000, Eqs. (8.23) and (8.25) show that

$$\psi(\epsilon) = \epsilon^{-2n} \tag{8.26}$$

i.e., from Eq. (8.9*e*)

$$\psi(\epsilon) = \epsilon^{-4.78} \tag{8.27}$$

For very low Reynolds numbers on the other hand, Eqs. (8.22) and (8.25) lead to the result

$$\psi(\epsilon) = \epsilon^{-n} \tag{8.28}$$

Therefore, from Eq. (8.9*a*)

$$\psi(\epsilon) = \epsilon^{-4.65} \tag{8.29}$$

A compromise between Eqs. (8.27) and (8.29) is[9]

$$\psi(\epsilon) = \epsilon^{-4.7} \tag{8.30}$$

The validity of Eq. (8.30) may be tested over the whole range of Reynolds numbers by raising Eq. (8.25) to the power $4.7/n$ and multiplying the result by Eq. (8.22). This gives

$$\mathrm{Re}_\infty^{(1-4.7/n)}(1 + 0.15\mathrm{Re}_\infty^{0.687}) = \mathrm{Re}_s^{(1-4.7/n)}(1 + 0.15\mathrm{Re}_s^{0.687}) \tag{8.31}$$

Evidently Eq. (8.31) cannot be satisfied exactly over a range of Reynolds numbers. However, one can require that the function of Reynolds number given by Eq. (8.31) should be approximately constant near the particular value under consideration. Differentiating Eq. (8.31) therefore leads to

$$\left(1 - \frac{4.7}{n}\right) \mathrm{Re}_\infty^{-4.7/n} + 0.15\left(1.687 - \frac{4.7}{n}\right) \mathrm{Re}_\infty^{(0.687-4.7/n)} = 0 \quad (8.32)$$

whence

$$n = 4.7 \frac{(1 + 0.15\mathrm{Re}_\infty^{0.687})}{(1 + 0.253\mathrm{Re}_\infty^{0.687})} \quad (8.33)$$

The values of n calculated this way are compared with the values of Richardson and Zaki in Table 8.1. The value for Re = 1000 is taken from Eqs. (8.26) and (8.30).

The results are sufficiently compatible for Eq. (8.30) to be regarded as a good approximation.

For spherical particles Eq. (8.15) can therefore be expressed as

$$-f_{fs} = \epsilon^{-4.7} C_{Ds} \frac{3}{4} \frac{\rho_f j_{f0}^2}{d} \quad (8.34)$$

and all the various forces can be calculated from known quantities.

For a particle in equilibrium under gravity, Eq. (8.20) becomes an explicit equation for the void fraction

$$\epsilon^{4.7} = \frac{3}{4} \frac{C_{Ds}\mathrm{Re}_s^2}{N_f^2} \quad (8.35)$$

In a system in which the particles are moving, the quantity j_{f0} is related to the relative velocity or the drift flux by

$$j_{f0} = \frac{j_{fs}}{1 - \epsilon} = v_{fs}\epsilon \quad (8.36)$$

In terms of the relative velocity, for example, Eq. (8.34) becomes

$$f_{fs} = -\epsilon^{-2.7} C_{Ds} \frac{3}{4} \frac{\rho_f v_{fs}^2}{d} \quad (8.37)$$

Table 8.1 Comparison between the values of n from Eqs. (8.33) and (8.9)

Re_∞	0	0.2	1	10	100	1000
Richardson and Zaki, Eqs. (8.9)	4.65	4.65	4.35	3.53	2.8	2.39
Eq. (8.33)	4.7	4.65	4.31	3.65	3.05	2.35

This is comparable to a similar expression, in which ϵ is raised to the power -2.5, derived by Zuber[52] for the laminar regime of dispersed two-phase flow.

8.3 PARTICULATE FLUIDIZATION

If some particles of a given size are put in a vertical vessel and a fluid of lower density is caused to flow upward through them with a sufficiently high velocity, the particles become fluidized; in other words, they no longer rest on one another but are supported by the fluid and are free to move about. Further increase in velocity causes the bed to expand and the void fraction to increase.

In *particulate* fluidization the particles are uniformly dispersed in the expanded mixture. Under some circumstances particulate fluidization cannot be achieved and the fluid is either channeled through regions of low resistance or forms bubbles that rise through the bed rather like gas bubbles in liquids.

THE MINIMUM FLUIDIZATION VELOCITY

The fluid volumetric flux at which the bed first becomes fluidized is known as the *minimum fluidization velocity*. At the minimum fluidization velocity the drag and pressure forces on the particles just equal their weight. Equation (8.35) is therefore valid at incipient fluidization and throughout the expansion of the bed as long as the particles remain uniformly dispersed.

The value of ϵ for spherical particles which are randomly packed is about 0.4. Therefore the value of $\epsilon^{4.7}$ is 0.0135 or $\frac{1}{74}$. The minimum fluidization velocity is then given implicitly by Eq. (8.35) as

$$(C_{Ds}\mathrm{Re}_s^2)_{mf} = 0.018N_f^2 \tag{8.38}$$

Knowing the value of $C_{Ds}\mathrm{Re}_s^2$, Re_s is found from Fig. 8.1, or Eqs. (8.6) and (8.7), and the minimum fluidization velocity is then calculated using Eq. (8.19).

The coefficient in Eq. (8.38) compares well with the value (0.0195) given by Rowe,[8] and also with the results of Pinchbeck and Popper[10] and van Herdens, et al.[11]

The value of ϵ at which nonspherical particles are in contact may be as high as 0.8. If this information is known, it can be used in Eq. (8.35) to estimate the minimum fluidization velocity.

PRESSURE DROP THROUGH A FLUIDIZED BED

Fixed bed The frictional pressure drop through a static pile of particles can be correlated in the usual way by defining a friction factor C_f' by the equation

$$-\left(\frac{dp}{dz}\right)_F = C'_f\, \tfrac{1}{2}\rho_f v_f^2 \frac{A_T}{\mathcal{V}_a} \tag{8.39}$$

A_T is the total surface area of the particles and $\mathcal{V}_a$ the available flow volume. For spherical particles of diameter d the ratio $\mathcal{V}_a/A_T$ is

$$\frac{\mathcal{V}_a}{A_T} = \frac{\epsilon}{1-\epsilon}\frac{d}{6} \tag{8.40}$$

For particles which are not spherical, a *shape factor* Φ is often defined by the equation

$$\frac{\mathcal{V}_a}{A_T} = \frac{1}{\Phi}\frac{\epsilon}{1-\epsilon}\frac{d}{6} \tag{8.41}$$

The average velocity through the bed is related to the volumetric flux as follows:

$$v_f = \frac{j_{f0}}{\epsilon} \tag{8.42}$$

Substituting Eqs. (8.40) and (8.42) into Eq. (8.39) and redefining a friction factor $C_f = \dfrac{3C'_f}{2}$ we obtain

$$-\left(\frac{dp}{dz}\right)_F = 2C_f \frac{\rho_f j_{f0}^2}{d}\frac{1-\epsilon}{\epsilon^3} \tag{8.43}$$

For nonspherical particles d is replaced by d/Φ.

The friction factor defined by Eq. (8.43) is usually correlated in terms of a Reynolds number which is defined in terms of the average fluid velocity [from Eq. (8.42)] and the hydraulic mean depth [from Eq. (8.40)]. Thus

$$\mathrm{Re}_f = \frac{\rho_f j_{f0} d}{(1-\epsilon)\mu_f} \tag{8.44}$$

Note that Re_f is not the same as the "superficial" Reynolds number in Eq. (8.19).

For low Reynolds numbers ($\mathrm{Re}_f < 10$) the Carman-Kozeny[12,13] equation for laminar flow gives

$$C_f = \frac{90}{\mathrm{Re}_f} \tag{8.45}$$

The coefficient is intermediate between the value of 75 in the Blake-Kozeny[14] equation, and the value of 100 recommended by Leva.[15]

The dependence of C_f on higher values of Re_f is shown in Fig. 8.3,

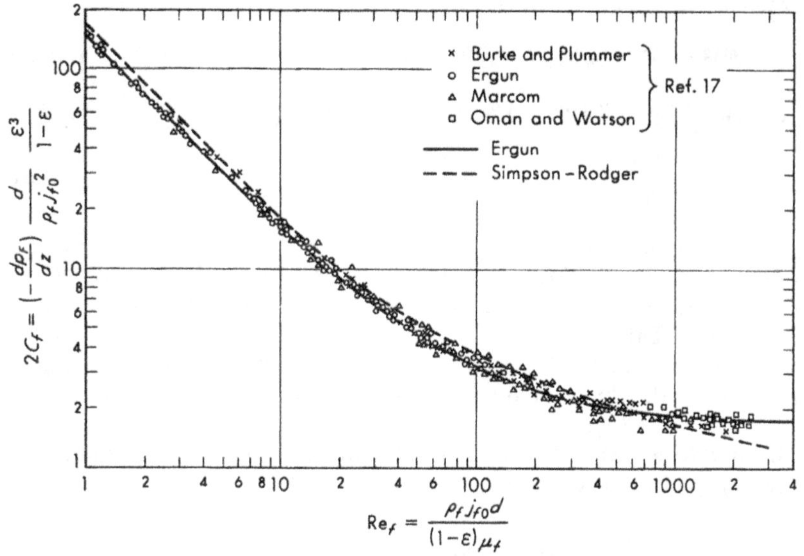

Fig. 8.3 Friction factors for flow in porous media compared with Ergun[17] [Eq. (8.46)] and Simpson-Rodger correlation.[16] (*Adapted from Ergun.*[17])

compared with the correlation of Simpson and Rodger[16] and an equation from Ergun[17] as follows:

$$C_f = \frac{75}{\mathrm{Re}_f} + 0.875 \tag{8.46}$$

Incipient fluidization At incipient fluidization the particles are just supported by the fluid. Therefore from Eqs. (8.14), (8.15), and (8.16)

$$-\left(\frac{dp}{dz}\right)_F = (1 - \epsilon)g(\rho_s - \rho_f) \tag{8.47}$$

Substituting Eqs. (8.44) and (8.45) into Eq. (8.43) we obtain

$$C_f\mathrm{Re}_f^2 = \frac{\epsilon^3}{2(1 - \epsilon)^2} N_f^2 \tag{8.48}$$

Equation (8.48) resembles Eq. (8.38) but has been arrived at by a completely different route. Using $\epsilon = 0.4$ for the value at incipient fluidization, Eq. (8.48) can be arranged to give

$$N_f^2 = 11.25C_f\mathrm{Re}_f^2 \tag{8.49}$$

whereas Eq. (8.38) becomes

$$N_f^2 = 55.5C_{Ds}\text{Re}_s^2 \tag{8.50}$$

In view of Eqs. (8.19) and (8.44) for $\epsilon = 0.4$,

$$\text{Re}_f = \frac{\text{Re}_s}{0.6} \tag{8.51}$$

For laminar flow, $C_{Ds} = 24/\text{Re}_s$, $C_f = 90/\text{Re}_f$ and the right side of Eq. (8.49) is equal to 1690Re_s, compared with 1330Re_s for Eq. (8.50). Comparison between the two methods over a wide range of Reynolds numbers is shown in Table 8.2.

The two methods are seen to be compatible within the range of accuracy with which the quantities C_f and C_D are usually known.

The fluidized state In the fluidized state the condition of equilibrium for a slice of the bed parallel to the horizontal is

$$-\frac{dp}{dz} = g[\epsilon\rho_f + (1 - \epsilon)\rho_s] \tag{8.52}$$

and therefore the pressure gradient can be calculated if ϵ is known.

For a fluidized bed in a straight vertical duct the total *frictional* pressure drop is constant and is simply equal to the submerged weight of all the particles in the bed, divided by the total cross-sectional area. If the total frictional pressure drop is plotted versus the fluid flux, curves such as those shown in Fig. 8.4 are obtained. The point where the curves flatten out corresponds to incipient fluidization.

Table 8.2 Comparison between values of N_f^2 at incipient fluidization predicted by Eqs. (8.49) and (8.50)

Re_s	*Eq.* (8.49)		*Eq.* (8.50)
	Simpson-Rodger[16]	*Ergun*[17]	
10^{-2}	16.9	14.1	13.3
1	1690	1410	1530
10	1.83×10^4	1.73×10^4	2.30×10^4
50	1.7×10^5	1.62×10^5	2.14×10^5
200	1.35×10^6	1.38×10^6	1.72×10^6
500	6.75×10^6	7.23×10^6	7.78×10^6
1000	2.20×10^7	2.8×10^7	2.44×10^7
2000	7.7×10^7	10.9×10^7	9.76×10^7

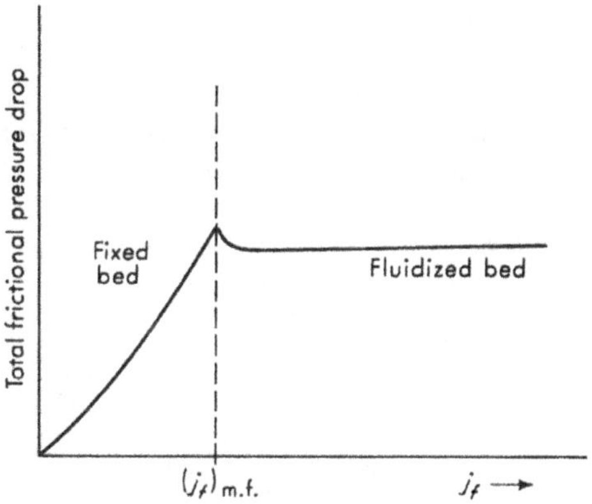

Fig. 8.4 Total frictional pressure drop versus fluid flux for a fluidized bed.

SUMMARY OF CALCULATION PROCEDURES FOR PARTICULATE FLUIDIZED BEDS

Stationary bed (no net particle motion) If the fluid and particle properties are known and the fluid volumetric flux is specified, the simplest calculation procedure for spherical particles is as follows.

1. Calculate Re_s from Eq. (8.19).
2. Substitute Re_s for Re_∞ in Eq. (8.6) or (8.7) to calculate C_{Ds}.
3. Calculate $C_{Ds}\mathrm{Re}_s^2$.
4. Using the value found in (3) calculate ϵ from Eq. (8.35). If ϵ lies between 0.4 and 1, then the bed is fluidized if the particles are spherical, otherwise additional information about the value of ϵ in the packed bed is required.
5. If the bed is fluidized, calculate the pressure gradient using Eq. (8.52). If the bed is not fluidized calculate the Reynolds number Re_f from Eq. (8.44) and use Fig. 8.3 to find C_f. Calculate $(dp/dz)_F$ from Eq. (8.43) and add the hydrostatic gradient $\rho_f g$ to find the net pressure gradient.

If the particles are not spherical, the form of the equations remains the same but correction factors should be applied as indicated by Heywood.[5]

Moving bed (net particle motion) The bed must be fluidized. It is easiest to work with Richardson and Zaki's correlation.

1. Evaluate $C_{D\infty}\mathrm{Re}_\infty^2$ from Eq. (8.5).
2. Calculate Re_∞ from Eqs. (8.6), (8.7), or Fig. 8.1.
3. Calculate v_∞ from Eq. (8.4).

4. Calculate n from Eq. (8.9).
5. Calculate j_{fs} from Eq. (8.2) and use Eq. (8.1) and the techniques of Chap. 4 to calculate ϵ. Calculate the pressure gradient using Eq. (8.52). If wall friction is thought to be important, estimate it from homogeneous flow theory and add to the gravitational pressure drop.

Example 8.1 Water at 20°C flows vertically upward in a tube of cross-sectional area 1 in.^{2} at a rate of 6.5×10^{-5} ft^3/sec. The tube contains 0.01-in.-diam copper shot. Is the bed fluidized? Calculate the pressure gradient and void fraction ϵ.

Solution Working in cgs units for convenience we have: $D = 2.87$ cm, $A = 6.45$ cm^2, $\mu_f = 10^{-2}$ g/(cm)(sec), $\rho_s = 8.92$ g/cm^3, $\rho_f = 1.00$ g/cm^2, $d = 0.0254$ cm, and $Q_f = 1.84$ cm^3/sec.

The liquid flux in the tube is

$$j_{f0} = \frac{Q_f}{A} = \frac{1.84}{6.45} = 0.285 \text{ cm/sec} \tag{a}$$

From Eq. (8.19)

$$\text{Re}_s = \frac{d j_{f0} \rho_f}{\mu_f} = \frac{(2.54 \times 10^{-2})(0.285)(1.0)}{0.01} = 0.725 \tag{b}$$

From Eq. (8.6)

$$C_{Ds} = 37.1 \tag{c}$$

Using Eqs. (*b*) and (*c*),

$$C_{Ds}\text{Re}_s^2 = (37.1)(0.725)^2 = 19.5 \tag{d}$$

From Eq. (8.35) and the definition of N_f

$$\epsilon^{4.7} = \frac{3}{4}\frac{C_{Ds}\text{Re}_s^2 \mu_f^2}{\rho_f d^3 g(\rho_s - \rho_f)} = \frac{(0.75)(19.5)(0.01)^2}{(10^{-6})(2.54)^3(981)(7.92)} = 0.0115 \tag{e}$$

Therefore the predicted value of ϵ is

$$\epsilon = (0.0115)^{1/4.7} = 0.385 \tag{f}$$

Since this value is less than the usual random packing value for spheres ($\epsilon_0 = 0.4$) the bed is just not fluidized.

Alternatively we can use Eq. (8.38) to calculate the value of $C_{Ds}\text{Re}_s^2$ which will just fluidize the bed. Thus

$$(C_{Ds}\text{Re}_s^2)_{mf} = 0.018\,\frac{\rho_f d^3 g(\rho_s - \rho_f)}{\mu_f^2} = \frac{(0.018)(1)(2.54 \times 10^{-2})^3(981)(7.92)}{10^{-4}} = 22.9 \tag{g}$$

Since the value given by Eq. (*d*) is less than this value the bed is not fluidized.

The void fraction is therefore the unfluidized void fraction and is about $\epsilon = 0.40$.

The pressure drop is calculated from Eq. (8.43). First we find the Reynolds number using Eq. (8.44):

$$\mathrm{Re}_f = \frac{d\rho_f j_{f0}}{(1-\epsilon)\mu_f} = \frac{(2.54 \times 10^{-2})(1)(0.285)}{0.6 \times 10^{-2}} = 1.21 \tag{h}$$

From Eq. (8.45)

$$C_f = \frac{90}{\mathrm{Re}_f} \tag{i}$$

Using Eq. (8.43)

$$-\left(\frac{dp}{dz}\right)_F = \frac{2C_f\rho_f j_{f0}^2(1-\epsilon)}{d\epsilon^3} = \frac{(2)(90)(1)(0.0813)(0.6)}{(1.21)(0.0254)0.064} = 4460\,\mathrm{g/(cm^2)(sec^2)} \tag{j}$$

Adding the hydrostatic gradient $\rho_f g$ we get the total pressure gradient

$$-\frac{dp}{dz} = 4400 + 981 = 5441\ \mathrm{dynes/cm^3} \qquad (0.202\ \mathrm{psi/in.}) \tag{k}$$

As a check we calculate what the pressure gradient would be if the bed were just fluidized. It is given by Eq. (8.52)

$$-\frac{dp}{dz} = g[\epsilon\rho_f + (1-\epsilon)\rho_s] = 981[(0.4)(1) + (0.6)(8.92)] = 5640\ \mathrm{dynes/cm^3} \tag{l}$$

Since the pressure gradient in Eq. (k) is less than the pressure gradient in Eq. (l), the bed is predicted to be just not fluidized, in agreement with the conclusions based on Eqs. (f) and (g).

8.4 UNSTEADY FLOW IN PARTICLE DISPERSIONS

PROPAGATION OF CONTINUITY WAVES

From Eq. (6.40) the continuity wave velocity in a dispersion of particles is

$$V_w = j + \frac{\partial}{\partial\epsilon}(j_{fs}) \tag{8.53}$$

In the case of a fluidized bed obeying Eq. (8.2), Eq. (8.53) becomes

$$V_w = j_f + \frac{\partial}{\partial\epsilon}[v_\infty\epsilon^n(1-\epsilon)] \tag{8.54}$$

This result has been confirmed experimentally by Slis, Willemse, and Kramers.[18]

In steady state the fluidization flux is related to the voidage by Eq. (8.24). Therefore, small variations about the steady state propagate with a velocity given by eliminating j_f or v_∞ from Eqs. (8.24) and (8.54),

$$V_w = n\frac{(1-\epsilon)}{\epsilon}j_{f0} \tag{8.55}$$

or,

$$V_w = n\epsilon^{n-1}(1-\epsilon)v_\infty \tag{8.56}$$

The continuity shock velocity is, from Eq. (6.41),

$$V_s = j + \frac{(j_{fs})_1 - (j_{fs})_2}{\epsilon_1 - \epsilon_2} \tag{8.57}$$

As long as inertia and compressibility effects are negligible, unsteady one-dimensional vertical flow is readily described in terms of the motion of continuity waves and shocks. However, an additional complication arises when the propagation of these waves causes a *density inversion* due to values of ϵ decreasing with height. This situation is unstable and tends to produce three-dimensional disturbances which break up the orderly flow structure.

8.5 BATCH SEDIMENTATION

Sedimentation provides an important example of the application of continuity wave theory. If there is no net flow the process is called *batch sedimentation*, whereas, if the solids are steadily removed, the operation is called *continuous thickening*. In this section the case of batch sedimentation will be considered in detail, but the method can be modified to suit processes in which the net solid or fluid flows are controlled in a specified way.[19,20]

Since the overall volumetric flux j is zero in batch sedimentation, Eqs. (8.53) and (8.57) reduce to

$$V_w = \frac{d}{d\epsilon}(j_{fs}) \tag{8.58}$$

$$V_s = \frac{(j_{fs})_1 - (j_{fs})_2}{\epsilon_1 - \epsilon_2} \tag{8.59}$$

It is usual to work in terms of the particle concentration rather than the void fraction. Since $\alpha = 1 - \epsilon$, Eqs. (8.58) and (8.59) become

$$V_w = -\frac{d}{d\alpha}(j_{fs}) \tag{8.60}$$

$$V_s = -\frac{(j_{fs})_1 - (j_{fs})_2}{\alpha_1 - \alpha_2} \tag{8.61}$$

Therefore, on a graph of j_{fs} versus α, the continuity wave velocity downward is represented by the slope of the tangent and the shock velocity downward by the slope of the chord.

A typical history of a suspension which is initially uniform is as follows: Initially the column contains a uniform two-phase mixture B, as shown in Fig. 8.5. As settling takes place, clear liquid A begins to appear at the top of the column and a dense sediment D at the bottom. In between B and D there is often a region C, in which the concentration

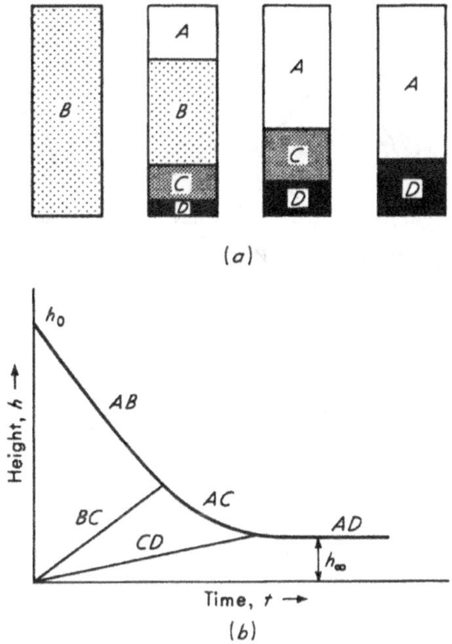

Fig. 8.5 Behavior in a typical batch sedimentation test. (*a*) Physical appearance. (*b*) Height of interfaces as a function of time.

is not uniform. If the particles are of fairly uniform size, a sharp discontinuity is formed between the layers A and B and this moves with the velocity of the settling particles. There may or may not be a distinct discontinuity between the regions B and C. Eventually the upper discontinuity meets the lower and the region B disappears altogether. Thereafter a slow compression or compaction of the regions C and D occurs until finally the sediment reaches its maximum density throughout.

Kynch[21] was apparently the first to formulate a comprehensive mathematical theory of sedimentation by using a plot of total solids flow rate against concentration which, in this case, is identical with the j_{fs}-versus-α relationship.

For the usual shape of this curve, three types of batch sedimentation can be identified as follows:

Type I If a direct shock is possible from the initial value of α (α_0) to the final, fully settled value α_∞ (Fig. 8.6), only one stage of settling is observed. The interface AB moves with a velocity given by the slope of the chord joining the points $\alpha = 0$, $\alpha = \alpha_0$, and the interface BD moves with a velocity given by the slope of the chord joining $\alpha = \alpha_0$ to $\alpha = \alpha_\infty$. Settling is complete when the two shocks meet and no compression occurs. For the curves of j_{fs} versus α shown in Fig. 8.6, type I sedimentation

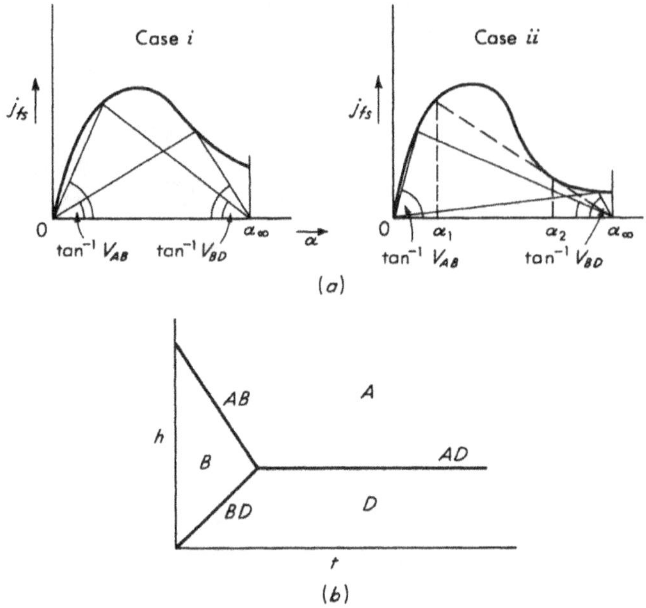

Fig. 8.6 Type I sedimentation. (*a*) Drift flux-concentration plane. (*b*) Wavelines in height-time plane.

occurs for all initial voidages in case *i*; in case *ii* it is only possible for $\alpha_0 < \alpha_1$ or $\alpha_0 > \alpha_2$.

When it is not possible for a direct shock to be propagated from α_0 to the fully packed state, more complicated processes occur.

Type II If the curve of j_{fs} is concave upward at α_0, the initial settling velocity V_{AB} is given by the slope of the chord $\alpha = 0$, $\alpha = \alpha_0$. At the bottom of the column the fully packed state α_∞ is propagated upwards with the velocity of the shock V_{CD} which is given by the tangent from the point $\alpha = \alpha_\infty$, $j_{fs} = 0$ to the curve. If this point of contact is at α_2, the various regions in the column contain voidages in the following ranges:

Region A: pure liquid, $\alpha = 0$
Region B: initial voidage, $\alpha = \alpha_0$
Region C: voidages from α_0 to α_2
Region D: final voidage, α_∞

Since there is no abrupt change in voidage at the boundary of regions B and C, the interface BC may tend to be overlooked in practice. When the top of region C, which moves with the velocity of the continuity wave corresponding to the voidage α_0, reaches the interface AB, the region B disappears and the shock begins to strengthen and slow down. At any subsequent time the value of α just below the shock is given by the continuity wave which has just reached the interface. This compression of region C continues until the AC and CD interfaces meet and settling is complete. The interface histories for this type of sedimentation are shown in Fig. 8.7 as well as the fan of continuity wave paths in region C.

Type III If the curve of j_{fs} is concave downward at α_0 (Fig. 8.8), a finite shock occurs between the regions B and C and three distinct interfaces are obtained. The voidage at the top of region C is determined by the point of contact of the tangent from the point α_0, since in this case the fastest continuity wave in region C is just able to keep pace with the shock BC, which is therefore neither strengthened nor eroded. When

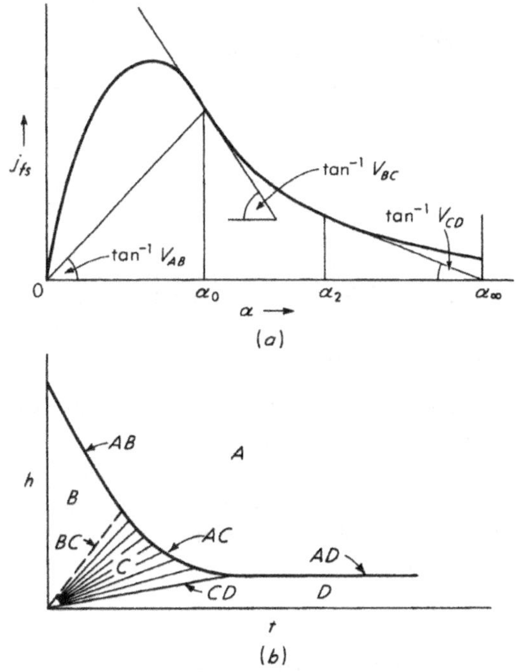

Fig. 8.7 Type II sedimentation. (*a*) Drift flux-concentration plane. (*b*) Wavelines.

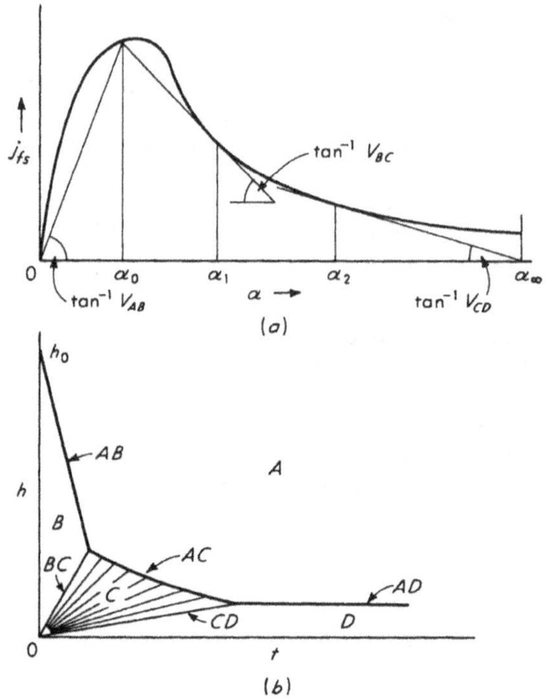

Fig. 8.8 Type III sedimentation. (*a*) Drift flux-concentration plane. (*b*) Wavelines.

the AB and BC interfaces meet, region B disappears and compression of region C continues until settling is complete. The interface history in this case is shown in Fig. 8.8.

A "GENERALIZED" REPRESENTATION OF BATCH SEDIMENTATION[7]

Let V_s denote the general velocity of the shock between clear liquid and the upper surface of the settling particles. It is represented on the curve of j_{fs} by the chord $\alpha = 0$, $\alpha = \alpha$ and moves with the velocity j_{fs}/α. At time t the position of a continuity wave moving up from the bottom of the sediment is given by

$$h = V_w t \tag{8.62}$$

which is represented on the ht plane by a waveline of slope V_w from the origin. Since V_s and V_w are both functions of α only, for a given fluid-particle combination, they may be regarded as functions of each other.

The slope of the curved part of the sedimentation curves therefore always has a particular slope V_s when it crosses a line of slope V_w from the origin. Hence all curves for various initial heights and concentrations have the same shape and may be made to coincide by multiplying both ordinate and abscissa by a suitable factor. The factor $1/h_0\alpha_0$ (h_0 being the original height) induces all curves to pass through the point $1/\alpha_\infty$ when sedimentation is complete, and is the one required. All the results for the compression stage will lie on this curve. The initial stage, in which the settling velocity is constant, is represented by a constant value of V_s, i.e., by a tangent to the curve that intersects the line $t = 0$ at the point $1/\alpha_0$. If the curve of j_{fs} contains an inflexion, V_w passes through a maximum and the generalized sedimentation curve contains two branches joined by a cusp. If α_0 is less than the value corresponding to the maximum value of V_s, the initial stage is represented by tangents to the lower branch and sedimentation of type III occurs.

If the line representing the initial stage cuts the horizontal line $1/\alpha_\infty$, which represents the final state, before intersecting the upper

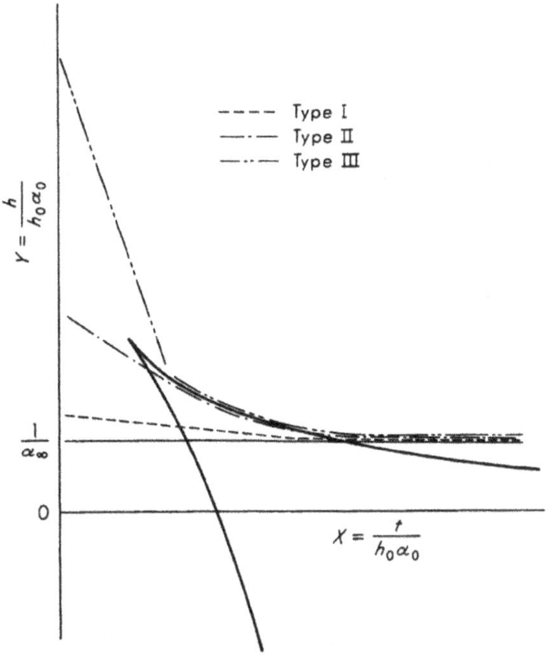

Fig. 8.9 Generalized representation of batch sedimentation.[7]

branch at all, type I sedimentation occurs. The three basic types of sedimentation are shown in Fig. 8.9 related to the general curve of $Y = h/h_0\alpha_0$ against $X = t/h_0\alpha_0$.

The complementary nature of the graph of j_{fs} against α and the graph of X against Y is shown by the following identities:

$$V_s = \frac{j_{fs}}{\alpha} = -\frac{dY}{dX} \tag{8.63}$$

$$V_w = -\frac{dj_{fs}}{d\alpha} = \frac{Y}{X} \tag{8.64}$$

The intercept of a tangent to the graph of X against Y with the Y axis is given by

$$Y - X\frac{dY}{dX} = \frac{1}{\alpha} \tag{8.65}$$

and hence from Eq. (8.63) the intercept with the X axis is $1/j_{fs}$.

Substitution of Eqs. (8.63) and (8.64) into Eq. (8.65) yields

$$X\left(\frac{j_{fs}}{\alpha} - \frac{dj_{fs}}{d\alpha}\right) = \frac{1}{\alpha} \tag{8.66}$$

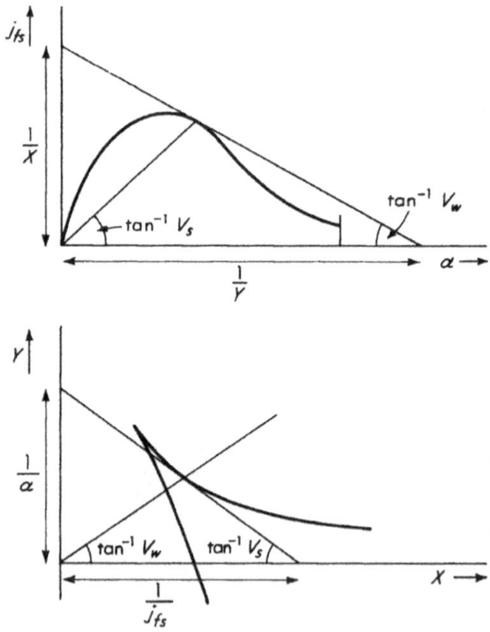

Fig. 8.10 Complementary relationships between graphs of X against Y, and j_{fs} against α.

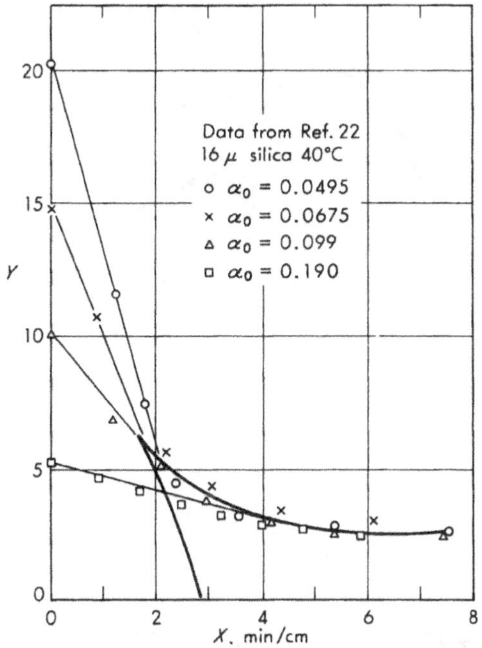

Fig. 8.11 Sedimentation for various initial concentrations and the same initial height represented in the XY plane.[7]

whence

$$\frac{1}{X} = j_{fs} - \alpha \frac{dj_{fs}}{d\alpha} \tag{8.67}$$

and therefore $1/X$ is given by the intercept of the tangent to the graph of j_{fs} with the line $\alpha = 0$. Equation (8.64) shows that the intercept with the α axis is $1/Y$.

These interesting complementary geometrical relationships are shown in Fig. 8.10 and make the task of deducing one curve from the other very simple. Each curve provides more than enough information to draw the other since both the coordinates of a point and the tangent at the point are defined.

If j_{fs} is given in the standard form of Eq. (8.2), the coordinates for the general sedimentation curve are, parametrically,

$$X = [nv_\infty \alpha^2 (1 - \alpha)^{n-1}]^{-1} \tag{8.68}$$

$$Y = \frac{(n + 1)\alpha - 1}{n\alpha^2} \tag{8.69}$$

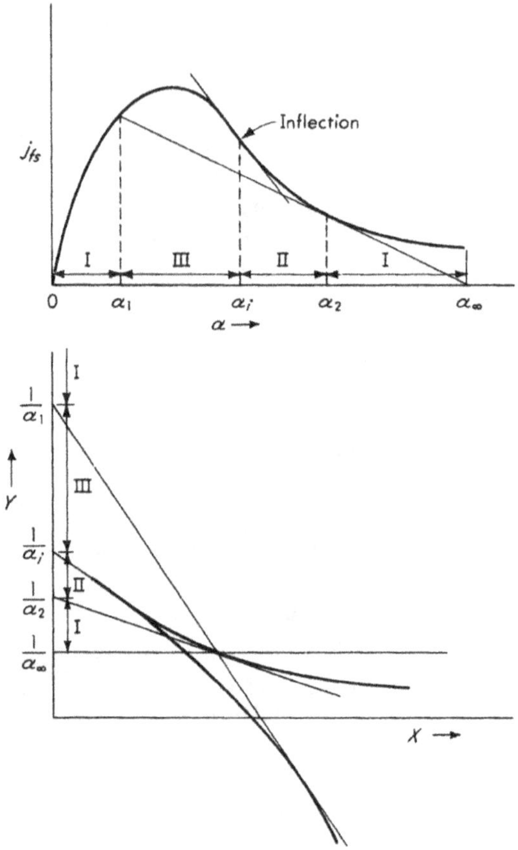

Fig. 8.12 The various types of sedimentation represented as a function of initial concentration on the $j_{fs}\alpha$ and XY planes.

For type III sedimentation it is often more accurate to draw the initial process as a straight line joining the points $1/\alpha_0$ and $1/j_{fs}$ on the axes if the point of contact of tangents to the lower branch is well below the X axis.

The above theory has been verified by Wallis[7] who plotted the results of Egolf and McCabe[22] on the basis of Y versus X. Typical results of this treatment are shown in Fig. 8.11. For the small irregular particles used in these experiments the value of n lay between 20 and 30, showing a considerable effect of flocculation.

The ranges of initial α values which promote the three types of sedimentation are shown on the αj_{fs} and XY curves in Fig. 8.12. If

Eq. (8.2) is obeyed, the inflection on the αj_{fs} curve and the cusp in the XY plot occur where

$$\alpha_i = \frac{2}{n+1} \tag{8.70}$$

whereas α_2 is given by

$$\alpha_2 = \frac{\alpha_\infty(1+n) + [\alpha_\infty^2(1+n)^2 - 4n\alpha_\infty]^{1/2}}{2n} \tag{8.71}$$

Example 8.2 Consider the settling of 0.01-in.-diameter spherical glass beads in 90% glycerol solution at 20°C in a 10-cm-diameter vertical tube. Describe the subsequent behavior for an initial height of 100 cm and initial concentrations of $\alpha_0 = 0.1$, 0.2, 0.3, 0.4, and 0.5. How long does settling take in each case before completion?

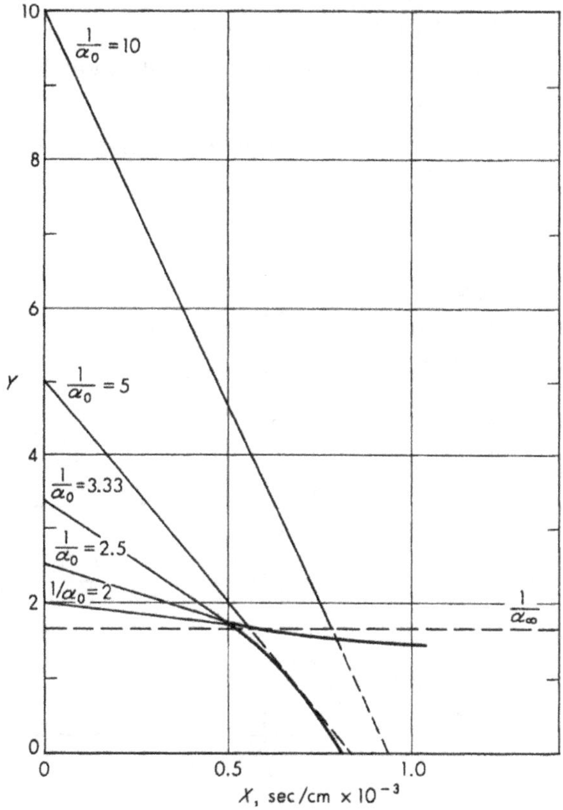

Fig. 8.13 XY plot for Example 8.2.

Solution To follow the standard procedure we first determine the $j_{fs}\alpha$ and XY relationships using the parameters v_∞ and n.

The following data are obtained from the "Handbook of Chemistry and Physics" (The Chemical Rubber Company, Cleveland, Ohio).

$$\mu_f = 2.19 \text{ g/(cm)(sec)} \qquad \rho_s = 2.32 \text{ g/cm}^3 \qquad \rho_f = 1.23 \text{ g/cm}^3$$

Using Eq. (8.5)

$$C_{D\infty}\mathrm{Re}_\infty{}^2 = \frac{4}{3}\frac{(2.54 \times 10^{-2})^3(1.23)(981)(1.09)}{(2.19)^2} = 0.006 \tag{a}$$

From Eq. (8.6) and the results of Eq. (*a*) we find

$$\mathrm{Re}_\infty = \frac{C_{D\infty}\mathrm{Re}_\infty{}^2}{24} = 2.5 \times 10^{-4} \tag{b}$$

From Eq. (8.4), therefore,

$$v_\infty = \frac{\mathrm{Re}_\infty \mu_f}{d\rho_f} = \frac{(2.5 \times 10^{-4})(2.19)}{(2.54 \times 10^{-2})(1.23)} = 0.01753 \text{ cm/sec} \tag{c}$$

From Eq. (8.9*a*) $\quad n = 4.65 + 19.5\, d/D = 4.70 \quad$ (*d*)

Combining the results of Eqs. (*c*) and (*d*) with Eq. (8.2) we get

$$j_{fs} = 0.01753\alpha(1 - \alpha)^{4.7} \text{ cm/sec} \tag{e}$$

X and Y are now found from Eqs. (8.68) and (8.69) and are used to draw the graph shown in Fig. 8.13.

The final settled value of α is about 0.6 for spheres. The value of $1/\alpha_\infty$ is therefore $1/0.6 = 1.667$.

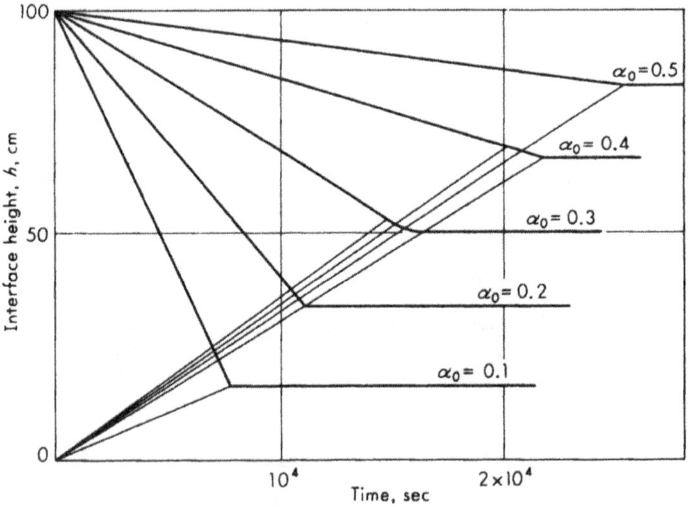

Fig. 8.14 Interface heights versus time for Example 8.2.

From Eqs. (8.70) and (8.71) α_i is found to be 0.35 and α_2 is 0.43. Therefore type II sedimentation occurs when $0.43 > \alpha_0 > 0.35$.

The lower bound of type III sedimentation is found by drawing the tangent to the lower branch of the XY curve which will just pass through the intersection between the upper branch and the $1/\alpha_\infty$ line. The range of type III sedimentation is then $0.22 < \alpha < 0.35$.

Type I sedimentation will therefore occur for $\alpha_0 = 0.1$, 0.2, and 0.5; type II for $\alpha_0 = 0.4$; and type III for $\alpha_0 = 0.3$.

To get the h-versus-t graphs, and hence the history of the interfaces, the axes in Fig. 8.13 are multiplied by the appropriate values of $h_0\alpha_0$. The resulting curves are shown in Fig. 8.14. The overall settling times are, in hours,

α_0	0.1	0.2	0.3	0.4	0.5
t, hr	2.18	3.06	4.55	6.07	7.00

8.6 PARTICLE-PARTICLE FORCES

In previous pages it has been assumed that sedimentation stops abruptly at the point where the particles rest on one another ($\alpha = \alpha_\infty$). In fact, if the particles (or the lattice in which they are arrayed) are able to distort under compressive stresses, a further compaction of the bed is possible. The force which resists this compression depends on the stress-strain relationship for the particle lattice. The *resultant* force on a particle in the lattice due to this compression will depend on the concentration gradient and perhaps other variables.

In order to characterize the particle-particle forces, consider a static experiment in which an array of particles is squeezed in a horizontal direction by a force F' acting over an area A. Assume that a suitable mean horizontal component of stress, σ_s, can be defined for the particles. Then, from a force balance across vertical planes we have

$$p\epsilon + \sigma_s(1 - \epsilon) = \frac{F'}{A} \tag{8.72}$$

which can be rearranged to the form

$$(\sigma_s - p)(1 - \epsilon) + p = \frac{F'}{A} \tag{8.73}$$

The first term in this equation can be used to define a *particle pressure* p_s, which is added to the fluid pressure in order to determine the overall force per unit area. Thus

$$p_s = (\sigma_s - p)(1 - \epsilon) \tag{8.74}$$

In a concentration gradient, the difference in the force per unit area which is carried by the particles is $-dp_s$ over a distance dz. However, the particles only occupy a fraction $(1 - \epsilon)$ of the volume. Therefore the contribution to f_s from the interparticle forces is $-\frac{1}{1-\epsilon}\frac{dp_s}{dz}$. In a horizontal, static test, if F_{fs} denotes the drag caused by fluid motion in the usual way, Eqs. (3.45) and (3.46) then become

$$0 = -\frac{dp}{dz} + \frac{F_{fs}}{\epsilon} \tag{8.75}$$

$$0 = -\frac{dp}{dz} - \frac{F_{fs}}{1-\epsilon} - \frac{1}{1-\epsilon}\frac{dp_s}{dz} \tag{8.76}$$

Eliminating F_{fs} we find

$$\frac{dp}{dz} + \frac{dp_s}{dz} = 0 \tag{8.77}$$

which could have been derived from Eqs. (8.73) and (8.74), since F'/A is constant.

Usually p_s is a function of ϵ, which can be determined by a compression test, and is independent of the fluid pressure for most particulate systems because only point contact occurs between the particles and the hydrostatic pressure surrounds them entirely. However, if the particles themselves are significantly compressible (gas bubbles, for example), p_s will depend on both ϵ and p.[87] For fibrous materials an empirical equation is often valid in the form

$$p_s^N = B(1 - \epsilon) \tag{8.78}$$

The last term in Eq. (8.76) can be conveniently rewritten as

$$-\frac{1}{1-\epsilon}\frac{dp_s}{dz} = -\frac{1}{1-\epsilon}\frac{\partial p_s}{\partial \epsilon}\frac{d\epsilon}{dz} = E\frac{d\epsilon}{dz} \tag{8.79}$$

where

$$E = -\frac{1}{1-\epsilon}\frac{\partial p_s}{\partial \epsilon} \tag{8.80}$$

is equivalent to $f_{s\nabla\epsilon}$ which was introduced in Chap. 6, and expresses the influence of concentration gradient on the force experienced by the particles.

Example 8.3 During filtration of a fibrous material, liquid flows through a "mat" which is held against a porous surface. How do the void fraction and pressure vary through the mat? Assume that the Carman-Kozeny equation applies, that $1 - \epsilon$ is small, and that the mat is built up so slowly that steady flow theory can be used.

Solution In terms of the fluid flux relative to stationary particles, j_{f0}, we have, from Example 3.5,

$$F_{fs} = -\frac{180\,\mu_f j_{f0}}{d^2}\left(\frac{1-\epsilon}{\epsilon}\right)^2 \tag{8.81}$$

Combining Eqs. (8.75) to (8.78) and (8.81) yields

$$\frac{dp_s}{dz} = \frac{180\mu_f j_{f0}}{d^2 B^2 \epsilon^3}\, p_s^{2N} \tag{8.82}$$

Since $1 - \epsilon$ is small, we can let $\epsilon = 1$ in Eq. (8.82) and integrate with $p_s = 0$ at $z = 0$ to get

$$p_s^{(1-2N)} = (1 - 2N)Cz \tag{8.83}$$

where

$$C = \frac{180\,\mu_f j_{f0}}{B^2 d^2} \tag{8.84}$$

The variation of void fraction with distance is obtained by using Eq. (8.78) in Eq. (8.83). Thus

$$(1 - \epsilon)^{(1-2N)/N} = \frac{(1 - 2N)Cz}{B^{(1-2N)/N}} \tag{8.85}$$

Since $N \approx \frac{1}{4}$ for fibers, the resulting pressure and void distribution will be as shown in Fig. 8.15.

From Eq. (8.85) the average particle concentration in the mat is readily found to be related to the concentration at the wall by the equation

$$\overline{(1 - \epsilon)} = \frac{1 - 2N}{1 - N}(1 - \epsilon)_w \tag{8.86}$$

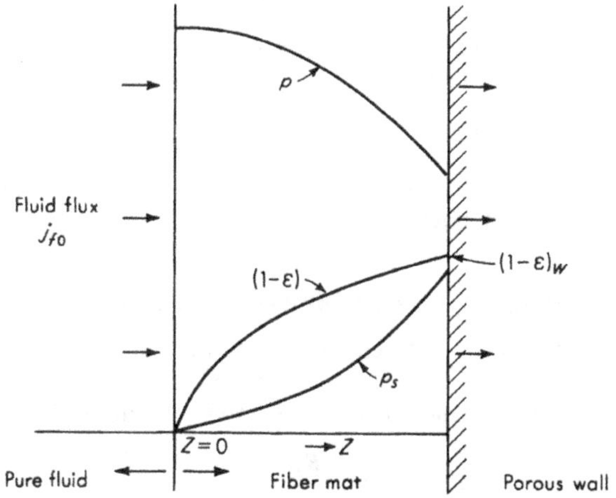

Fig. 8.15 Variation of density and pressure through a fiber mat during filtration from a dilute suspension.

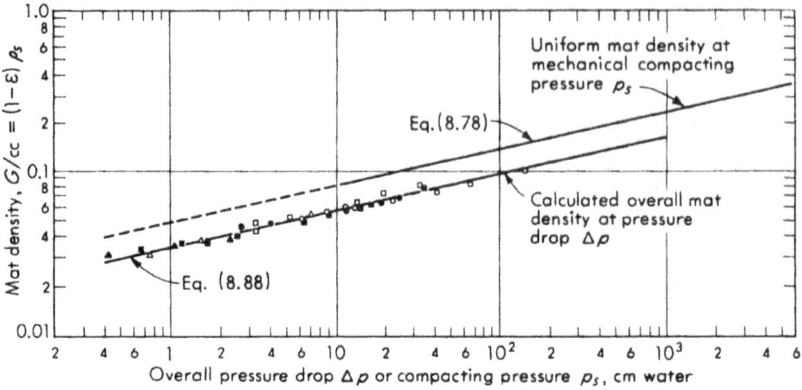

Fig. 8.16 Filtration through a fiber mat. (*Han and Ingmanson.*[23])

Therefore, in view of Eqs. (8.77) and (8.78) the overall pressure drop Δp through the mat is related to the mean density by the equation

$$B\overline{(1-\epsilon)} = \frac{1-2N}{1-N}\,\Delta p^N \tag{8.87}$$

The concentration $\overline{(1-\epsilon)}$ is proportional to the mat density. Thus the pressure-drop–density curve for such a mat is displaced from the compression stress-density curve by the factor $(1-2N)/(1-N)$. Figure 8.16 shows some typical data for which $N = 0.225$. The value of $(1-2N)/(1-N)$ is 0.71.

An alternative derivation by Han and Ingmanson[23] is based on Davis'[24] modification of the Carman-Kozeny equation for fiber mats which has $1-\epsilon$ raised to the power $3/2$ in Eq. (8.81). The result in this case is found by the same route to be

$$B\overline{(1-\epsilon)} = \frac{1-1.5N}{1-0.5N}\,\Delta p^N \tag{8.88}$$

For $N = 0.225$ we have $(1-1.5N)/(1-0.5N) = 0.75$, and the difference between the two theories can hardly be distinguished in Fig. 8.16.

8.7 UNSTEADY FLOW IN THE PRESENCE OF PARTICLE-PARTICLE FORCES

The behavior of a sediment which is being "compacted" to values of α greater than α_∞ can be predicted from Eqs. (3.45) and (3.46) by omitting the inertia terms but including the particle-particle forces. Thus

$$0 = -\frac{\partial p}{\partial z} - \rho_f g + \frac{F_{fs}}{\epsilon} \tag{8.89}$$

$$0 = -\frac{\partial p}{\partial z} - \rho_s g - \frac{F_{fs}}{1-\epsilon} + E\,\frac{\partial \epsilon}{\partial z} \tag{8.90}$$

Eliminating $\partial p/\partial z$ from these equations and making use of Eq. (8.15), we have

$$(\rho_s - \rho_f)g = E\frac{\partial \epsilon}{\partial z} - f_{fs} \tag{8.91}$$

f_{fs} is a function of the flow rates and the void fraction. Usually the value of j can be specified (for example it is zero in batch sedimentation or a soil compaction test) and a convenient form for the f_{fs} relationship is

$$f_{fs} = f_{fs}(j,j_f,\epsilon) \tag{8.92}$$

The continuity equations for the fluid and for the combined mixture (assumed to be incompressible) are

$$\frac{\partial \epsilon}{\partial t} + \frac{\partial j_f}{\partial z} = 0 \tag{8.93}$$

and

$$j = \text{const} \tag{8.94}$$

Equations (8.91) to (8.94) can now be solved simultaneously.

Differentiating Eq. (8.91) with respect to z and using Eqs. (8.92) and (8.94) we have

$$E\frac{\partial^2 \epsilon}{\partial z^2} + \frac{\partial E}{\partial \epsilon}\left(\frac{\partial \epsilon}{\partial z}\right)^2 - \left[\left(\frac{\partial f_{fs}}{\partial j_f}\right)_j \frac{\partial j_f}{\partial z} + \left(\frac{\partial f_{fs}}{\partial \epsilon}\right)_j \frac{\partial \epsilon}{\partial z}\right] = 0 \tag{8.95}$$

Eliminating $\partial j_f/\partial z$ between Eqs. (8.93) and (8.95) yields

$$E\frac{\partial^2 \epsilon}{\partial z^2} + \frac{\partial E}{\partial \epsilon}\left(\frac{\partial \epsilon}{\partial z}\right)^2 + \left(\frac{\partial f_{fs}}{\partial j_f}\right)_j \left[\frac{\partial \epsilon}{\partial t} - \frac{(\partial f_{fs}/\partial \epsilon)_j}{(\partial f_{fs}/\partial j_f)_j}\frac{\partial \epsilon}{\partial z}\right] = 0 \tag{8.96}$$

If E is zero or there is no particle-particle contact, the final term in Eq. (8.96) is easily shown to reduce to the equation governing continuity wave propagation during sedimentation.

If E is constant, Eq. (8.96) resembles a form of the diffusion equation, since $\partial f_{fs}/\partial j_f$ is negative, and can be compared with Terzaghi's well-known equation[59] for soil consolidation,

$$c\frac{\partial^2 \epsilon}{\partial z^2} = \frac{\partial \epsilon}{\partial t} \tag{8.97}$$

where c is a coefficient of consolidation. Evidently the Terzaghi theory is a reasonable approximation only if the diffusion of changes in ϵ dominates the motion of continuity waves.

8.8 STABILITY OF FLUIDIZED SYSTEMS

The stability of fluidized systems can be investigated by using the theory described in Sec. 6.4. The continuity wave velocity is given by Eq.

(8.53) and the dynamic wave velocity by Eq. (6.76). Note that unless the quantity $(-f_{\nabla\alpha})$, which could be thought of as inverse compressibility, is at least as great as the quantity

$$\frac{\rho_f \rho_s}{\alpha(1-\alpha)} \frac{(v_f - v_s)^2}{\rho_f/(1-\alpha) + \rho_s/\alpha}$$

dynamic waves are imaginary and the system is always unstable.

Unfortunately, insufficient data are available at present to enable the virtual compressibility of a fluidized bed[26] to be calculated with any confidence.

The usual method of analysis has been to neglect $f_{\nabla\alpha}$ completely. This leads to the conclusion that all fluidized systems are unstable, in the absence of mitigating circumstances.[26–29] However, the growth rate of instabilities is very dependent on the properties of the system.

The growth rate of instabilities can be determined as a function of wavelength by using Eqs. (6.162) and (6.163). The details of this calculation are reserved for Prob. 8.24. It is found that the shortest wavelengths grow most rapidly. Since waves must presumably be somewhat longer than individual particles, the shortest reasonable wavelength is chosen as 20 times the particle diameter and the distance L, which this wave must travel in order to grow by a factor of e, is computed. The results are shown in Table 8.3 and are comparable with the values computed by Jackson.[29] Also shown in the table is the value of the Froude

Table 8.3 Stability calculation for the data of Wilhelm and Kwauk.[30] L is the distance which a disturbance travels before growing by a factor of e (see Prob. 8.24)

Fluid	*Particle*	*d, cm*	$\frac{\rho_s}{\rho_f}$	*Fr*	$\frac{L}{10d}$	*L, cm*
Water	Glass beads	0.029	2.5	0.00052	342	99
	Glass beads	0.051	2.5	0.00067	266	136
	Socony beads	0.336	1.6	0.00088	12.0	40
	Socony beads	0.458	1.6	0.0099	10.6	49
	Glass beads	0.518	2.5	0.036	4.92	26
	Lead shot	0.128	10.8	0.13	2.28	29
Air	Glass beads	0.029	2000	1.1	0.66	0.2
	Glass beads	0.051	2000	2.03	0.54	0.3
	Socony beads	0.336	1300	10.0	0.40	1.3
	Socony beads	0.458	1300	13.0	0.39	1.8
	Glass beads	0.518	2000	40.0	0.36	1.9
	Lead shot	0.128	9000	85.0	0.34	0.4

number,

$$\mathrm{Fr} = \frac{j_{f0}^2}{gd} \tag{8.98}$$

which was suggested as a stability criterion by Wilhelm and Kwauk.[30]

If L is short, waves grow rapidly and soon generate three-dimensional disturbances as a result of density inversions. Bubbles of fluid are formed and rise through the bed rather like gas bubbles in a liquid.

8.9 COMPRESSIBLE FLOW OF PARTICLE SUSPENSIONS

The compressible flow of gases containing dispersed particles has received a great deal of attention as a result of the development of rocket technology. The high velocities which occur in rocket exhausts and the necessity for good accuracy in thrust prediction require theoretical methods for dealing with rapidly accelerating flows at high Mach number. Thus, by far the greater proportion of the published literature is concerned with nozzle flows in which supersonic flow is achieved.

The interested reader will find the review articles of Kliegel,[31] Hoglund,[32] and Soo[33] useful for obtaining a perspective on this topic.

ONE-DIMENSIONAL STEADY FLOW

One-dimensional steady flow is governed by the usual pairs of equations describing mass, momentum, and energy conservation. A completely general analysis is extremely complex, but it is often valid to assume that the following effects can be neglected:

1. Heat, mass, and momentum transfer with the nozzle walls
2. Particle-particle forces and particle thermal (brownian) motion
3. The effect of neighboring particles on drag forces
4. Virtual or apparent mass effects
5. Particle internal temperature gradients
6. Phase change and chemical reaction
7. Radiation
8. Dissociation, ionization, and "nonperfect gas" effects
9. The volume occupied by the particles
10. The contribution of the mean pressure gradient to the force on the particles (compared with the viscous drag)
11. Two- or three-dimensional effects
12. Variations in particle size and shape
13. Gravity

If any of the above assumptions prove to be invalid, the analysis should be modified accordingly.

The continuity equations are

$$\rho_g v_g A = W_g = \text{const} \tag{8.99}$$

$$W_s = \text{const} \tag{8.100}$$

The momentum equation for both components together is

$$W_g \frac{dv_g}{dz} + W_s \frac{dv_s}{dz} + A \frac{dp}{dz} = 0 \tag{8.101}$$

whereas the momentum equation for the particles alone takes the form

$$v_s \frac{dv_s}{dz} = \frac{3}{4} \frac{C_D \rho_g}{d \rho_s} (v_g - v_s)|v_g - v_s| \tag{8.102}$$

For the gas and the particles together the energy equation is

$$W_g(h_g + \tfrac{1}{2}v_g^2) + W_s(h_s + \tfrac{1}{2}v_s^2) = W_g h_{g0} + W_s h_{s0} \tag{8.103}$$

in which the subscript zero refers to stagnation conditions.

If the particles are incompressible, their thermal-energy equation is, from Example 3.10,

$$v_s \frac{dh_s}{dz} = \frac{6h}{\rho_s d} (T_g' - T_s) \tag{8.104}$$

The temperature T_g' represents the equivalent gas temperature for heat transfer to the particles and is not necessarily equal to the gas temperature because of the viscous dissipation in the gas layers surrounding the particle. If the relative velocity is large the value of T_g' may approach the gas stagnation temperature relative to the particles. In many cases of interest, however, this effect is negligible, and it can be assumed that $T_g' = T_g$.

For low particle Reynolds and Mach numbers relative to the surrounding gas the usual laminar-flow equations apply to the drag coefficient and heat-transfer coefficient. Equations (8.102) and (8.104) then become

$$v_s \frac{dv_s}{dz} = \frac{18\mu_g}{d^2 \rho_s} (v_g - v_s) \tag{8.105}$$

$$v_s c_s \frac{dT_s}{dz} = \frac{12 k_g}{d^2 \rho_s} (T_g - T_s) \tag{8.106}$$

c_s is the particle specific heat.

Substituting for enthalpies in terms of temperatures and assuming constant properties, Eq. (8.103) becomes

$$W_g\left(c_{pg}T_g + \frac{v_g^2}{2}\right) + W_s\left(c_sT_s + \frac{v_s^2}{2}\right) = W_g c_{pg} T_{g0} + W_s c_s T_{s0} \quad (8.107)$$

In addition, the equation of state for the gas relates changes in pressure, temperature, and density. Thus,

$$p = \rho_g R_g T_g \quad (8.108)$$

Equations (8.99), (8.101) to (8.104), and (8.108) constitute six equations for the six unknowns, T_g, T_s, v_s, v_g, p, *and* ρ_g (assuming that the flow rates and geometry are given), and they can be solved by a variety of methods.

HOMOGENEOUS EQUILIBRIUM FLOW

One of the simplest results is obtained if it can be assumed that momentum and energy are exchanged so rapidly that the component temperatures and velocities are everywhere equal. Equations (8.105) and (8.106) then reduce to the form

$$v_s = v_g \quad (8.109)$$

$$T_s = T_g \quad (8.110)$$

Substitution in the other equations reveals the conclusions which were already reached in the solution to Example 2.2. The mixture behaves as a pseudo gas with the following properties in terms of the relative mass fraction of the particles:

$$R' = \frac{R_g}{1 + m_s} \quad (8.111)$$

$$c'_p = \frac{c_{pg} + m_s c_s}{1 + m_s} \quad (8.112)$$

$$\gamma' = \frac{c_{pg} + m_s c_s}{c_{vg} + m_s c_s} \quad (8.113)$$

All of the conventional one-dimensional compressible flow equations now apply with the above parameters inserted in place of the gas properties. The sonic velocity is

$$c = \sqrt{\gamma' R' T} \quad (8.114)$$

and is less than the sound velocity in the gas alone. The pseudo Mach number is

$$M = \frac{V}{c} \quad (8.115)$$

LIMITING GASES OF NONEQUILIBRIUM FLOW

Three further simple results are obtained by assuming that heat and momentum transfer are either complete or nonexistent.[34] The mixture again behaves as a pseudo gas with appropriate properties, as outlined below.

Velocity equilibrium, thermal insulation Since the particles experience no heat transfer, their temperature stays constant. Thus

$$T_s = \text{const} \tag{8.116}$$

Because the virtual gas density is increased by the presence of the particles, but there are no other effects, the gas constant and the specific heat at constant pressure become

$$R' = \frac{R_g}{1 + m_s} \tag{8.117}$$

$$c' = \frac{c_{pg}}{1 + m_s} \tag{8.118}$$

The isentropic exponent is unchanged, and has the value for the gas alone

$$\gamma' = \gamma_g \tag{8.119}$$

The sonic velocity is determined by the gas temperature. Thus

$$c = \sqrt{\gamma' R' T_g} \tag{8.120}$$

Thermal equilibrium, velocity insulation In this case the particles move at a low velocity and act as a heat source for the gas. To make the solution reasonable, assumptions 9 and 10 should really be modified; however, the limiting condition which is approached is

$$v_s = \text{const} \tag{8.121}$$

The gas alone expands as if it had the properties

$$R' = R_g \tag{8.122}$$

$$c_p' = c_{pg} + m_s c_s \tag{8.123}$$

$$\gamma' = \frac{c_{pg} + m_s c_s}{c_{vg} + m_s c_s} \tag{8.124}$$

Thermal and velocity insulation The gas does not interact with the particles and expands as if it were "clean." Again, assumptions 9 and 10 should be modified to get a reasonable solution for the particle motion.

Evaluation of these limiting cases reveals that for many practical cases the predicted behavior is relatively insensitive to the assumptions which are chosen. However, when greater accuracy is required, more sophisticated methods of analysis are called for.

SIMILAR SOLUTIONS FOR CONSTANT FRACTIONAL LAG

A family of exact solutions to flows in the laminar regime of relative motion can be found by assuming that the ratios of the velocity and temperature changes of the components are constants. It is also assumed that both phases have a common stagnation temperature T_0, where they are at rest. Following Kliegel[35] we define the parameters

$$K = \frac{v_s}{v_g} \tag{8.125}$$

$$L = \frac{T_0 - T_s}{T_0 - T_g} \tag{8.126}$$

Equations (8.105) and (8.106) become

$$\frac{dv_g}{dz} = \frac{18\mu_g}{d^2\rho_s} \frac{1 - K}{K^2} \tag{8.127}$$

$$\frac{dT_g}{dz} = -\frac{12k_g(T_0 - T_g)}{v_g d^2 \rho_s c_s} \frac{1 - L}{KL} \tag{8.128}$$

The energy equation (8.107) is readily converted to the form

$$T_0 - T_g = \frac{v_g^2}{2} \frac{1 + m_s K^2}{c_{pg} + m_s c_s L} \tag{8.129}$$

Differentiating Eq. (8.129) we find that

$$\frac{1}{T_0 - T_g} \frac{dT_g}{dz} = -\frac{2}{v_g} \frac{dv_g}{dz} \tag{8.130}$$

Substituting Eq. (8.130) into Eq. (8.128) the result is

$$\frac{dv_g}{dz} = \frac{6k_g}{d^2\rho_s c_s} \frac{1 - L}{KL} \tag{8.131}$$

Both Eqs. (8.127) and (8.131) show that the gas velocity, and hence also the particle velocity, increases linearly with distance down the nozzle. Furthermore, since both expressions must be compatible

$$\frac{c_s \mu_g}{k_g} = \frac{1}{3} \frac{1/L - 1}{1/K - 1} \tag{8.132}$$

The left-hand side of Eq. (8.132) is a kind of pseudo Prandtl number in which the specific heat of the solid replaces the specific heat of the gas. This quantity determines the relationship between the thermal and velocity lags.

Further developments are possible by using the momentum and continuity equations. Using Eqs. (8.99), (8.125), and (8.126), Eq. (8.101) becomes

$$\rho_g v_g \frac{dv_g}{dz}(1 + m_s K) + \frac{dp}{dz} = 0 \tag{8.133}$$

which is just the one-dimensional momentum equation for a pseudo gas with density

$$\rho' = (1 + m_s K)\rho_g \tag{8.134}$$

This pseudo gas will obey a perfect-gas law equivalent to Eq. (8.108) if it has the temperature of the gas and the gas constant

$$R' = \frac{R_g}{1 + m_s K} \tag{8.135}$$

Equation (8.129) will be the energy equation for the pseudo gas if it has a specific heat given by

$$c_p' = \frac{c_{pg} + m_s c_s L}{1 + m_s K^2} \tag{8.136}$$

From Eqs. (8.135) and (8.136) the specific heat at constant volume is

$$c_v' = \frac{c_{pg} + m_s c_s L}{1 + m_s K^2} - \frac{R_g}{1 + m_s K} \tag{8.137}$$

Therefore the ratio of specific heats is

$$\gamma' = \frac{c_p'}{c_v'} = \left(1 - \frac{1 + m_s K^2}{1 + m_s K}\,\frac{R_g}{c_{pg} + m_s c_s L}\right)^{-1} \tag{8.138}$$

The two-component flow can now be treated as a pseudo-gas flow with the above properties and a total mass flow rate $W_s + W_g$. All of the standard formulas of one-dimensional gas dynamics can be applied in the usual way.

If a nozzle is to be designed on the above basis, the procedure might be as follows. First the value of K is chosen arbitrarily. The value of L follows from Eq. (8.132). All the pseudo-gas properties can then be determined and the area variation and velocities as a function of pressure determined, for given upstream conditions and flow rates. The nozzle shape, however, is not arbitrary since Eq. (8.127) enables the distance z down the nozzle to be calculated in terms of the other parameters. Thus a nozzle designed for specific upstream conditions, flow rates, and fractional lag will have a determinate geometry. A different choice of K will lead to a different geometry with smaller values of K leading to shorter nozzles for a given gas exit velocity.

PERTURBATION TECHNIQUES

A method for analyzing flows with small departures from equilibrium can be developed by formulating equations in terms of the deviations from homogeneous flow.[36,37]

The analysis starts from Eqs. (8.105) and (8.106) which govern the velocity and thermal lags. If a "characteristic" velocity, length, and temperature are defined as v_c, L_c, and T_c, respectively, we may write

$$\frac{d^2\rho_s v_c}{18\mu_g L_c}\frac{v_s}{v_c}\frac{d(v_s/v_c)}{d(z/L_c)} = \frac{v_g}{v_c} - \frac{v_s}{v_c} \tag{8.139}$$

and

$$\frac{d^2\rho_s c_s v_c}{12k_g L_c}\frac{v_s}{v_c}\frac{d(T_s/T_c)}{d(z/L_c)} = \frac{T_g}{T_c} - \frac{T_s}{T_c} \tag{8.140}$$

The quantities

$$\epsilon_v = \frac{d^2\rho_s v_c}{18\mu_g L_c} \tag{8.141}$$

and

$$\epsilon_T = \frac{d^2 c_s \rho_s v_c}{12k_g L_c} \tag{8.142}$$

are dimensionless parameters which characterize the deviations of the velocity and temperature from the homogeneous equilibrium values. If these quantities are small, they can be used as the basis for a perturbation expansion. Furthermore, their ratio

$$\frac{\epsilon_v}{\epsilon_T} = \frac{2}{3}\frac{k_g}{\mu_g c_s} \tag{8.143}$$

is usually close to unity. The sizes of ϵ_v and ϵ_T can also be used to estimate when the assumption of homogeneous equilibrium flow is reasonable.

SHOCK WAVES

Both normal and oblique shock waves occur in particle-gas mixtures. Moreover, the shock waves can be of (at least) two types depending on whether the velocity exceeds the homogeneous compressible wave velocity or the gas sonic velocity. The velocity of a homogeneous compressible wave is

$$c' = \sqrt{\gamma' R' T} \tag{8.144}$$

where γ' and R' are given by Eqs. (8.111) and (8.113). On the other hand the velocity of sound in the gas alone is

$$c_g = \sqrt{\gamma_g R_g T_g} \tag{8.145}$$

We may define Mach numbers based on either wave velocity as follows:

$$M' = \frac{v}{c'} \tag{8.146}$$

$$M_g = \frac{v}{c_g} \tag{8.147}$$

"Weak" shock waves are characterized by the condition $M' > 1$, $M_g < 1$ before the wave. In this case a rather gradual transition occurs through the wave. "Strong" shock waves can only exist if $M_g > 1$ before the wave front. The gas undergoes a normal shock of the usual single-phase type followed by a relaxation region in which the particles and the gas come to equilibrium. In each case the final state is governed by homogeneous theory.

Figure 8.17 shows a sketch of a strong normal shock wave. The condition before the wave is denoted by the subscript 1, the state after the gas shock by 2, and the final state by 3.

Across the gas shock the usual equations apply to the gas flow, whereas the particles are unchanged in either their velocity or temperature. In particular

$$M_{g2}^2 = \frac{M_{g1}^2 + 2/(\gamma_g - 1)}{[2\gamma_g/(\gamma_g - 1)]\, M_{g1}^2 - 1} \tag{8.148}$$

$$\frac{p_2}{p_1} = 1 + \frac{2\gamma_g}{\gamma_g + 1}(M_{g1}^2 - 1) \tag{8.149}$$

$$p_2 - p_1 = \frac{2(\rho_g)_1 v_1^2}{\gamma_g + 1}\left(1 - \frac{1}{M_{g1}^2}\right) \tag{8.150}$$

$$(v_g)_2 v_1 = \frac{2\gamma_g R_g T_1}{\gamma_g + 1} + \frac{\gamma_g - 1}{\gamma_g + 1} v_1^2 \tag{8.151}$$

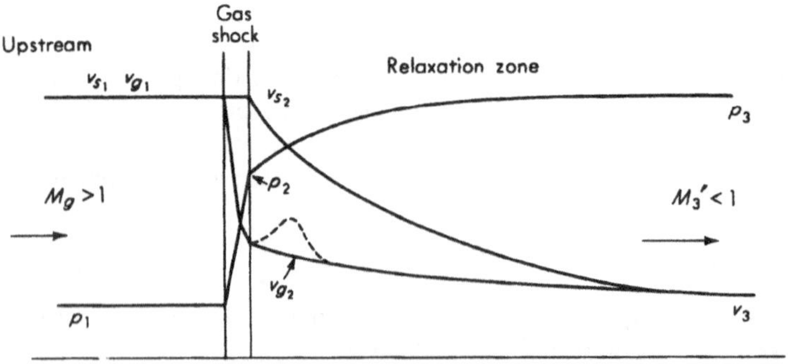

Fig. 8.17 Pressure and velocity variations through a "strong" two-phase shock wave.

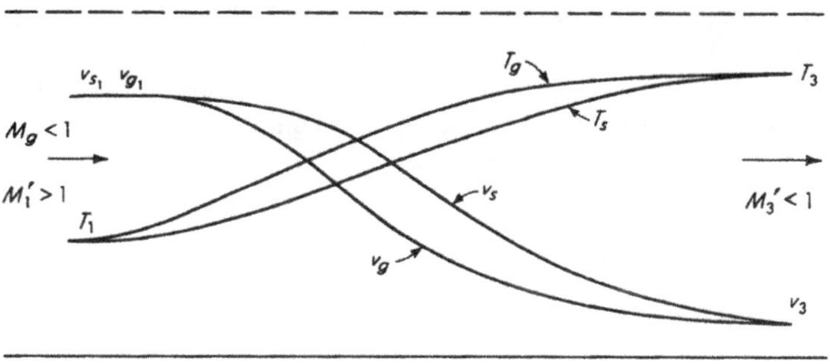

Fig. 8.18 Temperature and velocity variations through a "weak" two-phase shock wave.

or

$$\frac{(v_g)_2}{v_1} = 1 - \frac{2}{\gamma_g + 1}\left(1 - \frac{1}{M_{g1}^2}\right) \tag{8.152}$$

Exactly the same equations apply between states 1 and 3 as long as the Mach number is interpreted according to Eq. (8.146) and the gas constant and specific heat ratio are taken from Eqs. (8.111) and (8.113).

Since $\gamma_g > \gamma'$ and $M_1' > M_{g1}$, it is clear from Eqs. (8.150) and (8.152) that the final velocity at state 3 is less than the velocity at 2 and that the final pressure exceeds the pressure at 2. The deceleration of both the gas and the particles is brought about by a continuous rise in pressure across the relaxation zone. However, due to the drag from the faster moving particles the gas velocity may increase over the first part of the relaxation zone. A weak shock obeys the same equations between states 1 and 3 but is without the initial discontinuity in the gas properties (Fig. 8.18).

THE RELAXATION ZONE

In the relaxation zone of a shock wave the properties vary continuously,[3] whereas drag and heat transfer between the gas and the particles govern the rate of approach to equilibrium.[38-40]

Carrier's analysis[38] derives from taking the ratio of Eqs. (8.105) and (8.106) to eliminate the space coordinate and obtain

$$\frac{dT_s}{dv_s} = \frac{2k_g}{3c_s\mu_g}\,\frac{T_g - T_s}{v_g - v_s} \tag{8.153}$$

The coefficient $\frac{2}{3}(k_g/c_s\mu_g)$ is close to unity in many practical cases. Indeed, a relation of the same form as Eq. (8.153) is approximately valid over a wide range of Reynolds numbers, due to the interrelationship between heat and momentum transfer.

Eliminating the pressure and the gas density between Eqs. (8.99), (8.108), and (8.101), we get, since A is constant,

$$W_g \frac{dv_g}{dz} + W_s \frac{dv_s}{dz} + R_g W_g \frac{d}{dz} \frac{T_g}{v_g} = 0 \tag{8.154}$$

It follows by integration that

$$W_g v_g + W_s v_s + \frac{R_g W_g T_g}{v_g} = \text{const} \tag{8.155}$$

Equation (8.155) enables T_g to be expressed in terms of v_s and v_g. Use of the result in Eq. (8.107) gives T_s in terms of these velocities. Finally, substitution in Eq. (8.153) gives a differential equation, relating v_s and v_g, which can be solved numerically. Using either velocity as the independent variable, Eq. (8.105) is then used to find z, and all the other parameters can then be determined as functions of the position. For

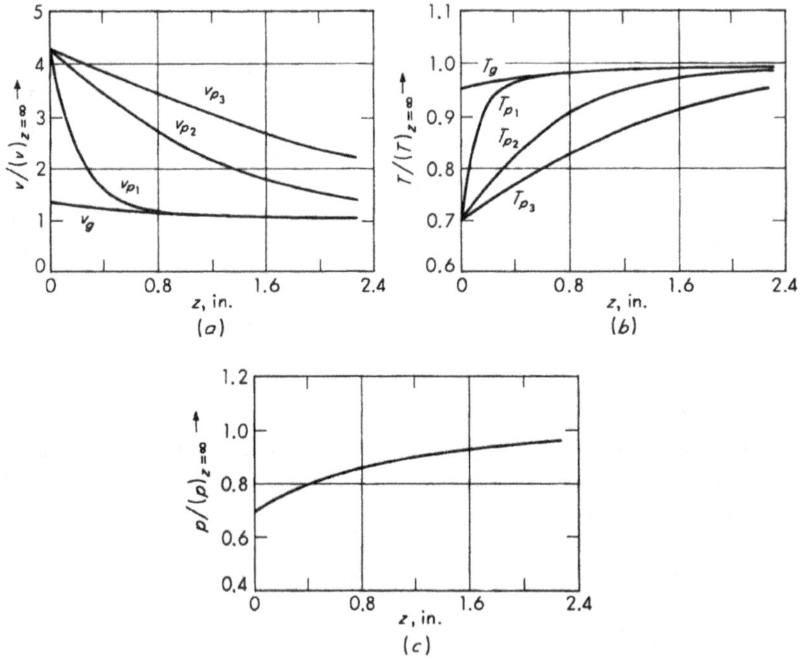

Fig. 8.19 Variation of gas and particle velocities and temperatures behind a strong shock. (*a*) Velocity versus z. (*b*) Temperature versus z. (*c*) Pressure versus z. Subscripts 1, 2, and 3 denote particles with radii 0.5, 1.5, and 2.5 μ. $M_{g1} = 2$. (*Kriebel.*[39])

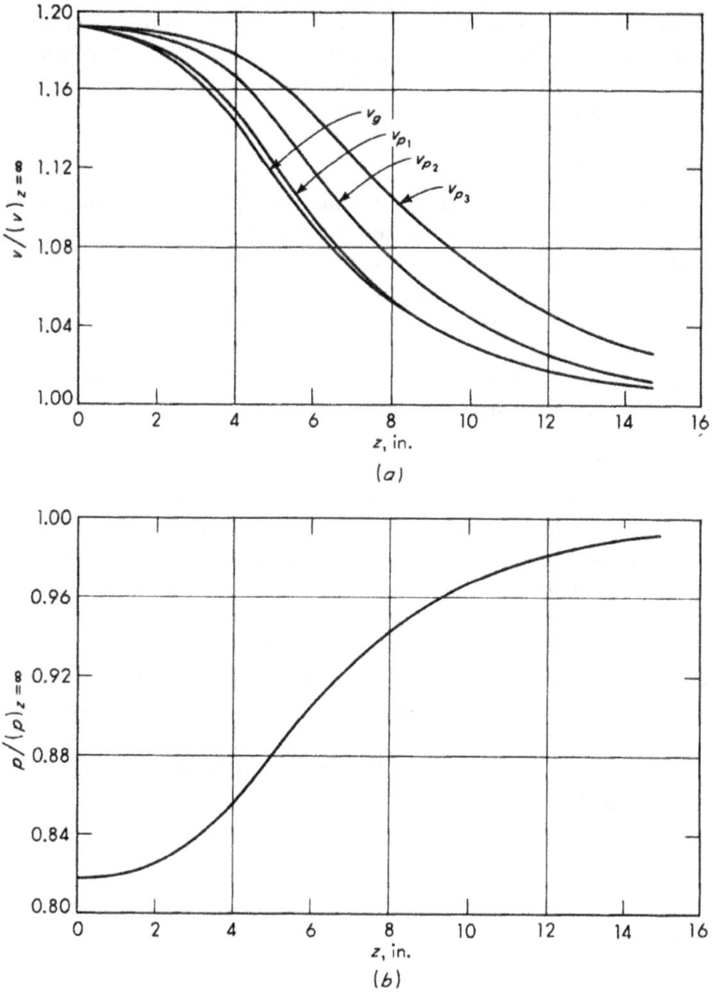

Fig. 8.20 Velocity and pressure variations through a "weak" shock. $M_{g1} = 0.91$, $M_1 = 1.1$. (*a*) Velocity versus *z*. (*b*) Pressure versus *z*. Subscripts 1, 2, and 3 denote particle radii 0.5, 1.5, and 2.5 μ. (*Kriebel.*[39])

example, Kriebel[39] gave a detailed description of his method of integration and solved several examples numerically. He also considered three groups of particle sizes such that 70 percent had the mean diameter, 12 percent were ⅓ of the mean diameter, and the remaining 15 percent were 5⁄3 of the mean diameter. These groups are denoted by subscripts 1, 2, and 3 in Figs. 8.19 and 8.20, which show examples of strong and weak shock relaxation zones. The particles were "typical of rocket propellants" with a mean diameter of 3 μ. Note that the relaxation zones are several inches thick. For particles 10 times as large the overall shock thickness could be several feet.

OBLIQUE SHOCKS

The initial and final states following an oblique shock can be derived in the usual way by resolving the motion along and perpendicular to the shock front. In the relaxation zone both the gas and particle streamlines are curved. In flow past a wedge, for example, there is no room for equilibrium to be established near the tip and the shock angle is determined by the gas flow alone. Farther away the equilibrium shock angle is approached. Just how the transition between these regions is achieved remains rather mysterious.

TWO- AND THREE-DIMENSIONAL EFFECTS

The conservation equations can be rewritten in vectorial form to describe multidimensional two-phase flow. Solution of these equations is possible numerically using the "characteristics" which provide a mesh for the numerical calculations.[41,42]

Qualitatively the particle thermal and velocity lags behave much the same as in the one-dimensional flow case, except that now the relative velocity must be interpreted vectorially. If the gas flow is turned, for example, the particles will tend to keep going straight until enough sideways relative velocity has been developed to produce a drag tending to bring them into line with the gas flow. Thus, in a converging-diverging nozzle the particles tend to leave the wall and move toward the axis,

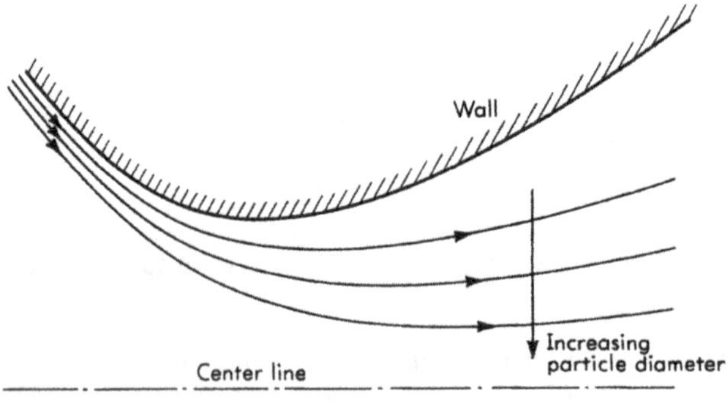

Fig. 8.21 Particle trajectories in a convergent-divergent nozzle, showing two-dimensional effects on limiting particle streamlines (for particles starting near the wall).

with the effect being more pronounced for larger particles (Fig. 8.21). Similar effects occur in two-dimensional shock relaxation, impinging jet flows, flows around bends, and so on.

OTHER EFFECTS

A vast number of additional effects occur if the assumptions outlined on page 207 are not all valid. Some useful references to these topics are given in the bibliography.[43]

8.10 ADDITIONAL FORCE COMPONENTS IN RAPIDLY ACCELERATING FLOWS

APPARENT MASS

When a particle is accelerated relative to a surrounding fluid it sets up a two-dimensional flow around it which possesses kinetic energy. Therefore work must be supplied to move the particle in addition to that which is required to accelerate it alone. This extra energy requirement shows up as an additional force on the particle. Thus, the acceleration a of a sphere in a stationary inviscid fluid of large extent requires a force

$$F' = \tfrac{1}{6}\pi d^3 \left(\rho_s + \frac{\rho_f}{2} \right) a \tag{8.156}$$

The sphere behaves as if it possessed an additional "apparent mass" equal to one-half of the fluid which it displaces.

Bodies of other shapes have different values of the apparent mass. In general we may say that the mass is increased in the ratio $1 + C\rho_f/\rho_s$ where $C = \tfrac{1}{2}$ for a sphere and is $\tfrac{1}{5}$ for an ellipsoid oriented with the flow with axes in the ratio 1:2 and 0.045 for axes in the ratio 1:6, for example.[44] One would also expect the presence of neighboring particles to influence the value of C which will therefore depend on the concentration α. For suspensions of spheres, Zuber[52] suggests that the apparent mass should be multiplied by the factor $(1 + 2\alpha)/(1 - \alpha)$.

The effect of apparent mass is that it introduces an additional component in the forces f that were introduced in Chap. 3. One might assume this force to have a form such as (see Appendix C)

$$f_s = -f_f \frac{\epsilon}{1 - \epsilon} = -C\rho_f \left[\frac{\partial}{\partial t} (v_s - v_f) + v_s \frac{\partial}{\partial z} (v_s - v_f) \right] \tag{8.157}$$

This force acts to reduce the velocity lag. The equation of motion of a particle moving upward in steady flow, for example, is, from Eqs. (3.45),

(8.15), (8.36), and (8.157),

$$v_s\rho_s \frac{dv_s}{dz} = -g\rho_s - \frac{dp}{dz} - \frac{3\epsilon^3(C_D)_\epsilon\rho_f(v_s - v_f)|v_s - v_f|}{4d} - C\rho_f v_s \frac{d(v_s - v_f)}{dz} \tag{8.158}$$

Under conditions of rapid acceleration, the final term will be significant compared with the term on the left-hand side if

$$\frac{C\rho_f}{\rho_s}\left(1 - \frac{dv_f}{dv_s}\right) \approx 0.1 \tag{8.159}$$

For particles suspended in gases the density ratio usually makes the effect negligible unless v_s is passing through a maximum or minimum. However, for liquid suspensions flowing through nozzles the effect should not be ignored.

Suppose that the inertia and pressure-drop terms dominate Eq. (8.158) entirely. Then we have

$$v_s\rho_s \frac{dv_s}{dz} = -\frac{dp}{dz} - C\rho_f v_s \frac{d(v_s - v_f)}{dz} \tag{8.160}$$

Using Eq. (8.157) the equivalent equation for the fluid is found to be

$$v_f\rho_f \frac{dv_f}{dz} = -\frac{dp}{dz} + \frac{1-\epsilon}{\epsilon} C\rho_f v_s \frac{d(v_s - v_f)}{dz} \tag{8.161}$$

Subtracting Eq. (8.161) from Eq. (8.160) to remove the pressure gradient we get

$$v_s\rho_s \frac{dv_s}{dz} - v_f\rho_f \frac{dv_f}{dz} = -\frac{C}{\epsilon}\rho_f v_s \frac{d(v_s - v_f)}{dz} \tag{8.162}$$

Collecting the terms in dv_s/dz and dv_f/dz and dividing by $\rho_f v_s$, we obtain

$$\left(\frac{\rho_s}{\rho_f} + \frac{C}{\epsilon}\right)\frac{dv_s}{dz} = \left(\frac{v_f}{v_s} + \frac{C}{\epsilon}\right)\frac{dv_f}{dz} \tag{8.163}$$

whence

$$\frac{dv_s}{dv_f} = \left(\frac{v_f/v_s + C/\epsilon}{\rho_s/\rho_f + C/\epsilon}\right) \tag{8.164}$$

If C and ϵ are constants, as in an incompressible flow with no distortion of the particles, Eq. (8.164) has the form

$$\frac{dv_s}{dv_f} = a\frac{v_f}{v_s} + b \tag{8.165}$$

Letting

$$n = \frac{v_s}{v_f} \tag{8.166}$$

Eq. (8.165) is transformed to

$$\frac{1}{v_f}\frac{dv_f}{dn} = \frac{-n}{n^2 - bn - a} \tag{8.167}$$

If the roots of the denominator of the right-hand side in Eq. (8.167) are n_1 and $-n_2$, expansion in partial fractions and integration eventually yields

$$v_f = (v_f)_0(n - n_1)^{-n_1/(n_1+n_2)}(n + n_2)^{-n_2/(n_1+n_2)} \tag{8.168}$$

where $(v_f)_0$ is a constant of integration.

Now, as v_f is increased during acceleration, Eq. (8.168) shows that n will be driven toward the value n_1, i.e.,

$$n \to n_1 = \frac{b + \sqrt{b^2 + 4a}}{2} \tag{8.169}$$

Equation (8.169) gives the maximum value of the velocity ratio during rapidly accelerating flows.

Example 8.4 What is the limiting velocity ratio during the accelerating nozzle flow of a dilute suspension of spherical gas bubbles of low density?

Solution The problem statement implies that

$$C = \tfrac{1}{2} \qquad \epsilon = 1 \qquad \frac{\rho_s}{\rho_f} \approx 0$$

Therefore, comparing Eqs. (8.164) and (8.165),

$$a = 2 \qquad b = 1$$

Using these values in Eq. (8.169) we obtain

$$n_1 = 2$$

which is the limiting *slip ratio* v_g/v_f. The drag term in Eq. (8.158) will act to keep the slip ratio below this limiting value.

THE BASSET FORCE

When a particle is accelerated relative to a fluid it establishes not only a potential flow field but also a viscous flow field. Since viscous effects, such as boundary layer growth, are governed by diffusion equations, the instantaneous flow field is a function of the entire previous history of the particle motion and is rather awkward to calculate. For laminar flow, Basset[45] obtained the result

$$f_s = -\frac{9}{\sqrt{\pi}}\frac{\sqrt{\rho_f\mu_f}}{d}\int_{t_0}^{t}(t - t')^{-1/2}\frac{d(v_s - v_f)}{dt'}\,dt' \tag{8.170}$$

where t' is a dummy variable which allows integration over the history of the particle between t_0 and the present time t. In particular, if $v_s - v_f$ varies linearly with time from t_0 to t, it is found (cf. Prob. 8.34) that the ratio of the Basset force to the steady flow drag force is $d/\sqrt{\pi \nu_f t}$. When the particle is part of a suspension, Zuber[52] suggests that this force should be increased by the factor $(1 - \alpha)^{-2.5}$, which probably overestimates the effect of neighboring particles.

8.11 FRICTION CHARACTERISTICS OF PARTICLE SUSPENSIONS

The friction characteristics of a fluid are governed by the way in which the shear stresses are related to an imposed rate of shearing strain. In the case of a newtonian fluid a simple linear relationship is valid in laminar parallel flow and this relationship defines the viscosity. Thus

$$\mu = \frac{\tau}{\partial v/\partial y} \tag{8.171}$$

Many suspensions of particles do not have a linear relationship between shear stress and rate of strain and have to be treated by rheological methods which are applicable to nonnewtonian fluids. For example, a typical shear diagram for a thorium oxide slurry is shown in Fig. 8.22. The laminar flow velocity profile can then be predicted by using a known shear stress distribution to deduce the velocity gradient and integrating across the duct, following the methods described in Chap. 5.

For low values of particle volumetric concentration, suspensions of spherical particles are approximately newtonian, and the ratio between the viscosity of the mixture and the viscosity of the fluid can be expressed as a function of α. For example, Einstein's equation[46] (for $\alpha < 0.05$) is

$$\mu = \mu_f(1 + 2.5\alpha) \tag{8.172}$$

which is equal to the limit of an equation proposed by Roscoe[83] and Brinkman,[86]

$$\mu = \mu_f(1 - \alpha)^{-2.5} \tag{8.173}$$

or, equivalently,

$$\mu = \mu_f \epsilon^{-2.5} \tag{8.174}$$

A rival equation which has a different coefficient was developed by Happel[49] as follows:

$$\mu = \mu_f(1 + 5.5\alpha) \tag{8.175}$$

and there is experimental evidence that the Einstein equation does tend to underestimate the effective viscosity.

Attempts have been made to obtain analogous results for particles which are not spherical[50] and to perform expansions in higher powers

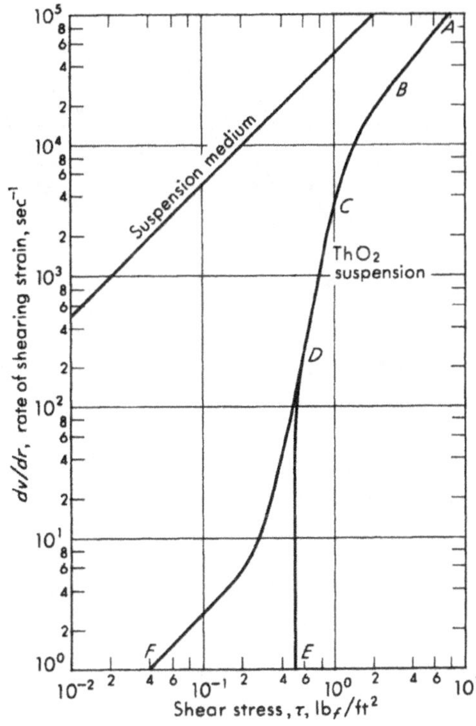

Fig. 8.22 Shear diagram for thorium oxide suspension. (*Thomas.*[47])

of α. There is a great variation in the value of the coefficients which are quoted by different authors.

Most nonnewtonian suspensions have a limiting viscosity η at high rates of shear, which can be simply related to the viscosity of the suspending fluid and the particle concentration.

The limiting viscosity of nonnewtonian suspensions is often correlated[51] by the equation

$$\eta = \mu_f e^{k\alpha} \tag{8.176}$$

An example of this dependence is shown in Fig. 8.23. The coefficient k is determined by the particle characteristics and is generally larger for smaller particles which flocculate and display an increased viscosity as a result of interparticle forces. Thomas[51] has found, for example, that k for suspensions in water can be expressed quite well by

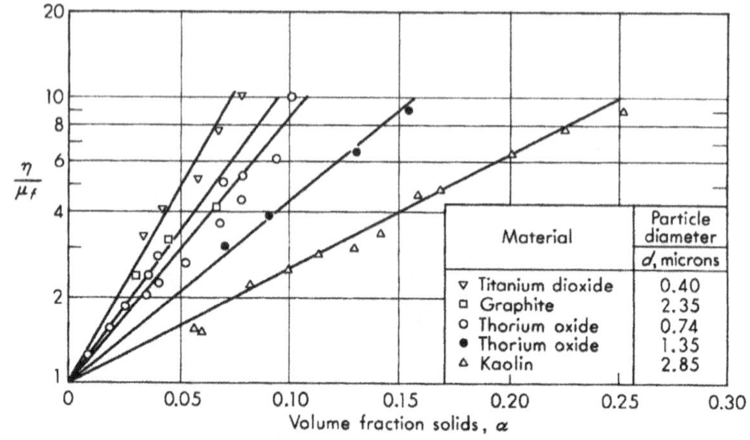

Fig. 8.23 Effect of volume fraction solids on limiting viscosity at high rates of shear. (*Thomas.*[51])

the equation

$$k = 2.5 + \frac{14}{\sqrt{d}}\Phi \tag{8.177}$$

where d is the particle diameter in microns and Φ is a shape factor which is equal to unity for spherical particles. For large particles and low concentrations, k becomes equal to 2.5 and Eqs. (8.172), (8.173), and (8.176) are approximately identical.

Some other models for the viscosity of suspensions are discussed by Zuber[52] and a comprehensive literature survey is given by Rutgers.[53,54]

LAMINAR FLOW

For laminar flow of a newtonian fluid in a round pipe the wall shear stress is related to the viscosity by the equation

$$\tau_w = \mu \frac{8V}{D} \tag{8.178}$$

and the pressure gradient is

$$-\left(\frac{dp}{dz}\right)_F = \frac{32\mu V}{D^2} \tag{8.179}$$

The equivalent relationships for a nonnewtonian fluid are dependent upon its particular rheological properties. However, it is possible to define an *effective viscosity* as

$$\mu_e = \frac{\tau_w}{8V/D} \tag{8.180}$$

Equation (8.171) is then replaced by a "pseudoshear" relationship which expresses the wall shear stress in terms of the rate of strain which would exist at the wall for the same overall flow rate in a newtonian fluid. Some typical pseudoshear results are shown in Fig. 8.24. When one changes the pipe diameter or velocity, the relationship between wall shear stress and effective rate of strain is unique for a given fluid-particle combination as long as the flow is laminar. Therefore the pseudoshear diagram can be used to predict laminar flow pressure drop.

Another way to proceed is to use a friction factor versus Reynolds number relationship in terms of the effective viscosity. The friction fac-

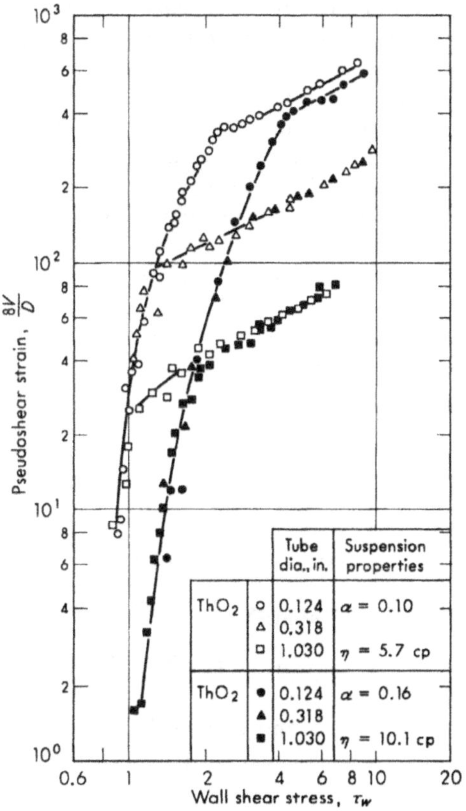

Fig. 8.24 Pseudoshear diagram for concentrated suspensions showing agreement of laminar and turbulent data, respectively, as tube diameter was varied. (*Thomas.*[48])

tor is defined as

$$C_f = \frac{\tau_w}{\frac{1}{2}\rho_m V^2} \tag{8.181}$$

Therefore, from Eq. (8.180),

$$C_f = \frac{16}{\text{Re}_m} \tag{8.182}$$

where Re_m is defined in terms of the effective viscosity

$$\text{Re}_m = \frac{\rho_m V D}{\mu_e} \tag{8.183}$$

In using this technique a pseudoshear diagram is still necessary in order to calculate μ_e. In cases where a quasi-newtonian approximation is valid, μ and μ_e are identical. Transition to turbulence occurs in round pipes when Re_m is greater than about 2500.

In principle, if the rheological relationships between stress and strain rate can be established, an accurate prediction of the pressure drop can be made as a function of the flow rate in laminar flow. The shear profile is used to deduce the velocity gradient profile and a double integration yields the flow rate. However, in practice, nonnewtonian characteristics display a whimsical and bewildering variety which defies any rational treatment from first principles.[50] One is always reduced to an attempt to fit experimental data with a suitable mathematical model. Since the engineering use of any theory is usually restricted to a limited range of shear rates, almost any reasonable curve fit will do.

The commonest models are:

(a) Power law:

$$\tau = k\left(-\frac{dv}{dr}\right)^n \tag{8.184}$$

(b) Bingham:[55]

$$\begin{aligned} \tau &= \tau_y + \mu\left(-\frac{dv}{dr}\right) \qquad \text{for } \tau > \tau_y \\ \frac{dv}{dr} &= 0 \qquad \text{for} \qquad \tau < \tau_y \end{aligned} \tag{8.185}$$

(c) Casson:[56]

$$\tau^{1/2} = \tau_y^{1/2} + \left[\mu\left(-\frac{dv}{dr}\right)\right]^{1/2} \text{for} \qquad \tau > \tau_y \tag{8.186}$$

(d) Ree and Eyring:[57]

$$\tau = C\left(-\frac{dv}{dr}\right) + \frac{1}{B}\sinh^{-1}\left(-\frac{1}{A}\frac{dv}{dr}\right) \tag{8.187}$$

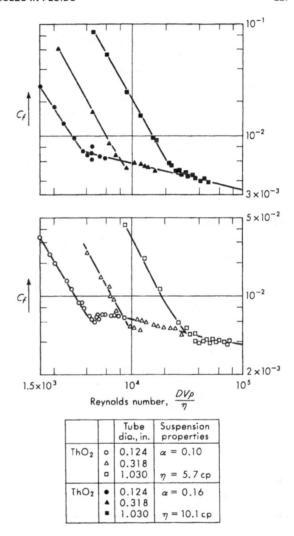

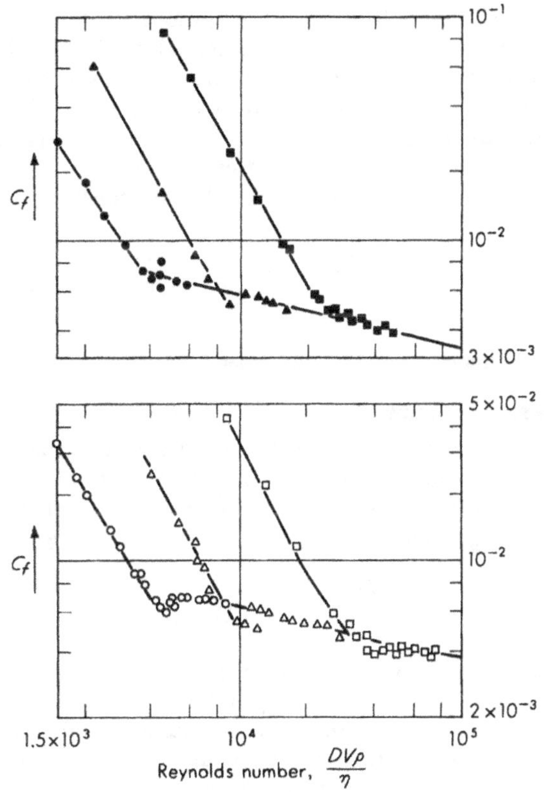

		Tube dia., in.	Suspension properties
ThO_2	○	0.124	$\alpha = 0.10$
	△	0.318	
	□	1.030	$\eta = 5.7$ cp
ThO_2	●	0.124	$\alpha = 0.16$
	▲	0.318	
	■	1.030	$\eta = 10.1$ cp

Fig. 8.25 Friction factor plot for concentrated suspensions showing agreement of turbulent data as tube diameter was varied. (*Thomas.*[48])

The Bingham model is appropriate for many plastic fluids. Casson's equation works well for blood,[58] inks,[56] and coal slurries;[59] it even works better on Thomas' thorium oxide suspensions than the models which Thomas originally tested.[47] The Ree-Eyring model often correlates data well (because it contains three adjustable constants), but is hard to deal with for analytical purposes.

An equation with three arbitrary constants which reduces to models a, b, and c as special cases is

$$\tau^m = \tau_y{}^m + k\left(-\frac{dv}{dr}\right)^n \tag{8.188}$$

and this equation will fit most data satisfactorily.

TURBULENT FLOW

Because of the high local rates of strain in turbulent flow, the effective viscosity does not give an accurate description of the force distribution in the flow field. A better approximation is usually obtained by using the limiting viscosity at high rates of shear in the Reynolds number. Thus

$$\mathrm{Re}_t = \frac{\rho_m VD}{\eta} \tag{8.189}$$

A friction factor–Reynolds number plot on this basis is shown in Fig. 8.25. Since the effective viscosity in laminar flow can be many times the limiting viscosity, the laminar flow lines are displaced on such a graph.

An interesting and important effect with suspensions of fibers or long chain molecules is the reduction of drag in turbulent flow.[48,61–64] Apparently the turbulence level is reduced by the damping action of the fibers on the eddy pattern. In practice it is sometimes economical to add fibers or polymers to fluids in order to reduce the pressure drop in pipelines.

PNEUMATIC TRANSPORT

The flow of particles of any significant size above a few microns which are suspended in gases is much less amenable to analysis than the flow of a liquid suspension. The particles "rattle around" in the pipe, there are many different flow regimes, and factors such as the coefficient of restitution for the particles are important. Numerous empirical correlations have been developed but few have any claim to general validity.

8.12 NONUNIFORM PARTICLE DISTRIBUTION

In previous sections of this chapter it was assumed that the particles were uniformly distributed over the flow field. Certain cases in which this assumption is no longer valid will now be discussed briefly. For convenience it is possible to define three general classes of nonuniform flows.

1. *Stratified flows* in which the particles concentrate in layers or annuli causing nonuniform distributions across the pipe.
2. *Periodic flows* in which the concentration varies along the pipe.
3. *Aggregative flows* in which the particles or the fluid tend to form many discrete areas of high concentration throughout the flow.

Often there exists a close analogy between these phenomena and similar phenomena which occur in gas-liquid flow.

STRATIFICATION

Gravitational effects One of the simplest ways in which stratification can occur is in horizontal flow where gravitational effects cause the heavier component to concentrate toward the bottom of the duct. The amount of stratification which occurs is governed by the balance between the buoyancy forces on the particle and the forces which are caused by its motion relative to the fluid. This balance is conveniently represented by taking the ratio of the terminal settling velocity v_∞ to the wall shear velocity u^* defined by Eq. (7.27). The concentration distribution is then often represented by expressing $\alpha/\langle\alpha\rangle$ as a function of v_∞/u^*.

According to Thomas,[65] the value $v_\infty/u^* \approx 0.2$ gives a convenient measure of the boundary between the regimes in which the solids are concentrated at the bottom of the pipe or flow predominantly in suspension. Corresponding values from the Reynolds flux theory (Prob. 7.16) are between 0.1 and 0.15. These simple expressions do not, however, give an adequate picture of the interaction between the particles and the velocity profile. Very small particles lie deep in the boundary layer and are exposed to velocities much less than u^*, whereas very large particles experience lift and drag forces which are scaled by the mean stream velocity. In general, particles which are small enough to lie inside the boundary layer at the condition of the minimum transport velocity obey the Stokes law ($\mathrm{Re}_\infty < 1$) and the critical value of u^* for dilute suspensions (u_0^*) is given by[65]

$$\frac{v_\infty}{u_0^*} = 0.01\left(\frac{d u_0^* \rho_f}{\mu_f}\right)^{2.71} \tag{8.190}$$

For larger particles ($\mathrm{Re}_\infty > 1$) a correlation is

$$\frac{v_\infty}{u_0^*} = 4.9\left(\frac{d u_0^* \rho_f}{\mu_f}\right)^{0.4}\left(\frac{d}{D}\right)^{0.6}\left(\frac{\rho_s - \rho_f}{\rho_f}\right)^{0.23} \tag{8.191}$$

which is remarkable for the large effect of the diameter and density ratios.

The above equations refer to the condition of infinite dilution, or $\alpha = 0$. When a significant concentration of particles is present, the wall shear velocity for minimum transport u_α^* can be found from the correlation[65]

$$\frac{u_\alpha^*}{u_0^*} = 1 + 2.8\left(\frac{v_\infty}{u_0^*}\right)^{1/3} \alpha^{1/2} \tag{8.192}$$

The effect of the density ratio in Eq. (8.191) is to make v_∞/u_0^* larger for gas-solids systems. This means that a small concentration of particles has a large effect in Eq. (8.192). In addition, the minimum transport velocity of the fluid becomes comparable with the terminal velocity of the particles, instead of being many times larger as in the case of liquid-particle systems. This is only possible if the particles "rattle around" in the pipe and acquire velocities comparable with the free stream. Once they settle out, they are difficult to dislodge, tend to fill up the pipe, and increase the pressure drop by an order of magnitude. This "point of saltation" is an important regime boundary in pneumatic conveying.[66]

Similar techniques are applicable to problems of erosion and mud and sand transportation in alluvial rivers.[67,68]

Symmetrical radial concentration variations Even when gravitational effects are absent or negligible it is possible for quite large variations to occur in particle concentration across a duct. For example, in viscous Poiseuille flow in a small diameter tube the particles tend to concentrate at about 0.6 of the tube radius from the axis and there is a fluid core that is relatively free of particles.[69]

Channeling or spouting in fluidized beds Nonuniform effects can be produced in fluidized beds by introducing the fluid in a suitable way. For example, if the fluid is introduced as a high-velocity jet in the center of a vertical column, a kind of annular countercurrent flow can be set up in which a dilute fluid-solids mixture flows up the middle while a denser concentration of solids circulates down near the wall. The behavior of such a system is described by Leva.[15]

PERIODIC FLOWS

Slugging Large axial variations in particle concentration can occur in a duct as a result of the growth of the instabilities which were discussed in Sec. 8.8 and also by the tendency of turbulence and other three-dimensional instabilities to break up the simple one-dimensional pattern.

Since many wavelengths are unstable, no simple regular pattern is observed in slugging. Individual slugs form, grow, collapse, and overtake each other continuously (Fig. 8.26). If the slugs develop a two-

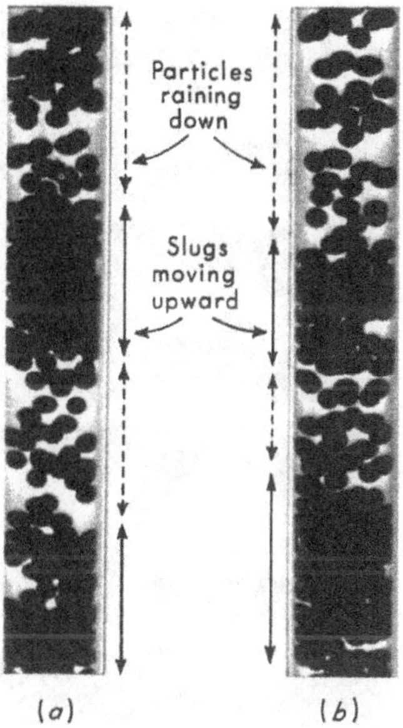

Fig. 8.26 Slugging in a fluidized bed. Glass spheres in a 1-in.-diam tube.

dimensional bubble shape, they can be treated by the methods to be described in Chap. 10.[70]

Wave formation in stratified flow In horizontal pipes in which dense particles lie on the bottom while fluid flows over them, it is possible to observe the periodic formation of waves or dunes which travel slowly in the direction of fluid motion (Fig. 8.27). These waves appear to be the result of a Helmholtz instability (due to the relative motion) and have been correlated by Thomas[71] using the equation

$$\frac{v_f^2}{g\lambda}\frac{\rho_f}{\rho_s - \rho_f} = \frac{1}{2\pi}\left[\frac{v_I^2}{4gH}\left(\frac{D}{H}\right)^{5/3}\left(\frac{D}{d}\right)^{2/3}\right]^{0.258} \tag{8.193}$$

for islands (particle groups) moving with velocity v_I, wavelength λ, height H, in a tube of diameter D, and composed of particles of density ρ_s and diameter d, beneath a fluid with velocity v_f and density ρ_f. Similar phenomena occur with alluvial rivers, sand dunes, and snowdrifts.

AGGREGATIVE FLOWS

Flocculation Very small particles (about 100 μ or less in diameter) or particles with irregular dendritic shapes tend to form clusters or *flocs* which retain their identity as they flow in the suspending fluid. The flocs are held together by interparticle forces. Because of the open

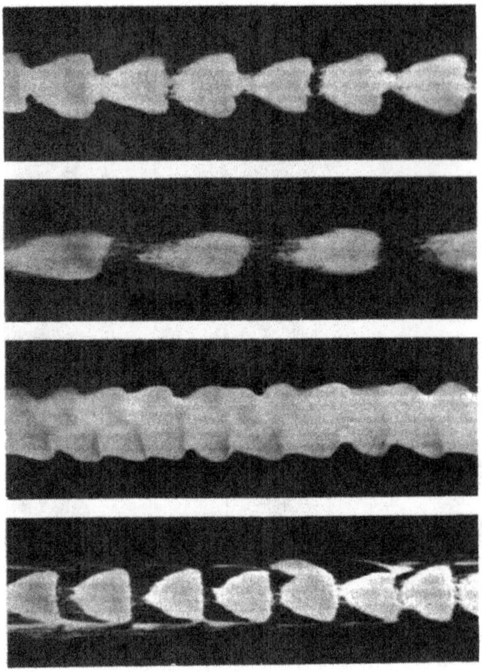

Fig. 8.27 Periodic phenomena observed with spherical particles in horizontal pipes. (*Thomas.*[71])

spatial structure of a floc, the suspending fluid can flow through it as well as around it. Flocculation is one cause of nonnewtonian flow characteristics of suspensions.

For flocculated suspensions the approximate value of n in Eq. (8.2) is often very high (e.g., a value of 20 to 30 in Ref. 7). Furthermore, n is not constant over the entire range of concentration since the floc size is not independent of concentration.

One effect of flocculation is to terminate settling at low values of α. The sediment contains a spongy mass of particles held together by interparticle forces in a loose open structure. Further compaction is only possible by the breaking of the floc bonds and the squeezing out of the liquid through the surrounding particles.

Bubbling Usually, when the suspending fluid is a gas and the particles are fairly large or dense, it is possible to observe the inverse of floc formation—namely, the formation of bubbles of fluid containing no particles.

The actual way in which these bubbles arise is a matter of debate. They may be the eventual three-dimensional result of the one-dimensional instability discussed on page 206.

A great deal of study has been devoted to the fascinating phenomenon of bubbling in large gas-fluidized beds.[72–78] The shape of such a bubble is shown in Fig. 8.28. The bubbles rise almost exactly like gas bubbles in a fluid with negligible viscosity and surface tension. In fact the rise velocity of a bubble in a bed which is just fluidized by gas of low density compared with the particle density is given by an equation similar to the equation of Davies and Taylor[79] for gas bubbles (the constant is slightly different), namely,[80]

$$v_b = 0.71 g^{1/2} \mathcal{V}_b^{1/6} \tag{8.194}$$

If the fluid density is significant, this equation is modified to the form

$$v_b = 0.71 g^{1/2} \mathcal{V}_b^{1/6} \left[\frac{1 - \epsilon_0}{\rho_s/(\rho_s - \rho_f) - \epsilon_0} \right]^{1/2} \tag{8.195}$$

The bubble surface is not impermeable to fluid and, in fact, there is a continual circulation, relative to the bubble, up through the bubble nose and round through the bubble wake. This circulation "holds up the roof" and prevents the bubble from collapsing.

Although a gas-fluidized bed usually bubbles, while a liquid-fluidized

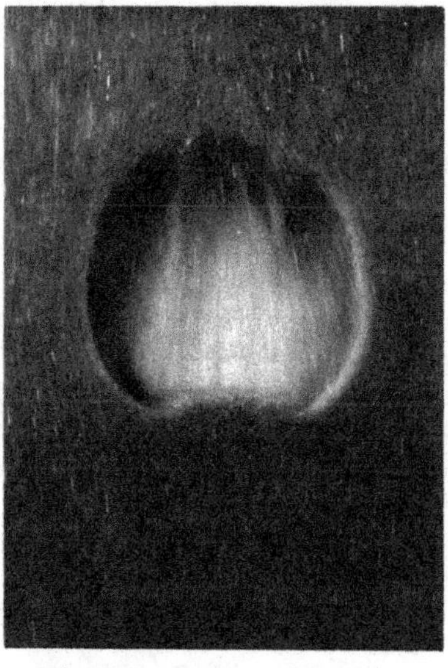

Fig. 8.28 Photograph of a two-dimensional bubble in a fluidized bed. (*Rowe, Partridge, and Lyall.*[78])

bed does not, there is a continuous transition from one form of behavior to the other.[16,81] Bubbling can be obtained with lead shot in water, and, on the other hand, gases under pressure will fluidize sufficiently light spheres quite uniformly.

A simple, but approximate, criterion for bubbling has been given by Wilhelm and Kwauk[30] in terms of the particle Froude number defined by Eq. (8.98). Uniform fluidization is predicted for a Froude number less than unity and bubbling at higher Froude numbers. Alternatively, Jackson[77] has suggested that the rate of growth of one-dimensional disturbances (Table 8.3) can be used as a measure of the tendency toward bubble formation. Similar analyses have also been carried out by Pigford[82] and Smith.[27]

Harrison, *et al.*[81] approach the boundary from the other side by considering the conditions under which a bubble can be destroyed. They reason that, if the circulation of fluid through the bottom of the bubble is rapid enough, it will entrain particles and fill the bubble from below. The criterion for this to happen is, approximately,

$$v_b > v_\infty \tag{8.196}$$

Since the bubble velocity is given by Eq. (8.194), Eq. (8.196) can be rearranged to give for unstable bubbles

$$\frac{v_\infty}{(gd)^{1/2}} < 0.71 \left[\frac{1 - \epsilon_0}{\rho_s/(\rho_s - \rho_f) - \epsilon_0}\right]^{1/2} \left(\frac{\mathcal{V}^{1/3}}{d}\right)^{1/2} \tag{8.197}$$

The Froude number is therefore the major factor determining the ratio of the diameter of the largest stable bubble to the particle diameter.

8.13 PERCOLATION THEORY

The flow of a fluid through a three-dimensional porous medium occurs in filtration equipment, leaching, settling tanks, and in many civil engineering applications such as the flow of ground water to wells, seepage through dams, and soil consolidation.

As usual, the formulation of the theory comprises the two continuity equations and equations of motion. In the simplest case, flow is steady, the fluid is incompressible, the porous matrix is homogeneous, and inertia terms are negligible. The continuity equation for the fluid is

$$\nabla \cdot \mathbf{j}_f = 0 \tag{8.198}$$

in which $\mathbf{j}_f$ is now a vector

Equation (3.45) in vector form becomes

$$-\nabla p - \rho_f g \mathbf{k} - \mathbf{f}_f = 0 \tag{8.199}$$

$\mathbf{k}$ is the unit vector in the upward direction.

The value of $\mathbf{f}_f$ will be determined by the value of $\mathbf{j}_f$ and the fluid and particle properties (since the matrix is stationary). In the case of laminar flow we have, from Example 8.3, for $\mathbf{j}_s = 0$,

$$\mathbf{f}_f = -\frac{180\mu_f}{d^2\epsilon^2}(1-\epsilon)\mathbf{j}_f \tag{8.200}$$

The Darcy permeability factor is defined as

$$K = \frac{g\rho_f d^2\epsilon^2}{180\mu_f(1-\epsilon)} \tag{8.201}$$

If we also define the nonhydrostatic pressure p' by the equation

$$p' = p + gz \tag{8.202}$$

the use of Eqs. (8.200) and (8.201) in Eq. (8.202) finally gives

$$\frac{\nabla p'}{g\rho_f} = -\frac{1}{K}\mathbf{j}_f \tag{8.203}$$

Combining this result with Eq. (8.198) we obtain

$$\nabla^2 p' = 0 \tag{8.204}$$

Therefore the pressure obeys Laplace's equation, whereas the fluid flux everywhere is proportional to the negative pressure gradient. All of the standard methods of potential flow theory can then be applied to solve specific problems. A good reference for the European literature is the book by Jaeger,[84] which also gives examples of many approximate methods.

For turbulent flow the best approach is usually a numerical one.

An approximate solution for unsteady flow was obtained by Terzaghi[25] who modified Eq. (8.204) to the form

$$c\nabla^2 p' = \frac{\partial p'}{\partial t} \tag{8.205}$$

where c was defined as the *coefficient of consolidation*. This theory is still repeated in most soil mechanics textbooks although, as pointed out in discussing Eq. (8.97), it is not entirely correct.

PROBLEMS

8.1. Compare the terminal velocities of "spent" duck shot in air and water if $d = 0.08$ in. Consider both lead shot ($\rho_s = 700$ lb/ft^3) and plastic shot ($\rho_s = 110$ lb/ft^3).

8.2. What are the values of n in Eq. (8.2) for the particle-fluid combinations in Prob. 8.1?

8.3. Rather than differentiating Eq. (8.31) use Eqs. (8.22) and (8.30) to find the best value of n for use in Richardson and Zaki's correlation as a function of Reynolds number.

8.4. Derive a general dimensionless drift-flux-versus-concentration relationship in terms of N_f, ϵ, and the drift-flux Reynolds number, $\text{Re}_{fs} = \rho_f j_{fs} d/\mu_f$.

8.5. Wen and Yu[9] derived a general correlation for the minimum fluidization velocity as follows:

$$(\text{Re}_s)_{\text{mf}} = [(33.7)^2 + 0.0408 N_f{}^2]^{1/2} - 33.7$$

Compare this result with the theories in the text.

8.6. Frantz[85] gives an equation for the fluid mass flux at minimum fluidization, as follows:

$$G_{\text{mf}} = 4.45 \times 10^5 \frac{d^2 \rho_f(\rho_s - \rho_f)}{\mu_f}$$

The various units are G, in pounds per hour per square foot; d, in feet; ρ, in pounds per cubic foot; μ, in pounds per hour per foot. Compare with the equations in the text.

8.7. Calculate the minimum-fluidization velocities for the following sizes of glass beads ($\rho_s = 2.5$ g/cm^3) in air or water and compare with Rowe's[8] measured values. Assume atmospheric conditions of 20°C and 1 atm.

Fluid	*d, cm*	*Measured* $(j_{f0})_{\text{mf}}$, *cm/sec*
Air	0.0083	0.823
Air	0.0273	7.01
Air	0.065	30.5
Water	0.065	0.597
Water	1.2	10.12

8.8. As air rises through a fluidized bed it expands. Show how a tapered bed can be designed so that the particles are just fluidized at all depths. Design such a bed for "stucco" with diameter 0.6 mm and density 2.6 g/cm^3 if the top is to be open to the atmosphere with an area 2 ft^2 and the depth is 3 ft.

8.9. Air at 70°F flows through a constant cross-section bed of copper shot with diameter 0.065 cm. The air velocity leaving the top of the bed is measured to be 1 m/sec at 14.7 psi. Estimate the depth to which the bed is fluidized.

8.10. Seeds ¼ in. in diameter are to be transported pneumatically at 50 psi and 100°F in a vertical duct 2 ft in diameter. It is determined empirically that the volume fraction of seeds in the duct must not exceed 0.05 if flow instabilities are to be avoided. Estimate the air velocity required to transport 100 tons of seeds per hour. One cubic foot of packed seeds weighs 50 lb.

8.11. A cylindrical filter in a compressed air line consists of sintered spheres each with a diameter 0.02 in. and having a void fraction of 0.3. It is ¼ in. thick, 6 in. in diameter, and 8 in. long. What is the pressure drop for an air flow of 40 scfm at 100 psi, 90°F inlet conditions? What is the variation in this pressure drop if the manufacturing tolerance in void fraction is ±10%?

8.12. Compare the predictions of Eqs. (8.48) and (8.35) for both low and high Reynolds numbers over the range of void fraction $0.3 < \epsilon < 0.6$. If the methods of Chap. 4 are used to make predictions in this range, how sensitive are the results to the particular equation which is chosen?

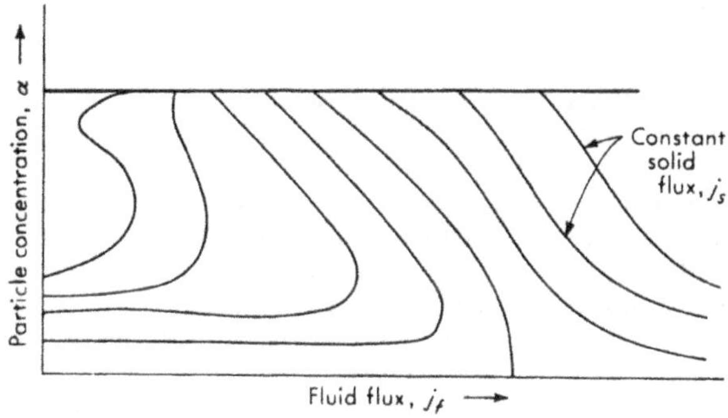

Fig. 8.29 Particle concentration characteristics for a fluidized system.

8.13. Ore with density $\rho_s = 140$ lb/ft³ and mean particle diameter 0.1 cm is washed by countercurrent flow of an acid [$\rho_f = 65$ lb/ft³, $\mu_f = 2$ lb/(hr)(ft)] in a vessel 6 ft in diameter. Evaluate the limiting relationship between the flow rates which is set by flooding.

8.14. Show that the void fraction of a fluidized system can be plotted versus the liquid flow rate for various upward or downward solids flow rates as shown in Fig. 8.29. Identify the direction of solids flow, the points corresponding to flooding, incipient fluidization, and the terminal particle velocity on the sketch.

8.15. Combine the results of Prob. 8.14 with packed-bed pressure-drop theory and homogeneous flow estimates for wall friction to generate the pressure-gradient-versus-flow-rate curves shown in Fig. 8.30. Identify the various features of the curves.

8.16. Thoria with a density of 9.5 g/cm³ is mixed into a slurry containing 500 g/liter with a mean size of 10 μ in water at 25°C. A liter of this mixture is put into a vertical cylinder 5 cm in diameter and allowed to settle. Determine the subsequent behavior.

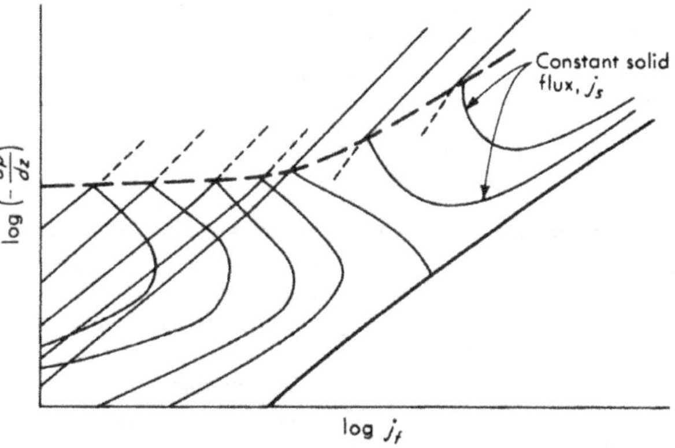

Fig. 8.30 Pressure gradient characteristics for a fluidized system.

8.17. A flocculated suspension is found to obey the equation

$$j_{fs} = 0.01\alpha(1 - \alpha)^{10} \qquad \text{fps}$$

and settling is complete at $\alpha_\infty = 0.5$. Predict the batch sedimentation behavior of this suspension for $\alpha_0 = 0.01, 0.05, 0.1, 0.15$, and 0.2.

8.18. Derive Eq. (8.71).

8.19. A fluidized bed containing catalyst beads ($d = 0.4$ cm, $\rho_s = 1.6$ g/cm³) is operated steadily in a 10-cm-diam tube with $\epsilon = 0.8$ using a flow of water at 60°F. If the flow of water is suddenly cut in half, describe the subsequent behavior.

8.20. Integrate Eq. (8.82) if $1 - \epsilon$ is not small. If $N = 0.25$, what errors are introduced by assuming $(1 - \epsilon)$ to be small for the following values of $1 - \epsilon$: 0.002, 0.01, 0.05, 0.1?

8.21. A mat of fibers is formed on a filter, and the pressure drop across the mat is 1 psi. If the stress-strain relationship for the fibers is $p_s = 8000\ (1 - \epsilon)^4$ psi, what is the mean volumetric density of fibers in the mat?

8.22. Show that, if there is no particle contact, Eq. (8.96) represents continuity waves. If the Carman-Kozeny equation applies, show that the continuity wave velocity is three times the liquid velocity.

8.23. If E is constant in Eq. (8.96), what conditions are necessary for Eq. (8.97) to be reasonably valid?

8.24. Deduce the results shown in Table 8.3 assuming that $f_{\nabla\alpha} = 0$, $n = 3$, $f_{fs} \propto (v_f - v_s)^{1.5}$, and $\epsilon = 0.4$.

8.25. Design a sandblast nozzle to provide a flow of 1 lb/min of sand ($d = 0.02$ in., $\rho_s = 150$ lb/ft³) with a velocity of 400 fps if a compressed air supply is available with $p_{0g} = 100$ psi, $T_{0g} = 70$°F.

8.26. One of the sand particles mentioned in Prob. 8.25 is dropped into an air stream with velocity 1000 fps at 200°F and 20 psia. How far does it go before attaining a velocity of 800 fps.

8.27. A plasma spray gun is to use nitrogen at 5000°K ($\mu_g \approx 10^{-3}$ g/(cm)(sec), $\rho_g \approx 3 \times 10^{-5}$ g/cm³) and nickel particles ($\rho_s = 8$ g/cm³). If it is designed for constant fractional velocity lag with $K = \frac{1}{2}$, what distance is required to accelerate 10-μ-diam particles to 2000 fps? What is the time spent in the nozzle by the particles in accelerating from 20 to 2000 fps? Estimate their change in enthalpy in this period if $k_g \approx 10^{-3}$ cal/(cm)(sec)(°K)? Do they evaporate?

8.28. If the particles in Prob. 8.27 are large so that their velocity is low and the plasma velocity is constant, show that their enthalpy change after going a distance z is

$$\Delta h = \frac{4k_g\overline{\Delta T}}{\mu_g}\left(\frac{z}{d}\frac{\mu_g}{\rho_s d v_g}\right)^{\frac{1}{2}}$$

where $\overline{\Delta T}$ is an average temperature difference between the particles and the plasma.

8.29. A 1-ft-long nozzle is to operate using air at 500°F and velocities of about 1000 fps and is to carry a flow of aluminum particles with density 170 lb/ft³ [and specific heat 0.2 Btu/(lb)(°F)]. What must the particle size be if ϵ_v and ϵ_T are to be "small" in Eqs. (8.141) and (8.142)?

8.30. A stationary shock wave occurs in dusty air. The upstream conditions are $v = 2000$ fps, $m_s = 1$, $T = 400$°F, $\rho_s = 200$ lb/ft³, $c_s = 0.2$ Btu/(lb)(°F), and $d = 10$ μ. Discuss the shock structure.

8.31. Solve Prob. 8.30 if $v = 1200$ fps.

8.32. What is the effect of apparent mass on dynamic waves in two-phase flow? Estimate the maximum possible departure of the compressibility wave velocity from homogeneous theory for an air-water mixture in which $\alpha = \frac{1}{2}$ and $C = \frac{1}{2}$.

8.33. Under what conditions can the final term in Eq. (8.158) be larger than the drag term? Evaluate typical cases numerically.

8.34. Prove that, if the relative velocity increases linearly with time for a coordinate system moving with a particle, the ratio of the Basset force to the steady-flow drag force in laminar flow is $d/\sqrt{\pi \nu_f t}$. Explain this result physically, using simple ideas of diffusion. For what size of particles in atmospheric air will the Basset force be significant if $t = 0.01$ sec?

8.35. A coal slurry consisting of 40% coal, with density 84.5 lb/ft³ in water at 70°F, is found to obey Casson's equation with $\tau_y = 9 \times 10^{-2}$ lb/ft² and $\eta = 5 \times 10^{-3}$ lb/(ft)(sec). Derive the pseudoshear diagram for laminar flow. What is the frictional pressure drop for flow with a mean velocity of 1 fps in a pipe 1 in. in diameter and 100 ft long?

8.36. The coal slurry in Prob. 8.35 flows in a pipe for which single-phase turbulent-flow data were correlated by the equation $C_f = 0.026\ \mathrm{Re}^{-0.12}$. What is the pressure drop for a velocity of 10 fps in a 1-in. pipe which is 100 ft long?

8.37. A flocculated suspension is transported in water at 200°F in a chemical treatment plant. If the particle diameter is 1-μ, the density ρ_s is 400 lb/ft³, and the slurry contains 0.3 lb of solids per pound of water, what is the pressure drop for an upward flow of 2×10^4 lb/hr in a 100-ft length of vertical 1-in. pipe?

8.38. What is the minimum transport velocity for pneumatically conveying grain ($\rho = 75$ lb/ft³, $d = \frac{1}{4}$ in.) in a 12-in. pipe at 200 psia, 70°F, (*a*) at low concentrations, (*b*) if $\alpha = 0.1$?

8.39. An alluvial river flows over a bed consisting of mud particles for which $\rho_s = 2$ g/cm³ and $d = 100\,\mu$. What velocity of the river will cause erosion? If the river is lined with pebbles 1 cm in diameter, what velocity is tolerable during the spring floods?

8.40. A suggested correlation for the minimum transport velocity of large particles in large pipes is

$$v = 4\left[\frac{gd(\rho_s - \rho_f)}{\rho_f}\right]^{1/2}$$

Compare this with the theory in the text.

8.41. The circulation pattern of gas in a bubble rising through a fluidized bed depends on the ratio v_b/j_{f0}. What is this ratio for lead shot, $d = 0.01$ in., fluidized by atmospheric air if $\mathcal{V}_b = 100$ in.³?

8.42. How do apparent mass effects alter the theory of entropy generation which was presented in Chap. 3? Show how Eq. (3.87) can be modified to account for the kinetic energy which is associated with the three-dimensional flow around particles which are moving relative to a fluid. This kinetic energy is not dissipated and can be recovered. What are the correct forms of Eqs. (3.106), (3.107), and (3.108) in this case?

REFERENCES

1. Lapidus, L., and J. C. Elgin: *A. I. Ch. E. J.*, vol. 3, p. 63, 1957.
2. Lapidus, L., J. A. Quinn, and J. C. Elgin: *A. I. Ch. E. J.*, vol. 7, p. 260, 1961.
3. Rowe, P. N.: *Trans. Inst. Chem. Engrs.*, vol. 39, p. 175, 1961.
4. Schiller, L., and A. Neumann: *Z. Ver. Deutsch. Ing.*, vol. 77, p. 318, 1935.

5. Heywood, H.: *Symp. Interaction Fluids Particles*, Inst. Chem. Engrs., London, pp. 1–8, 1962.
6. Richardson, J. F., and W. N. Zaki: *Trans. Inst. Chem. Engrs.*, vol. 32, pp. 35–53, 1954.
7. Wallis, G. B.: *Symp. Interaction Fluids Particles*, Inst. Chem. Engrs., London, pp. 9–16, 1962.
8. Rowe, P. N., and G. A. Henwood: *Trans. Inst. Chem. Engrs.*, vol. 39, pp. 43–54, 1961.
9. Wen, C. Y., and Y. H. Yu: *Chem. Eng. Progr. Symp.*, ser. 62, vol. 62, pp. 100–111, 1966.
10. Pinchbeck, P. H., and F. Popper: *Chem. Eng. Sci.*, vol. 1, p. 57, 1956.
11. van Heerden, D., A. P. P. Nobel, and D. W. van Krevelen: *Chem. Eng. Sci.*, vol. 1, p. 63, 1951.
12. Carman, P. C.: "Flow of Gases through Porous Media," Butterworth & Co. (Publishers), Ltd., London, 1956.
13. Carman, P. C.: Fluid Flow through Granular Beds, *Trans. Inst. Chem. Engrs.*, vol. 15, pp. 150–166, 1937.
14. Blake, F. E.: *Trans. Am. Inst. Chem. Engrs.*, vol. 14, p. 415, 1922.
15. Leva, M.: "Fluidization," pp. 42–77, McGraw-Hill Book Company, 1959.
16. Simpson, H. C., and B. W. Rodger: *Chem. Eng. Sci.*, vol. 16, p. 179, 1962.
17. Ergun, S.: *Chem. Eng. Progr.*, vol. 48, p. 93, 1952.
18. Slis, P. L., Th. H. Willemse, and H. Kramers: *Appl. Sci. Res.*, vol. A8, p. 209, 1959.
19. Shannon, P. T., and E. M. Tory: *Ind. Eng. Chem.*, vol. 57, pp. 18–25, 1965.
20. Montcrieff, A. G.: *Trans. Instn. Mining Metal*, vol. 73, part 10, pp. 729–759, 1963–1964.
21. Kynch, G. J.: *Trans. Faraday Soc.*, vol. 48, pp. 166–176, 1952.
22. Egolf, C. B., and W. L. McCabe: *Trans. Am. Inst. Chem. Engrs.*, vol. 33, p. 630, 1937.
23. Han, S. T., and W. L. Ingmanson: *TAPPI, 21st Eng. Conf.*, Boston, Mass., vol. 50, no. 4, pp. 176–180, April, 1967.
24. Davis, C. N.: *Proc. Inst. Mech. Engrs.*, vol. B1, p. 185, 1952.
25. Terzaghi, K.: "Theoretical Soil Mechanics," chap. 13, John Wiley & Sons, Inc., New York, 1943.
26. Wallis, G. B.: One-dimensional Waves in Two-component Flow, UKAEA Rept. AEEW-R162, 1962.
27. Smith, J. L.: Massachusetts Institute of Technology, 1963, unpublished work.
28. Murray, J.: *J. Fluid Mech.*, vol. 21, pp. 465–493, 1965.
29. Jackson, R.: *Trans. Inst. Chem. Engrs.*, vol. 41, pp. 13–21, 1963.
30. Wilhelm, R. H., and M. Kwauk: *Chem. Eng. Progr.*, vol. 44, p. 201, 1948.
31. Kliegel, J. R.: *Intern. Symp. Combust.*, 9th, pp. 811–826, Academic Press, Inc., New York, 1963.
32. Hoglund, R. F.: Recent Advances in Gas-particle Nozzle Flows, *ARS J.*, pp. 662–672, May, 1962.
33. Soo, S. L.: *A. I. Ch. E. J.*, vol. 7, no. 3, pp. 384–391, 1961.
34. Altman, D., and J. M. Carter: "Combustion Processes," sec. B, edited by B. Lewis, R. N. Pease, and H. S. Taylor, Princeton University Press, Princeton, N.J., 1952.
35. Kliegel, J. R.: IAS paper no. 60–65, 1960.
36. Rannie, W. D.: Detonation and Two-phase Flow, in "Progress in Astronautics and Rocketry," vol. 6, S. S. Penner and F. A. Williams (eds.), Academic Press, Inc., New York, 1962.

37. Marble, F. E.: *AIAA J.*, vol. 1, no. 12, pp. 2793–2801, 1963.
38. Carrier, G. F.: *J. Fluid Mech.*, vol. 4, pp. 376–382, 1958.
39. Kriebel, A. R.: *ASME Trans. J. Basic Eng.*, vol. 86, no. 4, pp. 655–665, December, 1964.
40. Rudinger, G.: *Phys. Fluids*, vol. 7, pp. 658–663, 1964.
41. Kliegel, J. T., and G. R. Nickerson: Detonation and Two-phase Flow, "Progress in Astronautics and Rocketry," vol. 6, S. S. Penner and F. A. Williams (eds.), Academic Press, Inc., New York, 1962.
42. Nickerson, G. R., and J. R. Kliegel: TRW Rept. 6120-8345-MU000, May, 1962.
43. Boyer, M. H., and R. Grandey: Theoretical Treatment of Detonation Behavior of Composite Propellants, Detonation and Two-phase Flow, in "Progress in Astronautics and Rocketry," vol. 6, S. S. Penner and F. A. Williams (eds.), Academic Press, Inc., New York, pp. 75–98, 1962.
Also Williams, F. A.: Detonations in Dilute Sprays, *ibid.*, pp. 99–114.
Morgenthaler, J. H.: Analysis of Two-phase Flow in Supersonic Exhausts, *Ibid.*, pp. 145–172.
Marble, F. E.: *Phys. Fluids*, vol. 7, no. 8, pp. 1270–1282, 1964.
Bailey, W. S., E. N. Nilson, R. A. Serra, and T. F. Zupnik: *Am. Rocket Soc. J.*, vol. 31, no. 6, pp. 793–798, June, 1961.
Rudinger, G.: *AIAA J.*, vol. 3, pp. 1217–1222, 1965.
Rudinger, G., and A. Chang: *Phys. Fluids*, vol. 7, pp. 1747–1754, 1964.
Gilbert, M., J. Allport, and R. Dunlap: *Am. Rocket Soc. J.*, pp. 1929–1930, December, 1962.
44. Prandtl, L.: "Essentials of Fluid Dynamics," p. 342, Blackie & Son, Ltd., Glasgow, 1952.
45. Basset, A. B.: "Hydrodynamics," p. 270, Dover Publications, Inc., New York, 1961.
46. Einstein, A.: *Ann. Phys.*, vol. 4, p. 289, 1906.
47. Thomas, D. G.: *A. I. Ch. E. J.*, vol. 6, no. 4, pp. 631–639, 1960.
48. Thomas, D. G.: *A. I. Ch. E. J.*, vol. 8, pp. 266–278, 1962.
49. Happel, J.: *J. Appl. Phys.*, vol. 28, pp. 1288–1292, 1957.
50. Eirich, F. R.: "Rheology," 3 vols., Academic Press, Inc., New York, 1956.
51. Thomas, D. G.: *A. I. Ch. E. J.*, vol. 7, no. 3, pp. 431–437, 1961.
52. Zuber, N.: *Chem. Eng. Sci.*, vol. 19, pp. 897–917, 1964.
53. Rutgers, R.: *Rheol. Acta*, vol. 2, no. 4, pp. 305–348, 1962.
54. Rutgers, R.: *Rheol. Acta*, vol. 2, no. 3, pp. 202–210, 1962.
55. Bingham, E. C.: "Fluidity and Plasticity," McGraw-Hill Book Company, New York, 1922.
56. Casson, N.: "Rheology of Disperse Systems," C. C. Mill (ed.), Pergamon Press, New York, 1959.
57. Ree, T., and H. Eyring: chap. 3 in Ref. 50.
58. Charm, S. E., G. S. Kurland, and S. L. Brown: *ASME Biomed. Fluid Mech. Symp.*, Denver, Colo., pp. 89–93, 1966.
59. Huff, W. R., J. H. Holden, and J. A. Phillips: U.S. Bureau of Mines Rept. RI6706, 1965.
60. Thomas, D. G.: Paper no. 64, *Progr. Intern. Res. Thermodyn. Transport Properties*, ASME, 1962.
61. Daily, J. W., and G. Bugliarello: *TAPPI*, vol. 44, pp. 497–512, 1961.
62. Bugliarello, G., and J. W. Daily: *TAPPI*, vol. 44, pp. 881–893, 1961.
63. Mih, W., and J. Parker: *TAPPI, 21st Eng. Conf.*, vol. 50, no. 5, p. 237–246, Boston, Mass., May, 1967.

64. Savins, J. G.: *Soc. Pet. Eng. J.*, vol. 4, no. 3, pp. 203–214, 1964.
65. Thomas, D. G.: Transport Characteristics of Suspensions, part VI, *A. I. Ch. E. J.*, vol. 8, pp. 373–378, 1962.
66. Zenz, F. A.: *Ind. Eng. Chem.*, vol. 41, pp. 2801–2806, 1949.
67. Chow, V. T.: "Open Channel Hydraulics," pp. 164–179, McGraw-Hill Book Company, New York, 1959.
68. Kennedy, J. F.: *J. Boston Soc. Civil Engrs.*, vol. 52, pp. 247–266, 1965.
69. Segre, J., and A. Silberberg: *J. Fluid Mech.*, vol. 14, pp. 136–157, 1962.
70. Ormiston, R. M., F. R. G. Mitchell, and J. F. Davidson: *Trans. Inst. Chem. Engrs.*, vol. 43, pp. 209–216, 1965.
71. Thomas, D. G.: *Science*, vol. 144, pp. 534–536, 1964.
72. Rowe, P. N.: *Chem. Eng. Progr. Symp.*, vol. 58, p. 42, 1962.
73. Rowe, P. N., and B. A. Partridge: UKAEA Rept. AERE-R4660, 1964.
74. Harrison, D., and L. S. Leung: *Trans. Inst. Chem. Engrs.*, vol. 40, p. 146, 1962.
75. Davidson, J. F., R. C. Paul, M. J. S. Smith, and H. A. Duxbury: *Trans. Inst. Chem. Engrs.*, vol. 37, p. 323, 1959.
76. Murray, J. D.: Harvard University, Cambridge, Mass., NSF Grant GP-2226, 1963.
77. Jackson, R.: *Trans. Inst. Chem. Engrs.*, vol. 41, pp. 22–28, 1963.
78. Rowe, P. N., B. A. Partridge, and E. Lyall: UKAEA Rept. AERE-R4543, 1964.
79. Davies, R. M., and G. I. Taylor: *Proc. Roy. Soc.* (London), vol. 200, p. 375, 1950.
80. Davidson, J. F., and D. Harrison: "Fluidised Particles," pp. 33 and 37, Cambridge University Press, London, 1963.
81. Harrison, D., J. R. Davidson, and J. W. DeKock: *Trans. Inst. Chem. Engrs.*, vol. 39, p. 202, 1961.
82. Pigford, R. L.: Massachusetts Institute of Technology, unpublished work, 1959.
83. Roscoe, R.: *Brit. J. Appl. Phys.*, vol. 3, p. 267, 1952.
84. Jaeger, C.: "Engineering Fluid Mechanics," St Martin's Press, Inc., New York, 1957.
85. Frantz, J. F.: *Chem. Eng. Proc. Symp.*, ser. 62, vol. 62, p. 29, 1966.
86. Brinkman, H. C.: *J. Chem. Phys.*, vol. 20, p. 571, 1952.
87. Taub, P. A.: *J. Fluid Mech.*, vol. 27, pp. 561–580, 1967.

9
Bubbly Flow

9.1 INTRODUCTION

The bubbly flow pattern is characterized by a suspension of discrete bubbles in a continuous liquid. There are numerous regimes of bubbly flow. Void fractions range from the extreme case of a single isolated bubble in a large container to the quasi-continuum flow of a foam, containing less than 1 percent of liquid by volume. Interactions between the forces that are due to surface tension, viscosity, inertia, and buoyancy produce a variety of effects which are quite often evidenced by different bubble shapes and trajectories. The regime in which bubbles are so large that they assume a cylindrical shape and almost fill the duct in which they are flowing is important enough to warrant a separate name, *slug flow*, and will be discussed in the next chapter.

Engineering applications include bubble columns for promoting mass transfer, high-pressure evaporators, flash distillation, fire-fighting foams, pumping of beer, champagne, and ice cream, cryogenics, foam "stripping" of impurities, sewers to handle the efflux from washing machines, instant lather for shaving, underwater breathing, and control of wave action in harbors.

9.2 BUBBLE FORMATION

It is only rarely that bubbly flow is the final stable equilibrium flow regime in a given duct. Gas bubbles suspended in fluids usually tend to agglomerate and lose their identity, and when evaporation or condensation occur the existence of small bubbles is only transitory. For these reasons it is particularly important to understand the ways in which bubbles can be formed. Moreover, the bubble size has an influence on the dynamics of a bubbly mixture and must often be specified in terms of the mechanism of bubble generation. In spite of the wealth of available literature only a few equations for predicting bubble size have received general acceptance. Some of the simplest of these will now be described.

BUBBLE FORMATION AT AN ORIFICE

One of the simplest ways to form a bubble is at a circular orifice facing upward in a stationary fluid. Assuming an approximately spherical bubble of radius R_b attached to an orifice of radius R_0 by a cylindrical neck we have, for the largest bubble which can be in static equilibrium,

$$\tfrac{4}{3}\pi R_b^3 g(\rho_f - \rho_g) = 2\pi R_0 \sigma \tag{9.1}$$

The radius of a bubble which will be formed by blowing through a small orifice at low flow rates is therefore given approximately by

$$R_b \approx \left[\frac{3\sigma R_0}{2g(\rho_f - \rho_g)}\right]^{1/3} \tag{9.2}$$

According to Kutateladze and Styrikovich[1] a more accurate version of Eq. (9.2) which is derived from experimental data is

$$R_b = 1.0\left[\frac{\sigma R_0}{g(\rho_f - \rho_g)}\right]^{1/3} \tag{9.3}$$

Equation (9.3) ceases to be valid when the orifice diameter is comparable with the bubble radius, that is, when

$$R_0 > 0.5\left[\frac{\sigma}{g(\rho_f - \rho_g)}\right]^{1/2} \tag{9.4}$$

When a bubble is formed at a finite rate many other factors are important including all the liquid and gas properties and the details of the orifice design and gas supply. The literature in this field has been reviewed by Jackson,[2] and the general phenomenon of bubble formation is shown to be surprisingly complicated.

When the gas flow rate through the orifice is increased, the bubble size first increases due to the fact that the bubble takes a finite time to break from the orifice after reaching the size given by Eq. (9.3). For a system in which the flow rate through the orifice is carefully maintained constant, the bubble size at departure can be predicted by knowing the time for which the bubble remains attached to the orifice.[3] Davidson and Harrison[4] have calculated this time from the equation of motion of the rising bubble and deduced the following result for the volume of a bubble at detachment in an inviscid liquid:

$$\mathcal{V}_b = 1.138 \frac{Q_g^{6/5}}{g^{3/5}} \tag{9.5}$$

where Q_g is the volumetric gas flow rate through the orifice. This equation is compared with data for a wide variety of orifice sizes in water in Fig. 9.1. The deviations from the theory at lower gas flow rates are due to approaching the quasi-static limit given by Eq. (9.3). A more general solution for bubbles growing as a power of time is considered in Probs. 9.3 to 9.6. Davidson and Schüler[5] have also made similar calculations for viscous liquids and obtained the result

$$\mathcal{V}_b = \left(\frac{4\pi}{3}\right)^{1/4} \left[\frac{15\mu_f Q_g}{2g(\rho_f - \rho_g)}\right]^{3/4} \tag{9.6}$$

When the gas velocity is large, the momentum flux from the orifice becomes significant and the bubble volume is given by solving

$$(36\pi)^{1/3} = \frac{2g(\rho_f - \rho_g)}{5\mu_f Q_g} \mathcal{V}_b^{4/3} + \frac{Q_g \rho_g}{A_0 \mu_f} \mathcal{V}_b^{1/3} \tag{9.7}$$

A_0 is the area of the orifice.

If the gas velocity through the orifice is made sufficiently high, bubbles are no longer formed individually, but the gas leaves the orifice in the form of a jet which eventually breaks into individual bubbles. According to Kutateladze and Styrikovich,[1] the condition for the formation of a gas jet is

$$\frac{v_g \sqrt{\rho_g}}{[g\sigma(\rho_f - \rho_g)]^{1/4}} > 1.25 \left[\frac{\sigma}{g(\rho_f - \rho_g)R_0^2}\right]^{1/12} \tag{9.8}$$

where v_g is the gas velocity through the orifice. Bubbles which are formed in this way have a radius which is about twice the radius of the orifice.

In commercial applications bubbles are not usually formed at a

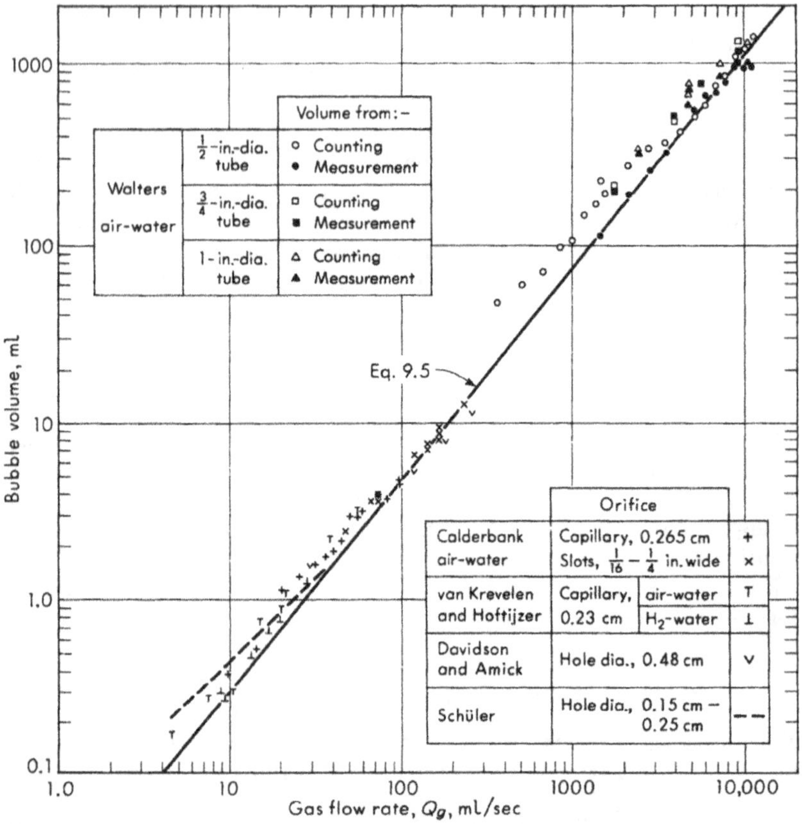

Fig. 9.1 Volume of bubbles produced by blowing from submerged orifices and nozzles in inviscid fluids. (*Davidson and Harrison.*[4])

single orifice but by a group of orifices or a porous plate. In this case the single-orifice theory is useful only as a first approximation.

FORMATION OF BUBBLES BY TAYLOR INSTABILITY

Under some circumstances bubbles can be formed by detachment from a blanket of gas or vapor which is maintained over a porous or heated surface. The continual formation of these bubbles is not identical with the classical case of "Taylor instability"[6] of a fluid below a denser fluid, but the bubble size is scaled by the same dimensionless parameter. Therefore

$$R_b \approx \left[\frac{\sigma}{g(\rho_f - \rho_g)}\right]^{1/2} \tag{9.9}$$

The theory of Taylor instability has been extended to the case of continuous bubble production by Zuber[7] and is particularly important for describing film boiling.

FORMATION OF BUBBLES BY EVAPORATION OR MASS TRANSFER

Bubbles can also be produced by the evaporation of the surrounding liquid and by the release of gases which are dissolved in the liquid (e.g., beer, soda, champagne). These bubbles almost invariably form around nucleation centers which are either impurities suspended in the fluid or pits, scratches, and cavities on the walls of the containing vessel. Of particular interest is the formula derived by Fritz[8] for the equivalent diameter (diameter of a sphere having the same total volume) of a bubble which is just large enough to break away from a horizontal surface. If β is the contact angle in degrees, the equivalent diameter is

$$d = 0.0208\beta \left[\frac{\sigma}{g(\rho_f - \rho_g)}\right]^{1/2} \tag{9.10}$$

This equation is valid only in the quasi-static case and does not adequately describe bubbles which are formed rapidly during boiling, for example.

THE INFLUENCE OF SHEAR STRESSES ON BUBBLE SIZE

In forced convection or mechanically agitated systems the bubble size is determined by shear stresses. These stresses influence both the size of bubbles which are torn away from their point of formation and also the maximum bubble size which is stable in the flow field. One would expect the critical bubble size in both situations to be governed by the balance between surface-tension forces and fluid stresses, i.e., by a suitably chosen Weber number. Little authoritative work seems to have been done in this field but a formula from Hinze[9] may be useful for estimating bubble size, namely,

$$d = 0.725 \left(\frac{\sigma}{\rho_f}\right)^{3/5} \left(\frac{P}{M}\right)^{-2/5} \tag{9.11}$$

The quantity P/M represents the mechanical power dissipated per unit mass.

9.3 ONE-DIMENSIONAL VERTICAL FLOW OF A BUBBLY MIXTURE WITHOUT WALL SHEAR

As long as the wall shear stresses are small and the concentration and velocity profiles are approximately uniform, the techniques described in Chap. 4 can be applied to analyze bubbly flow in a vertical pipe.

The key to a successful understanding is to find an expression for the drift flux j_{gf} in terms of basic quantities. For most practical purposes the empirical equation (4.14) is a good approximation. Thus

$$j_{gf} = v_\infty \alpha(1 - \alpha)^n \tag{9.12}$$

and it remains to express v_∞ and n in terms of the fluid properties and bubble size.

THE RISE VELOCITY OF SINGLE BUBBLES

The dependence of the terminal rise velocity of a single bubble, v_∞, upon fluid properties has been determined experimentally by Peebles and Garber,[10] Haberman and Morton,[11] and by numerous other investigators. The dependence of rise velocity on bubble volume for air bubbles in water is shown in Fig. 9.2. For the smallest bubbles, which are approximately perfect spheres because of the dominant effect of surface tension on their shape, Stokes' solution[12] provides a reasonably accurate description,

$$v_\infty = \frac{1}{18} \frac{d^2 g(\rho_f - \rho_g)}{\mu_f} \tag{9.13}$$

The equation is valid for *solid* spheres and it is assumed that the liquid velocity goes to zero at the bubble surface. For *fluid* spheres contain-

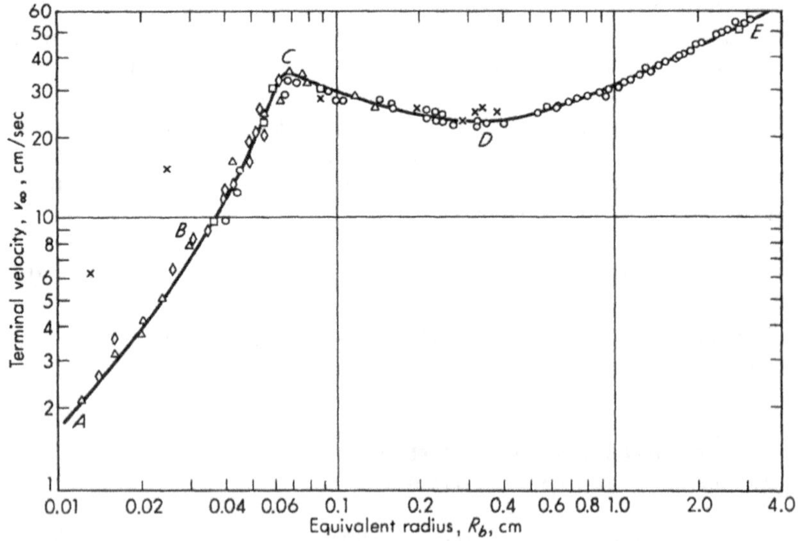

Fig. 9.2 Terminal velocity of air bubbles in filtered or distilled water as function of bubble size. (*Haberman and Morton.*[11])

ing liquid with viscosity μ_g and having a completely nonrigid surface, Hadamard[13] and Rybczynski[14] obtained the equation

$$v_\infty = \frac{d^2 g(\rho_f - \rho_g)}{18\mu_f} \frac{3\mu_g + 3\mu_f}{3\mu_g + 2\mu_f} \tag{9.14}$$

If $\mu_g \ll \mu_f$ this reduces to

$$v_\infty = \frac{d^2 g(\rho_f - \rho_g)}{12\mu_f} \tag{9.15}$$

In the complete absence of impurities, which tend to collect at the bubble surface and give it a certain resistance to shear stress, it is possible to obtain the result predicted by Eq. (9.15). However, in most practical cases some contamination is present and the bubble rise velocity is found to lie between the values which are given by Eqs. (9.13) and (9.15).

At the other extreme, when the bubbles are very large, the effects of surface tension and viscosity are negligible and the rise velocity is given by the equation of Davies and Taylor,[15]

$$v_\infty = \tfrac{2}{3} \sqrt{gR_c} \tag{9.16}$$

where R_c is the radius of curvature in the region of the bubble's nose. The shape of the bubble is approximately a spherical cap with an included angle of about 100° and a relatively flat tail.

Equation (9.16) can be expressed in terms of the volume of gas in the bubble, $\mathcal{V}_b$. Thus

$$v_\infty = 0.79(g\mathcal{V}_b^{1/3})^{1/2} \tag{9.17}$$

Alternatively, we can define an equivalent radius R_b, which is the radius which the bubble would have if it were spherical

$$\mathcal{V}_b = \tfrac{4}{3}\pi R_b^3 \tag{9.18}$$

The bubble rise velocity is then

$$v_\infty = 1.02 \sqrt{gR_b} \tag{9.19}$$

For bubbles of intermediate size, both the effects of liquid inertia, surface tension, viscosity, and cleanliness are important, as well as whether the bubbles rise in straight lines, oscillate, or describe a spiral path. Many correlations exist in the literature of which perhaps the most comprehensive are from Peebles and Garber,[10] who suggest the equations shown in Table 9.1 (for a gas density negligible compared with liquid density). The range of applicability of each equation is determined in

Table 9.1 Terminal velocity of single gas bubbles in liquids (according to Peebles and Garber)[10]

	Terminal velocity	*Range of applicability*
Region 1	$v_\infty = \dfrac{2R_b^2(\rho_f - \rho_g)g}{9\mu_f}$	$\mathrm{Re}_b < 2$
Region 2	$v_\infty = 0.33g^{0.76}\left(\dfrac{\rho_f}{\mu_f}\right)^{0.52} R_b^{1.28}$	$2 < \mathrm{Re}_b < 4.02G_1^{-0.214}$
Region 3	$v_\infty = 1.35\left(\dfrac{\sigma}{\rho_f R_b}\right)^{0.50}$	$4.02G_1^{-0.214} < \mathrm{Re}_b < 3.10G_1^{-0.25}$ or $16.32G_1^{0.144} < G_2 < 5.75$
Region 4	$v_\infty = 1.18\left(\dfrac{g\sigma}{\rho_f}\right)^{0.25}$	$3.10G_1^{-0.25} < \mathrm{Re}_b$ $5.75 < G_2$

terms of the following dimensionless groups:

$$\mathrm{Re}_b = \frac{2\rho_f v_\infty R_b}{\mu_f} \tag{9.20}$$

$$G_1 = \frac{g\mu_f^4}{\rho_f \sigma^3} \tag{9.21}$$

$$G_2 = \frac{g R_b^4 v_\infty^4 \rho_f^3}{\sigma^3} \tag{9.22}$$

It is noteworthy that in region 4 the bubble rise velocity is independent of size. Harmathy[16] suggests that a better value for the constant (1.18) is 1.53 in this region. The upper limit of region 4 is reached when the rise velocity is comparable with the value given by Eq. (9.19), i.e., for

$$R_b \geq 2\left(\frac{\sigma}{g\rho_f}\right)^{1/2} \tag{9.23}$$

which defines a further region 5 in which Eq. (9.19) is valid.

A more complete representation of Table 9.1 is given by Wallis.[51]

Example 9.1 Calculate the rise velocity of air bubbles of equivalent radii 0.2, 0.5, and 2 cm in water if $\sigma = 70$ dynes/cm, $\rho_f = 1$ g/cm³, $\mu_f = 0.01$ poise, and $g = 981$ cm/sec².

Solution From Eq. (9.21) and the given property values

$$G_1 = \frac{(981)(0.01)^4}{(1)(70)^3} = 2.86 \times 10^{-11}$$

Therefore $4.02G_1^{-0.214} = 715$ and $3.1G_1^{-1/4} = 1340$.

From Eq. (9.23), moreover,

$$2\left(\frac{\sigma}{g\rho_f}\right)^{1/2} = 0.51 \text{ cm}$$

Rather than iterating to determine the various regions in Table 9.1, we use Fig. 9.2 to obtain a first approximation to the bubble velocity and hence the Reynolds number given by Eq. (9.20). The results are

R_b, cm	0.2	0.5	2
v_b, cm/sec	25	25	40
Re_b	1000	2500	16,000
Region	3	4	5

The predicted bubble velocities, using Table 9.1 and Eq. (9.19), are then

$$R_b = 0.2 \text{ cm} \qquad v_\infty = 1.35\left(\frac{70}{0.2}\right)^{1/2} = 25.2 \text{ cm/sec}$$

$$R_b = 0.5 \text{ cm} \qquad v_\infty = \left.\begin{matrix} 1.18 \\ \text{or} \\ 1.53 \end{matrix}\right\} [(981)(70)]^{1/4} \quad \begin{matrix} 19.1 \text{ cm/sec} \\ \\ 24.8 \text{ cm/sec} \end{matrix}$$

$$R_b = 2 \text{ cm} \qquad v_\infty = 1.00[(981)(2)]^{1/2} = 44.3 \text{ cm/sec}$$

Comparison with Fig. 9.2 indicates agreement with experiment and favors the value of the constant given by Harmathy in region 4.

THE INFLUENCE OF CONTAINING WALLS

When a bubble rises in a finite vessel, its velocity is generally lower than the value predicted by Table 9.1. In a tube of diameter D, the ratio of the bubble velocity v_b to the velocity in an infinite medium can be expressed as a function of d/D, where $d = 2R_b$.

Collins[17] studied large inviscid bubbles corresponding to region 5. A good approximation to his results is given by

$$\frac{d}{D} < 0.125 \qquad \frac{v_b}{v_\infty} = 1 \tag{9.24}$$

$$0.125 < \frac{d}{D} < 0.6 \qquad \frac{v_b}{v_\infty} = 1.13e^{-d/D} \tag{9.25}$$

$$0.6 < \frac{d}{D} \qquad \frac{v_b}{v_\infty} = 0.496\left(\frac{d}{D}\right)^{-1/2} \tag{9.26}$$

Equation (9.26) is equivalent to the equation governing the velocity of rise of slug flow bubbles in an inviscid fluid [Eq. (10.5)].

Similar corrections can be made for bubbles in viscous fluids. If the bubbles behave as solid spheres, Ladenburg[18] derived the result

$$\frac{v_b}{v_\infty} = \left(1 + \frac{2.4d}{D}\right)^{-1} \tag{9.27}$$

whereas for fluid spheres with $\mu_g \ll \mu_f$, Edgar[19] obtained

$$\frac{v_b}{v_\infty} = \left(1 + \frac{1.6d}{D}\right)^{-1} \tag{9.28}$$

If d/D exceeds about 0.6, the bubbles behave as slug flow bubbles and obey the equation

$$0.6 < \frac{d}{D} \qquad \frac{v_b}{v_\infty} = 0.12 \left(\frac{d}{D}\right)^{-2} \tag{9.29}$$

A reasonable fit to Eq. (9.28) which is tangential to Eq. (9.29) at $d/D = 0.6$ is

$$\frac{v_b}{v_\infty} = 1 - \frac{d/D}{0.9} \tag{9.30}$$

and this may be used to estimate bubble velocities for $d/D < 0.6$.

INFLUENCE OF VIBRATIONS

If a bubble is placed in a vibrating vertical column of liquid, it experiences a downward force which opposes gravity. Under suitable circumstances it can be made to oscillate stably about a stationary mean position or even move downward.[20,21,50]

THE INFLUENCE OF VOID FRACTION

The influence of void fraction is conveniently represented by Eq. (9.12). It is necessary merely to obtain suitable values for the index n.

Wallis[22] made the simple assumption that the relative velocity varied linearly with concentration and recommended a value of n equal to 2. Using the value of bubble rise velocity given by Peebles and Garber's region 4 (Table 9.1) led to the result

$$j_{gf} = 1.18\alpha(1 - \alpha)^2 \rho_f^{-1/2}[\sigma g(\rho_f - \rho_g)]^{1/4} \tag{9.31}$$

If Harmathy's recommendation is followed the value of the constant in Eq. (9.31) should be changed to give

$$j_{gf} = 1.53\alpha(1 - \alpha)^2 \rho_f^{-1/2}[\sigma g(\rho_f - \rho_g)]^{1/4} \tag{9.32}$$

A comparison between the predictions of Eqs. (9.31) and (9.32) and experimental results of Shulman and Molstad[23] for countercurrent flow of air and water is shown in Fig. 9.3. Again Harmathy's value of the constant is superior to Peebles and Garber's and is recommended for use with gas-liquid systems. For liquid-liquid systems Eq. (9.31) is superior.

The method of plotting is the one indicated in Fig. 4.1 and enables steady-state operating values to be indicated by points, whereas the observed flooding points are indicated by tangents. The departure of the points from the theoretical bubbly flow curve at approximately $\alpha = 0.3$ is due to flooding and a consequent change in flow pattern. The new flow pattern contains very large bubbles and the total gas-liquid

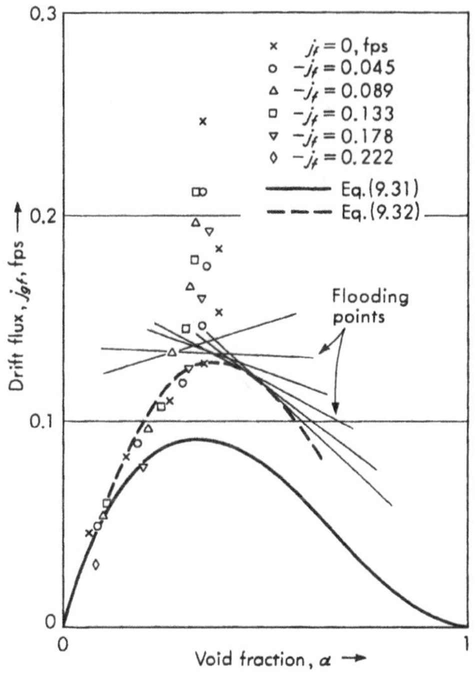

Fig. 9.3 Results of Shulman and Molstad[23] for an air-water system in vertical bubbly flow plotted according to the methods of Fig. 4.1.

interfacial area is markedly reduced above the flooding point. Thus Shulman and Molstad[23] found a sudden reduction in mass-transfer coefficients at the flooding points which are shown in the figure.

In many applications a gas (or light fluid) is bubbled through a stagnant liquid. In this case Eqs. (9.31) and (4.2) lead to the result

$$j_{g0} = 1.18\alpha(1 - \alpha)\rho_f^{-1/2}[\sigma g(\rho_f - \rho_g)]^{1/4} \tag{9.33}$$

Comparison between this equation and the results of Kutateladze[24] bubbling water through mercury are shown in Fig. 9.4. Flooding and a change in flow regime occur at $\alpha = 0.4$ near the maximum in the curve.

There are indications that the value of n is not always exactly 2, although at low values of α in the main region of application of bubbly flow theory this variation is not particularly significant. Gaylor, Roberts, and Pratt[25] working with liquid-liquid systems obtained a value $n = 2$ in the region 1 given in Table 9.1. Miles, Shedlovsky, and Ross[26] working with stable foams at high values of α obtained results in the range $n = 1.6$ to 1.9. Zuber and Hench[27] present further analytical and experimental results and recommend that the following values of n are

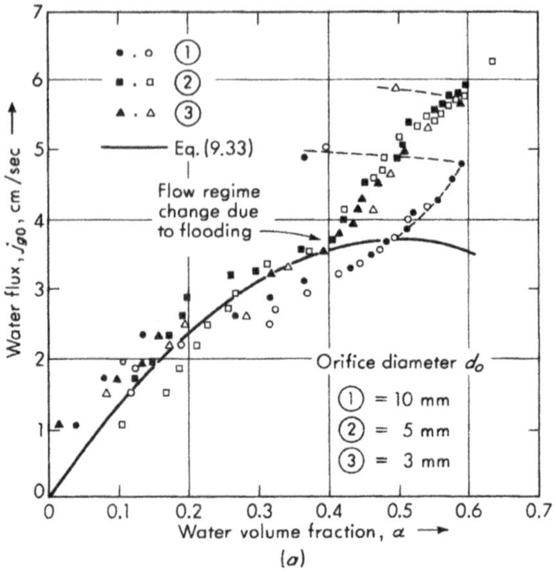

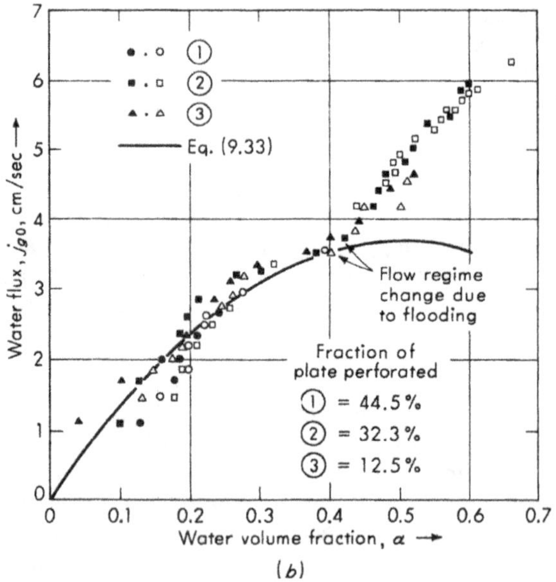

Fig. 9.4 Results of Kutateladze and Moskvicheva[24] bubbling water through mercury from a variety of orifices. *a* Fraction of plate perforated = 33 percent, h_0 = 55 mm. *b* Orifice diameter d_0 = 5 mm, initial height h_0 = 155 cm.

applicable in the various regimes defined in Table 9.1:

Region 1 $n = 2$
Region 2 $n = 1.75$
Region 4 $n = 1.5$

When the bubble size is larger than the value given by Eq. (9.23), three-dimensional effects become important and there is significant entrainment of bubbles in each other's wakes. The result of this "streaming" or "channeling" is an increase in the relative velocity with an increase in the number of bubbles present. The index n in Eq. (9.12) therefore becomes less than unity. The flow pattern becomes agitated and unsteady. This regime, which is called *churn-turbulent* by Zuber, appears to represent a transition region between "ideal bubbly flow," in which bubbles rise uniformly and steadily, and slug flow in which large bubbles fill the tube and flow entirely in each other's wakes. Agglomeration is particularly significant in the churn-turbulent regime since a bubble which is flowing in the wake of another tends to rise faster than its predecessor and eventually coalesces with it. Steady state is not reached until coalescence is complete, and this may require a considerable length of duct.

According to Zuber,[27] in the churn-turbulent regime bubbles rise with the velocity characteristic of region 4 in Table 9.1 relative to the volumetric average velocity of the mixture. The drift velocity of the bubbles is therefore

$$v_{gj} = 1.53 \left[\frac{\sigma g(\rho_f - \rho_g)}{\rho_f^2}\right]^{1/4} \tag{9.34}$$

The corresponding value of the drift flux is

$$j_{gf} = 1.53\alpha \left[\frac{\sigma g(\rho_f - \rho_g)}{\rho_f^2}\right]^{1/4} \tag{9.35}$$

and the value of n is zero.

MODIFICATIONS TO THE SIMPLE THEORY TO TAKE ACCOUNT OF VARIATIONS IN CONCENTRATION AND VELOCITY

Variations in bubble concentration and velocity can be taken into account by the method of Zuber and Findlay[28] which was described in Chap. 4. Making use of the flow distribution parameter C_0, Eq. (4.24), for example, becomes

$$v_g = C_0 j + 1.53 \left[\frac{\sigma g(\rho_f - \rho_g)}{\rho_f^2}\right]^{1/4} \tag{9.36}$$

for the churn-turbulent bubbly regime. For this regime the value of C_0 sufficiently far from the point of bubble injection usually lies between

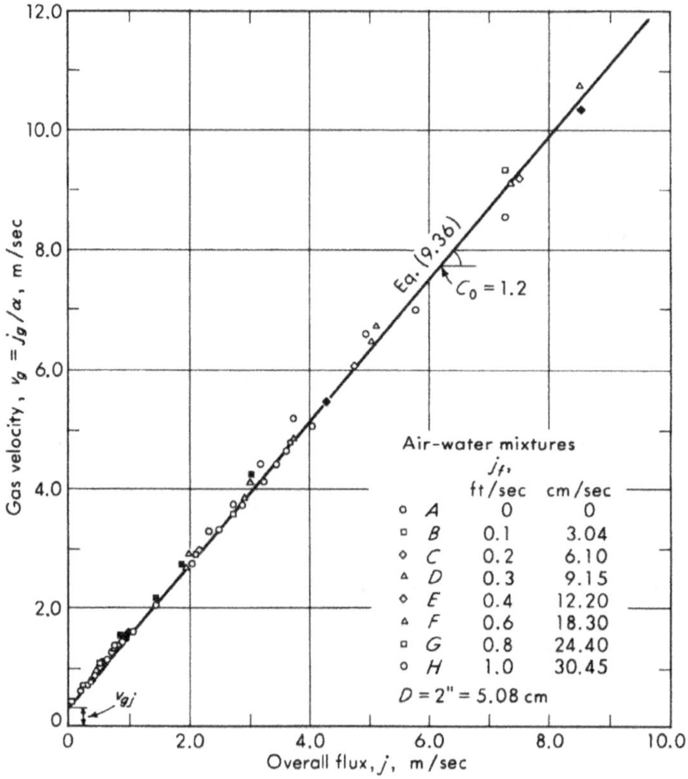

Fig. 9.5 Comparison between Eq. (9.36) and experimental data. (*Zuber and Findlay.*[28])

1.0 and 1.5 with a most probable value of about 1.2. Comparison between this theory and some data is shown in Fig. 9.5. The method of plotting shown in the figure gives C_0 as the slope of the curve and v_{gj} as the intercept on the v_g axis.

9.4 UNSTEADY FLOW

Unsteady flows are conveniently analyzed by combining continuity wave theory with the appropriate expression for the drift flux. The methods are exactly parallel to those which were used in the analysis of sedimentation in the previous chapter.

For example, consider the drainage of liquid from an initially uniform foam in a vertical tube. Because α is close to unity it is convenient to work in terms of the liquid fraction ϵ.

Let the initial value of ϵ be ϵ_0. As the foam starts to drain, waves will begin to propagate from the top and the bottom where the end conditions are specified.

Figure 9.6 shows the situation on the drift flux-concentration graph and on the time-distance plane.

The conditions for a stable shock are satisfied at the bottom of the foam, and an interface is formed which rises with a velocity v_s, the fluid below the interface being pure liquid. No shock can occur at the top of the foam because of the upward curvature of the j_{gf} curve. Continuity waves corresponding to the spectrum of voidages $\epsilon = 0$, to $\epsilon = \epsilon_0$ will therefore propagate down into the foam, each with its own particular velocity. The foam drainage proceeds in two stages as shown in the figure.

Stage 1 Initially a shock between $\epsilon = \epsilon_0$ and $\epsilon = 1$ propagates up from the bottom while continuity waves propagate down through the

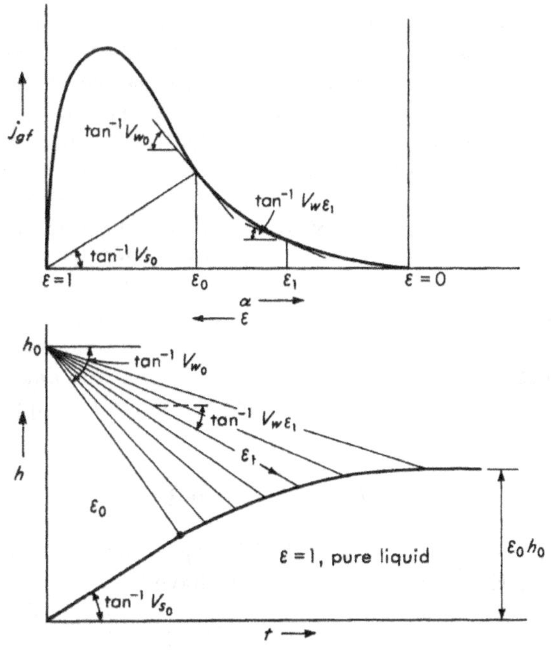

Fig. 9.6 Drainage of an initially uniform foam represented on the drift flux-concentration and time-distance planes.

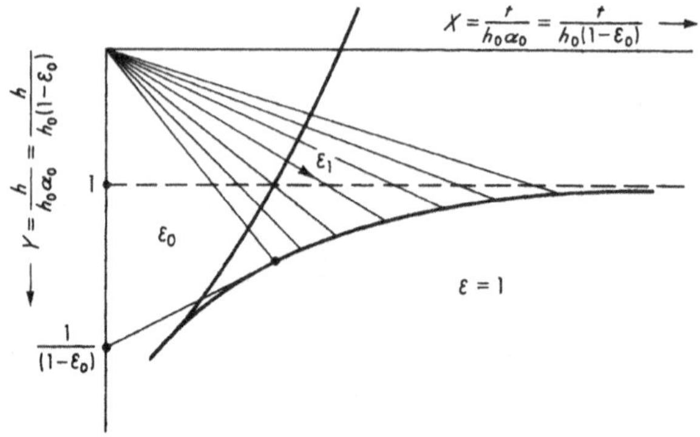

Fig. 9.7 Foam drainage represented on the XY plot. For convenient comparison with Fig. 9.6 the direction of Y is shown downward.

foam from the top. Throughout this stage the rate of drainage of liquid from the foam is given by the shock velocity and is constant.

Stage 2 When the first continuity wave corresponding to a liquid concentration $\epsilon_0 - \delta\epsilon$ reaches the interface, the voidage above the shock begins to increase and the shock strengthens and slows down. At any subsequent time the liquid concentration above the interface is given by the value which has just had sufficient time to propagate down from the top.

This series of events resembles an upside-down version of type III sedimentation and can be represented by the universal XY plot as shown in Fig. 9.7 and described by Eqs. (8.68) and (8.69). The value of α_∞ is essentially unity when drainage is complete.

If the height of the shock at any time during stage 2 is h and the tube has a cross-sectional area A, then the amount of liquid which has been drained is $A(h_0 - h)$ and the amount of liquid remaining in the foam is

$$Ah_0\epsilon_0 - A(h_0 - h) = Ah_0\alpha_0\left(\frac{h}{h_0\alpha_0} - 1\right) = Ah_0\alpha_0(Y - 1) \qquad (9.37)$$

The quantity $Ah_0\alpha_0$ is the initial gas content of the entire foam. Rewriting Eq. (8.69) in terms of ϵ we have

$$(Y - 1) = \frac{n - (n + 1)\epsilon}{n(1 - \epsilon)^2} - 1 = \frac{\epsilon(n - 1) - n\epsilon^2}{n(1 - \epsilon)^2} \qquad (9.38)$$

If ϵ is small, an adequate approximation for the amount of liquid left in the foam when ϵ is the concentration just above the shock is therefore, from Eqs. (9.37) and (8.38),

$$\mathcal{V}_f = \frac{(Ah_0\alpha_0)\epsilon(n-1)}{n} \tag{9.39}$$

This occurs after a time t which is derived from the parameter X, as follows:

$$t = X\alpha_0 h_0 = \frac{\alpha_0 h_0}{n(1-\epsilon)^2\epsilon^{n-1}v_\infty} \approx \frac{\alpha_0 h_0}{n\epsilon^{n-1}v_\infty} \tag{9.40}$$

Stage 2 starts when the value of ϵ in these equations is equal to ϵ_0. Stage 1 is represented by a straight line between the initial conditions and the start of stage 2 on the ht plane.

Example 9.2 Miles, Shedlovsky, and Ross[26] performed experiments in a vertical column of diameter 3.2 cm with a uniform foam which was produced from a 0.24% sodium lauryl sulfate solution. Steady flow experiments for liquid draining through stationary foam gave results which correlated with the equation

$$y = 0.24x^{1.8} \tag{9.41}$$

in which y was the rate of liquid flow through the foam in cubic centimeters per minute and x was the volume of liquid in a total 295 cm³ of foam.

They also studied unsteady flow by suddenly closing the valve in the liquid supply line and plotting the amount of liquid remaining in the foam as a function of time. Predict their results if at the beginning of an experiment 6.9 cm³ of liquid were uniformly dispersed in a total 295 cm³ of foam.

Solution With the present notation Eq. (9.41) becomes

$$-j_{f0} = 13.9\epsilon^{1.8} \tag{9.42}$$

if velocities are measured in centimeters per second. Substituting Eq. (9.42) into Eq. (4.2) with $j_g = 0$ gives

$$j_{gf} = 13.9\epsilon^{1.8}(1-\epsilon) \tag{9.43}$$

Therefore $v_\infty = 13.9$ cm/sec and $n = 1.8$. $A_0\alpha_0 h_0$ is equal to $295 - 6.9 = 288.1$ cm³ and $\alpha_0 h_0$ is $288.1/8.05 = 35.8$ cm. Using these values in Eqs. (9.39) and (9.40) we get

$$\mathcal{V}_f = 128\epsilon \quad \text{cm}^3 \tag{9.44}$$

$$t = 1.432\epsilon^{-0.8} \quad \text{sec} \tag{9.45}$$

Substituting $\epsilon_0 = 6.9/295 = 0.0234$ into Eqs. (9.44) and (9.45) the end of stage 1 is predicted to occur when

$$\mathcal{V}_f = 3.0 \text{ cm}^3,\ t = 28.8 \text{ sec}$$

Stage 1 is represented by a straight line joining the initial condition to this point.

The predictions for both stages in the process are compared with the experimental data in Fig. 9.8.

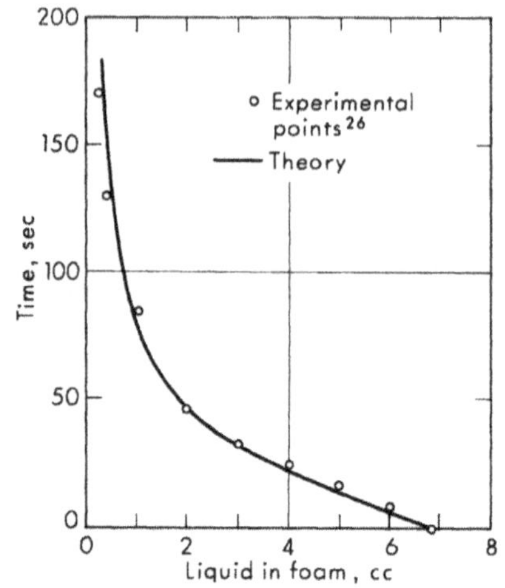

Fig. 9.8 Foam drainage results (for Example 9.2).

9.5 SPECIAL PROBLEMS ASSOCIATED WITH THE BUBBLY FLOW REGIME

BUBBLE SIZE

A considerable problem in practice is the determination of bubble size since the simple equations given in 9.2 do not apply to the complex geometries of most engineering equipment. However, the bubble size is an important variable for determining bubble rise velocity from Table 9.1 and the corresponding value of n in Eq. (9.12) as well as influencing whether operation is in the ideal bubbling or churn-turbulent regime. Figure 9.9, for example, shows a striking example of the way in which completely different results can be obtained for the same values of overall flow rates when air is blown into a column of water from various perforated plates.

AGGLOMERATION AND FRACTURE OF BUBBLES

Unless special precautions are taken, such as the addition of surface-active agents to the fluid, bubbles tend to coalesce as they touch each other. If the bubbles which are introduced into a system are smaller than the maximum stable size for the prevailing shear field, which can be estimated from Eq. (9.11), they will eventually agglomerate. The bubble size, and hence dependent variables such as void fraction, is then a function not only of the way in which bubbles are produced but also of the distance which they travel from the point of injection. For example, if air is bubbled through stagnant tap water the mean void fraction depends on the depth of water in the vessel (Fig. 9.10). Similarly, quite different results may be obtained in the same apparatus if the fluid purity, and hence resistance to agglomeration, is altered (Fig. 9.11). An approach to the agglomeration problem has been made by Radovcich and Moissis.[29]

BUBBLE GROWTH AND COLLAPSE

Bubble size is time dependent when boiling, flashing, or condensation occur, when gas bubbles are released from solution or dissolved, and when bubbles move through significant pressure differences and expand or collapse. Some effects which result from these phenomena will be introduced in the problems at the end of this chapter.

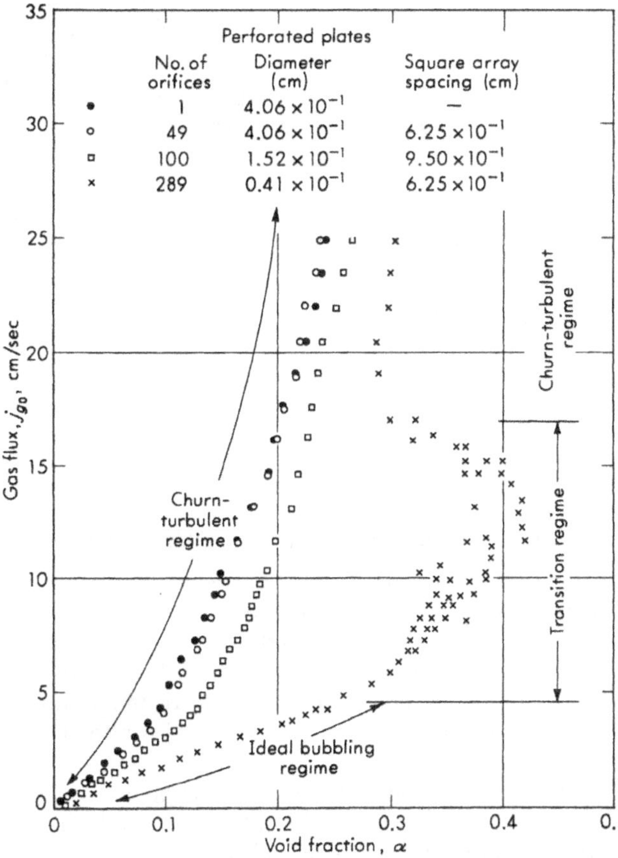

Fig. 9.9 The effect of inlet conditions on void fraction in bubbly flow. (*Zuber and Hench.*[27])

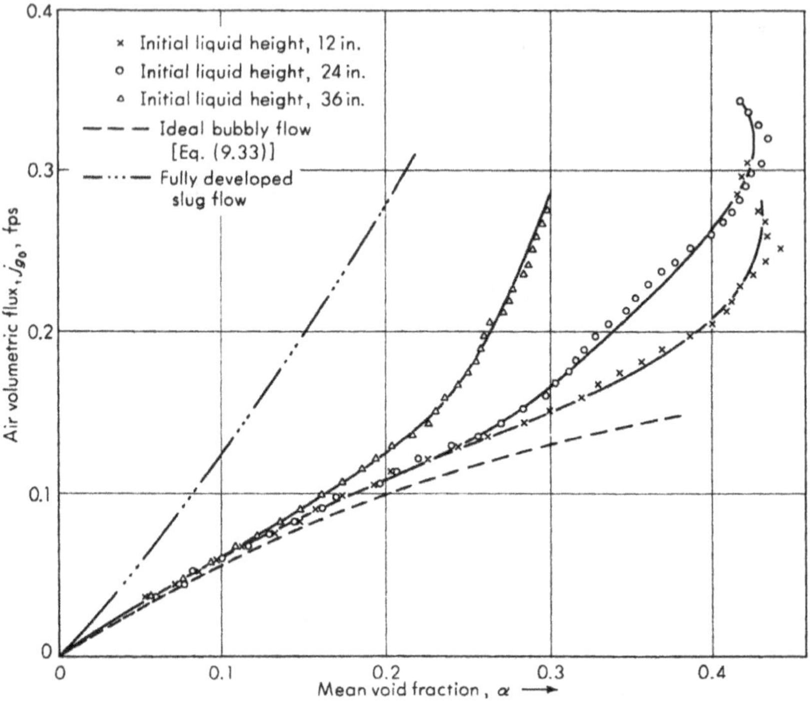

Fig. 9.10 Agglomeration effects on mean void fraction when bubbling through different depths of tap water. (*Wallis.*[22])

9.6 FRICTION AND MOMENTUM FLUX IN BUBBLY FLOW

Although the simple theory gives an adequate description of vertical bubbly flow at low velocities, it is obvious that in general one cannot neglect the effects of wall shear stress and momentum changes. Fortunately it turns out that homogeneous theory is a good approximation in most cases where friction and momentum effects are important, whereas at low velocities, where homogeneous theory is inaccurate, the effects are usually negligible.

The usual method of correlating frictional pressure drop is to define a friction factor C_f which is the ratio of the wall shear stress τ_w to the dynamic pressure of the stream, $Gj/2$.

$$C_f = \frac{2\tau_w}{Gj} \tag{9.46}$$

This friction factor is then plotted versus the Reynolds number

$$\mathrm{Re} = \frac{GD}{\mu} \tag{9.47}$$

Bubbly mixtures in laminar flow are newtonian at low values of α and the effective viscosity is given by Eq. (2.58)

$$\mu = \mu_f(1 + \alpha) \tag{9.48}$$

If the liquid is contaminated the influence of α is enhanced due to the tendency of the bubbles to behave as solid spheres.

Unfortunately, Eq. (9.48) is valid only for values of α below about 0.05. At higher bubble concentrations the mixture rapidly becomes non-newtonian, exhibiting a yield stress,[30,31] a decreasing apparent viscosity with increasing shear rate,[32] and even electrical effects.[33] Foams exhibit considerable rigidity at high void fractions, and the bubbles behave like the atoms in a crystal. Observations of the flow of frothy beer in plastic

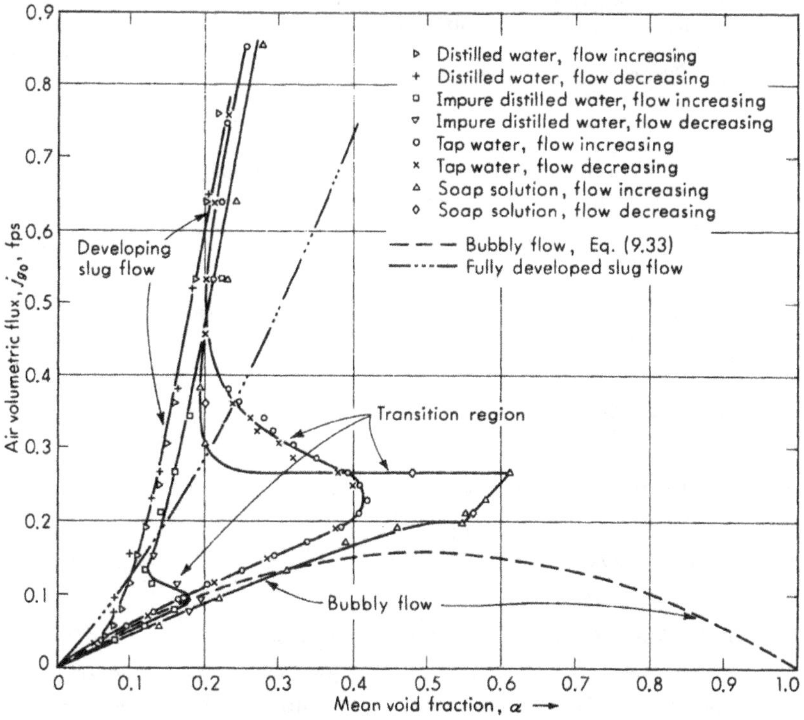

Fig. 9.11 Effect of liquid purity on void fraction in bubbly flow. (*Wallis.*[22])

pipelines at college picnics, for example, show all of the bubbles moving at the same velocity, while shear strain is confined to a thin liquid film on the tube wall.

In turbulent flow it is usually adequate to use the liquid viscosity in the Reynolds number and employ single-phase flow correlations. Up to Reynolds numbers of 10^5 a good approximation is $C_f \approx 0.005$. For example, Meyer[30] found that the data of Rose and Griffith[34] scattered randomly around this value with maximum deviations of 25 percent. Momentum fluxes can also be estimated within about 20 percent accuracy by using homogeneous flow theory. If the component fluxes are not very large compared with the drift flux, a better estimate for vertical flow is given by the solution to Prob. 4.1 as

$$j\left(\frac{G_1}{1 + j_{21}/j_1} + \frac{G_2}{1 - j_{21}/j_2}\right)$$

Example 9.3 Water and air flow upward in a vertical duct. The various values of fluxes and densities are: $\rho_f = 1$ g/cm³, $\rho_g = 0.002$ g/cm², $j_f = 20$ cm/sec, $j_g = 50$ cm/sec. The drift flux is given by

$$j_{gf} = 20\alpha(1 - \alpha)^2 \qquad \text{cm/sec}$$

What correction factor should multiply the homogeneous theory momentum flux to get the correct value?

Solution The value of momentum flux predicted by homogeneous theory is jG. Therefore the required correction factor is

$$\frac{G_f}{G(1 + j_{gf}/j_f)} + \frac{G_g}{G(1 - j_{gf}/j_g)}$$

Using Eq. (4.2) and the given values of j_f and j_g, the value of α is found to be 0.25, and the corresponding value of j_{gf} is 2.8 cm/sec. The values of the mass fluxes are

$$G_f = \rho_f j_f = 20 \text{ g/(cm}^2\text{)(sec)}$$
$$G_g = \rho_g j_g = 0.1 \text{ g/(cm}^2\text{)(sec)}$$

The required correction factor is therefore

$$\frac{20}{(20.1)(1 + 0.14)} + \frac{0.1}{(20.1)(1 - 0.056)} = 0.878$$

9.7 THE VELOCITY OF SOUND IN BUBBLY MIXTURES

The velocity of sound in a bubbly mixture can be calculated from the equations derived in Chap. 6. For small bubbles with a density much lower than the liquid, the homogeneous assumption is approximately valid and the appropriate equation is (6.110). This equation is the same as the equation given by Wood[35] and is

$$c^2 = \frac{1}{[\alpha\rho_g + (1 - \alpha)\rho_f][\alpha/\rho_g c_g^2 + (1 - \alpha)/\rho_f c_f^2]} \tag{9.49}$$

For an air-water mixture at 60°F and an adiabatic process, $c_g = 1117$ fps, $c_f = 4800$ fps, $\rho_g/\rho_f = 0.0012$, and for values of α greater than 10^{-3} a good approximation is

$$c \approx \frac{c_g}{[\alpha(1 - \alpha)\rho_f/\rho_g]^{1/2}} \tag{9.50}$$

The minimum value of c occurs at $\alpha = 0.5$.

Because of transient heat-transfer effects which occur during the passage of a wave, it is not obvious which path one should use to calculate the quantity $\partial p/\partial \rho$ for the gas. For a rapid compression and expansion one would expect the gas to behave adiabatically, in which case

$$c_g^2 = \gamma_g R_g T \tag{9.51}$$

whereas, for slow changes one would expect approximately isothermal behavior, i.e.,

$$c_g^2 = R_g T \tag{9.52}$$

The adiabatic result should be approached at high frequencies and the isothermal prediction at low frequencies.[36,37]

The predictions of Eqs. (9.51) and (9.52) for atmospheric pressure and moderate frequencies are compared with some data in Fig. 9.12.

If the bubbles are small, their virtual compressibility is altered by the effects of surface tension. The sonic velocity is then obtained[38] by multiplying Eq. (9.50) by the factor

$$\left(1 + \frac{2\sigma}{3pR_b + 4\sigma}\right)^{-1/2}$$

R_b is the bubble radius and p the pressure. For very small bubbles this factor approaches a limiting value of 0.82. Some resonance effects occur at high frequencies (above about 1 kc) which are comparable with the natural frequency of pulsation of the bubbles[39] given[50] by the expression $(3\gamma p/\rho_f R_b^2)^{1/2}$.

9.8 THE LIMITS OF THE BUBBLY FLOW REGIME

There is no lower limit to the void fraction at which bubbly flow can occur in cocurrent two-phase flow. The bubbly pattern breaks down in practice for one of two reasons.

1. Coalescence of the bubbles either as they are being formed or as they collide while flowing along the duct.
2. The characteristics of the process of injection or production of the gas or vapor and its interaction with the flow dynamics of the channel.

Both of the above processes are rather whimsical. The speed with which coalescence occurs is particularly sensitive to impurities, even in minute quantities (see Fig. 9.11). Velocity gradients and turbulence tend to increase the rate at which small bubbles collide, thus promoting agglomeration, but also have the effect of tearing apart the bigger bubbles. A very approximate "rule of thumb" which is sometimes used by engineers

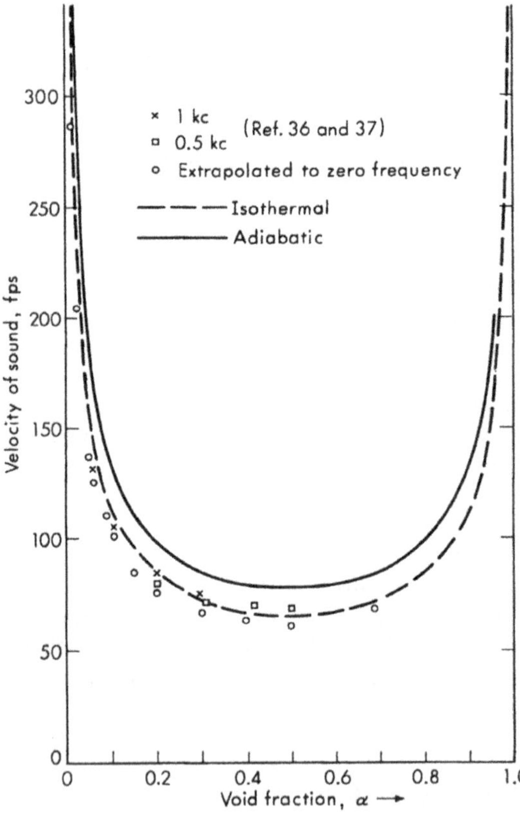

Fig. 9.12 Velocity of sound in a bubbly mixture of air and water under atmospheric conditions.[36,37]

places the transition from bubbly to slug flow at 10 percent void fraction for "pure" liquids although Rose[34] has reported bubbly flow at 60 percent void fraction in tap water. At the other extreme there exist foaming agents which allow bubbly flow to persist up to virtually 100 percent void fraction.

The second mode of breakdown of bubbly flow often cannot be analyzed without reference to the entire system characteristics. For example, in a boiler tube bubbly flow may be converted to slug flow by suppressing a sufficient number of nucleation sites so that the vapor is formed as several large bubbles instead of numerous small ones. Similarly, if gas is introduced into a liquid stream through a porous surface, there exists a transition at the injector surface from the formation of small bubbles to the formation of a gas "blanket" which breaks off to form individual slug-flow bubbles.[22] Only in certain special cases where the limiting process can be identified (it might perhaps be a flooding phenomenon) has rationalization of this transition been possible.

From an academic standpoint it should be possible to determine exactly which bubble size distribution (if any) will be ultimately stable in a very long channel far away from all inlets. However, in the majority of practical cases the bubbly flow pattern never becomes "fully developed" and entrance effects predominate.

Example 9.4 Air and water flow upward in a 10-ft-long, 2-in.-diameter vertical pipe and discharge into an environment at 14.7 psia. Assuming bubbly flow and a temperature of 70°F, calculate the inlet pressure for the following volumetric fluxes measured at atmospheric pressure and temperature,

j_f, fps	0.5	1	10	15	30	32	34	36	38	40
j_g, fps	1	2	2	10	30	32	34	36	38	40

Solution The bubble size being unspecified we have a choice of the regimes in Table 9.1 and the corresponding values of n in Eq. (9.12). Assuming that no particular precautions are taken to produce small bubbles, the most likely regime is probably churn-turbulent and the appropriate equation is (9.35). Substituting values for air and water at 70°F we find

$$v_{gj} = 0.82 \text{ fps} \qquad (a)$$

$$\text{Therefore } j_{gf} = 0.82\alpha \qquad (b)$$

Using Eq. (4.5) and solving for α we have

$$\alpha = \frac{j_g}{j_f + j_g + v_{gj}} \qquad (c)$$

Since the pressure changes down the pipe, the value of j_g will change. Assuming isothermal expansion we obtain

$$j_g = \frac{p_a}{p}(j_g)_{p_a} \tag{d}$$

where p_a is the atmospheric pressure and $(j_g)_{p_a}$ the given gas flux.

A further complication will arise if the flow becomes sonic at the duct exit. This corresponds to choking, and in this case the exit pressure is not necessarily the same as the pressure of the surroundings.

The three components of pressure gradient are as follows.

Acceleration

$$-\left(\frac{dp}{dz}\right)_A = G_g\frac{dv_g}{dz} + G_f\frac{dv_f}{dz} \tag{e}$$

v_g and v_f are found by using Eq. (*c*) in Eqs. (1.22) and (1.23)

$$v_g = j_f + j_g + v_{gj} \tag{f}$$

$$v_f = j_f\frac{j_f + j_g + v_{gj}}{j_f + v_{gj}} \tag{g}$$

Only j_g changes down the duct. Therefore,

$$\frac{dv_g}{dz} = \frac{dj_g}{dz} \tag{h}$$

$$\frac{dv_f}{dz} = \frac{dj_g}{dz}\frac{j_f}{j_f + v_{gj}} \tag{i}$$

Equation (*e*) therefore becomes, with the use of Eqs. (*h*) and (*i*),

$$-\left(\frac{dp}{dz}\right)_A = \left(G_g + G_f\frac{j_f}{j_f + v_{gj}}\right)\frac{dj_g}{dz} \tag{j}$$

If the interval is taken small enough, dj_g/dz can be found by differentiating Eq. (*d*). Thus

$$\frac{dj_g}{dz} = -(j_g)_{p_a}\frac{p_a}{p^2}\frac{dp}{dz} \tag{k}$$

Combining Eqs. (*j*) and (*k*),

$$\left(\frac{dp}{dz}\right)_A = \left(G_g + G_f\frac{j_f}{j_f + v_{gj}}\right)(j_g)_{p_a}\frac{p_a}{p^2}\frac{dp}{dz} \tag{l}$$

Friction

$$-\left(\frac{dp}{dz}\right)_F = \frac{2C_fGj}{D} \tag{m}$$

The appropriate values are $C_f = 0.005$, $D = \frac{1}{6}$ ft,

$$G = \rho_g j_g + \rho_f j_f \tag{n}$$

$$j = j_f + j_g \tag{o}$$

Gravity

$$-\left(\frac{dp}{dz}\right)_G = g[\alpha\rho_g + (1 - \alpha)\rho_f] \tag{p}$$

Table 9.2 The components of pressure drop and inlet and exit pressures predicted by the solution to Example 9.4

j_f, fps	j_g, fps	Δp_{accel}, psi	Δp_{fric}, psi	Δp_{grav}, psi	p_e, psi	p_i, psi
0.5	1	0.0004	0.003	2.546	14.7	17.25
1	2	0.0019	0.016	2.137	14.7	16.85
10	2	0.061	0.937	3.736	14.7	19.43
15	10	0.568	2.78	2.84	14.7	20.89
30	30	6.96	11.32	2.81	14.7	35.79
32	32	8.32	12.61	2.87	14.7	38.49
34	34	9.30	13.92	2.93	15.2	41.36
36	36	9.83	15.29	2.99	16.2	44.31
38	38	10.36	16.72	3.05	17.2	47.33
40	40	11.39	18.21	3.10	17.7	50.40

Combining Eqs. (*l*), (*m*), and (*p*) and solving for dp/dz gives the usual form of result

$$-\frac{dp}{dz} = \frac{2C_fGj/D + g[\alpha\rho_g + (1-\alpha)\rho_f]}{1 - (G_g + G_f j_f/(j_f + v_{gj}))(j_g)_{pa}\, p_a/p^2} \tag{q}$$

The second term in the denominator of Eq. (*q*) has the same significance as the square of the Mach number in Eq. (2.45). If this term exceeds unity, the acceleration pressure drop becomes negative and the flow must be supersonic. Normally this will not be permissible and the exit pressure will adjust until choking is reached with a Mach number of unity. The condition for choking is therefore

$$p_e^2 = p_a(j_g)_{pa}\left(G_g + G_f\frac{j_f}{j_f + v_{gj}}\right) \tag{r}$$

where p_e is the exit pressure.

The method of solution is now as follows. The pipe is divided into intervals and the pressure is calculated by working back from the pipe exit. First Eq. (*r*) is used to calculate the exit pressure which would produce choking. If this is less than p_a, the exit pressure is atmospheric, whereas, if it exceeds p_a, the exit is choked and the exit pressure follows from Eq. (*r*). α is calculated from Eq. (*c*) and $-dp/dz$ from Eq. (*q*). Stepping back up the pipe new values of p, α, j_g, and j are found and the procedure is repeated until the inlet is reached.

This problem was solved on the Dartmouth College computer by Gary Grulich and Andrew Porteous. The results are shown in Table 9.2. Note how the gravitational pressure drop dominates at low flow rates while acceleration and friction take over at the higher flow rates. p_i is the inlet pressure.

9.9 ISOTHERMAL HOMOGENEOUS FLOW OF GAS-LIQUID MIXTURES IN STRAIGHT PIPES

Rather than resorting to numerical methods, as in Example 9.4, it is often convenient to have explicit analytical expressions for the dependent variables. These equations can be derived by using homogeneous theory and treating the bubbly mixture as a pseudo gas with appropriate properties.

Fortunately, since bubbly flow does not exist at high void fractions and the liquid density is usually much higher than the gas density, the liquid makes up almost all of the mass flow rate and the resulting flow is essentially isothermal. Thus, although it is interesting academically to develop exact equations for the general case, the assumption of constant temperature is usually quite adequate.

At this stage we shall neglect the effects of phase change, remembering that this will complicate matters under some circumstances.

In the usual way, the pressure gradient in a straight pipe at an angle θ to the vertical is then

$$-\frac{dp}{dz} = \frac{2C_f jG/D + (G/j)g\cos\theta}{1 - j_g G/p} \tag{9.53}$$

The isothermal Mach number is

$$M^2 = \frac{j_g G}{p} \tag{9.54}$$

Denote conditions at a Mach number of unity by the * superscript. From Eq. (9.54), therefore,

$$p^* = j_g^* G \tag{9.55}$$

The isothermal expansion law for the gas, if there is no area change, is

$$\frac{p}{p^*} = \frac{j_g^*}{j_g} \tag{9.56}$$

Combining Eqs. (9.54) to (9.56) we find

$$p = \frac{p^*}{M} \tag{9.57}$$

$$j_g = Mj_g^* \tag{9.58}$$

Differentiation of Eq. (9.57) yields

$$\frac{1}{p}\frac{dp}{dz} = -\frac{1}{M}\frac{dM}{dz} \tag{9.59}$$

Substituting for all the variables j, p, j_g, dp/dz in Eq. (9.53) in terms of the Mach number we eventually find

$$\frac{dM}{dz}\frac{1 - M^2}{M^2} = \frac{2C_f}{D}\frac{\delta^* M + 1}{\delta^*} + \frac{g\cos\theta}{j_f{}^2\delta^*(1 + M\delta^*)} \tag{9.60}$$

where δ is defined as j_g/j_f (Example 6.7) and

$$\delta^* = \frac{j_g^*}{j_f} \tag{9.61}$$

Equation (9.60) shows that friction and gravity in upflow act to drive the Mach number toward unity since dM/dz is positive when $M < 1$ and negative when $M > 1$. In downflow, however, there is apparently an interaction between friction and gravity which can lead to a smooth transition through $M = 1$ when

$$\frac{2C_f}{D}(1 + \delta^*)^2 = \frac{-g \cos \theta}{j_f^2} \tag{9.62}$$

Since, under these conditions, the right-hand side of Eq. (9.60) is positive for $M > 1$ and negative for $M < 1$, the flow must go from supersonic to subsonic and the Mach number decreases continuously.

Equation (9.60) can be integrated to give a rather long explicit equation for the variation of Mach number with distance. The pressure and gas flow rate everywhere are then found from Eqs. (9.57) and (9.58).

In horizontal flow the integration is simpler and results in the "Fanno line" equation for a bubbly mixture. Arranging in partial fractions we get

$$\frac{2C_f}{D}\,dz = dM\left[\frac{\delta^*}{M^2} - \frac{\delta^{*2}}{M} - \frac{\delta^*(1 - \delta^{*2})}{1 + M\delta^*}\right] \tag{9.63}$$

which integrates to give

$$2C_f\,\frac{z^* - z}{D} = \delta^*\left(\frac{1}{M} - 1\right) + \delta^{*2} \ln M - (1 - \delta^{*2}) \ln \frac{1 + \delta^*}{1 + M\delta^*} \tag{9.64}$$

and is compatible with the results of Huey.[40,41]

9.10 ISOTHERMAL HOMOGENEOUS FLOW WITH AREA CHANGE ONLY

If only the momentum terms are significant in the equation of motion we have

$$-\frac{dp}{dz} = G\,\frac{dj}{dz} \tag{9.65}$$

Now, in most cases in which bubbly flow occurs almost all of the mass flux is due to the liquid and we may write

$$G \approx G_f = \rho_f j(1 - \alpha) = \frac{\rho_f j}{1 + \delta} \tag{9.66}$$

Furthermore, for isothermal flow from stagnation conditions δ_0, p_0,

$$\frac{\delta}{\delta_0} = \frac{p_0}{p} \tag{9.67}$$

Substituting Eq. (9.67) into Eq. (9.66) and using the result in Eq. (9.65), we obtain

$$\frac{dp}{dz}\left(1 + \delta_0 \frac{p_0}{p}\right) = -\rho_f j \frac{dj}{dz} \tag{9.68}$$

which may be integrated from the stagnation conditions to give an explicit expression for the velocity in terms of the pressure ratio, as shown by Tangren, Dodge, and Seifert,[42]

$$\frac{\rho_f j^2}{2p_0} = 1 - \left(\frac{p}{p_0}\right) - \delta_0 \ln \frac{p}{p_0} \tag{9.69}$$

Equations (9.66) and (9.67) can also be used in Eq. (9.54) to show that

$$M^2 = \frac{\rho_f j^2}{p_0 \delta_0}\left(1 + \frac{p}{p_0 \delta_0}\right)^{-2} \tag{9.70}$$

Eliminating j^2 between Eqs. (9.69) and (9.70) gives an equation for the Mach number in terms of the pressure ratio, as follows:

$$M^2 = \frac{2}{\delta_0}\left(1 - \frac{p}{p_0} - \delta_0 \ln \frac{p}{p_0}\right)\left(1 + \frac{p}{p_0 \delta_0}\right)^{-2} \tag{9.71}$$

When the Mach number is unity, $p = p^*$ and the ratio p^*/p_0 can be found from Eq. (9.71). Some results are given in Table 9.3.

Knowing the pressure ratio as a function of Mach number, other useful variables follow from relationships which are easily derived from the condition of isothermal flow, such as

$$M = \frac{j_g}{j_g^*} = \frac{\delta}{\delta^*}\frac{A^*}{A} = \frac{p^*}{p}\frac{A^*}{A} \tag{9.72}$$

Figure 9.13 shows the resulting dependence of Mach number, pressure ratio, velocity ratio, and density ratio as functions of nozzle geometry.

The throat area is found in terms of the upstream conditions from Eq. (9.54) with $M = 1$.

$$1 = \frac{j_g G^*}{p} = \frac{Q_g^* W}{A^{*2} p^*} = \frac{(Q_g)_0 W p_0}{A^{*2} p^{*2}} \tag{9.73}$$

Table 9.3 Critical pressure ratio for bubbly flow in a nozzle as a function of stagnation gas-liquid ratio and void fraction[42]

α_0	0.67	0.5	0.333	0.2	0.091	0.05
δ_0	2	1	0.5	0.25	0.1	0.05
p^*/p_0	0.55	0.52	0.46	0.39	0.30	0.25

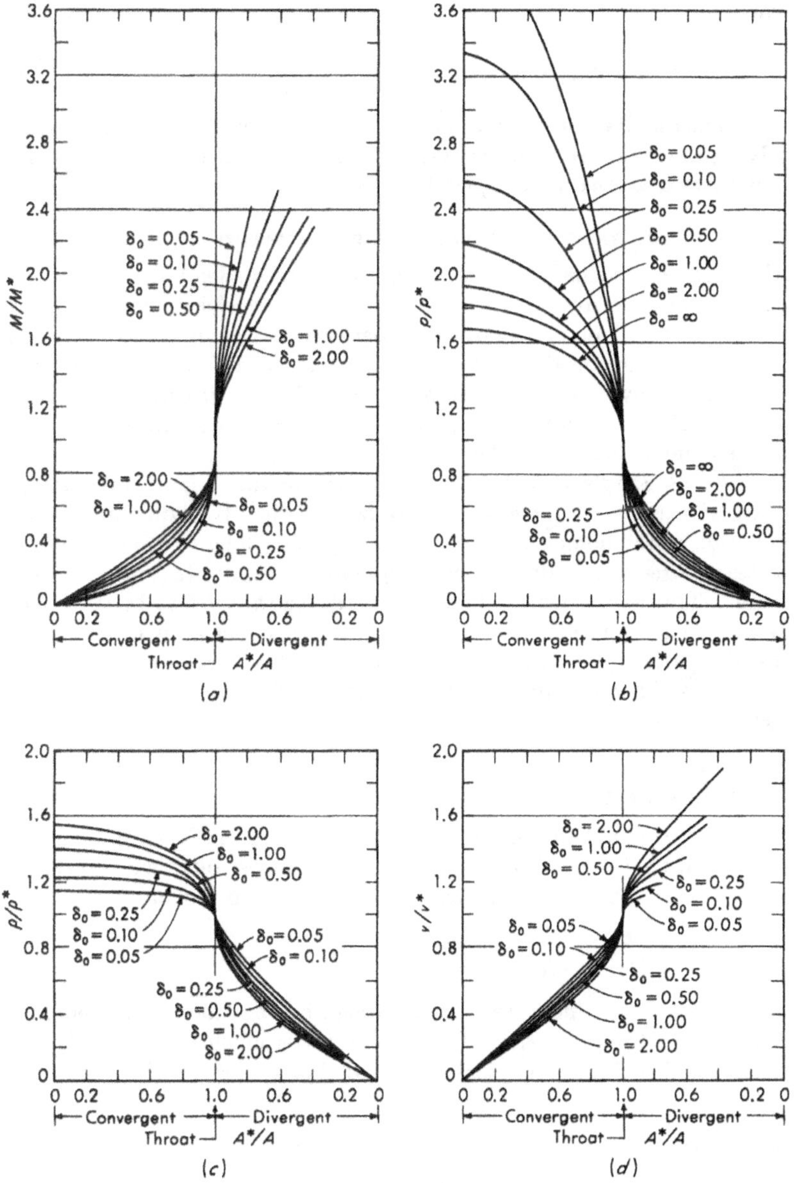

Fig. 9.13 Nozzle characteristics for isothermal frictionless bubbly flow. (*Tangren, Dodge, and Seifert.*[42])

whence

$$A^* = \frac{p_0}{p^*}\left[\frac{(Q_g)_0 W}{p_0}\right]^{1/2} \tag{9.74}$$

The qualitative behavior of bubbly nozzle flows and the methods for solving practical problems are very similar to the analogous single-phase flow cases discussed by Shapiro.[43]

USE OF THE EQUATIONS OF MOTION FOR BOTH COMPONENTS

Although the assumption of homogeneous flow is adequate for most purposes it can be erroneous when a body force or accelerational field causes significant relative velocity or "slip" between the components. We have already seen how the drift velocity model can account for slip in low velocity vertical flow. When acceleration is significant (it can be very many g's in nozzles), it is necessary to consider the separate equations of motion of the components, Eqs. (3.45) and (3.46).

Unless the f's are substantial it is evident that a given pressure gradient will accelerate the gas far more rapidly than the liquid and cause a violation of the homogeneous flow assumption. The f's are composed of at least three parts: drag forces, apparent mass effects, and the Basset force. The way in which these forces interact has yet to be resolved in detail. The apparent mass effects impose an upper limit of about 2 on the slip ratio v_g/v_f, as shown in Example 8.4. For example, practical maximum slip ratios in high-speed nozzle flows measured by Muir and Eichhorn[44] were in the range 1.1 to 1.8.

9.11 SHOCK WAVES

Eddington[45] found that both normal and oblique shock waves in bubbly flow could be described by the isothermal homogeneous theory presented in Sec. 6.6. The normal shock relations are simply, from Eq. (6.175),

$$M_1^2 = \frac{1}{M_2^2} = \frac{(j_g)_1}{(j_g)_2} = \frac{p_2}{p_1} = \frac{\delta_1}{\delta_2} \tag{9.75}$$

Shock-wave thicknesses are governed by the relaxation phenomena discussed in Chap. 8. Equilibration proceeds much more rapidly than in gas-particle flows. Typical measured thicknesses are about 1 in.

PROBLEMS

9.1. An air bubble is formed very slowly in water at 20°C. What is its diameter when it detaches if (*a*) it is formed on upward-facing nonwettable orifices with radii 0.1, 0.5, 1, and 2 mm; (*b*) the orifice is a hole in a wettable surface with contact angle β and β can be 30, 60, 90, or 120°?

9.2. Estimate the equilibrium bubble size in a milk-shake machine which contains one liter of fluid and is run by a $\frac{1}{8}$th-horsepower motor.

9.3. By considering apparent mass effects show that the equation of motion of a spherical bubble growing freely in an inviscid liquid is[4]

$$\mathcal{V}_b(\rho_f - \rho_g)g = \frac{d}{dt}\left[\mathcal{V}_b(\rho_g + \tfrac{1}{2}\rho_f)\frac{dz}{dt}\right]$$

Explain why the fact that the bubble is growing introduces an additional component into the force f_g.

9.4. If the bubble in Prob. 9.3 grows according to the equation $\mathcal{V}_b = At^a$ where A and a are constants, and $\rho_g \ll \rho_f$, show that the distance its center has gone after time t is

$$z = \frac{1}{a+1}gt^2$$

9.5. If the bubble in Prob. 9.4 is grown from a nozzle, which exerts no force on it, and breaks away when $z = R_b$, show that the bubble volume at departure is

$$\mathcal{V}_b = \left[\left(\frac{3}{4\pi}\right)^a\left(\frac{a+1}{g}\right)^{3a}A^6\right]^{1/(6-a)}$$

and that this agrees with Eq. (9.5) when $a = 1$. What happens if $a = 6$?

9.6. When bubbles grow as a result of transient heat or mass transfer, their radii increase in proportion to the square root of the time. A bubble is nucleated in a superheated liquid by passing a current pulse through a hot wire. The bubble grows according to the equation $R_b = 2t^{1/2}$ where R_b is in centimeters and t in seconds. What is the bubble volume when it breaks from the wire? Use Probs 9.4 and 9.5.

9.7. Solve Probs. 9.3 and 9.4 for a bubble growing in a very viscous liquid. Assume that the bubble always moves with its terminal velocity. Show that Eq. (9.6) results if $a = 1$.

9.8. What is the maximum bubble volume which can be blown from a given orifice in a viscous liquid if Eq. (9.7) is valid? Evaluate this volume for air blowing through a $\frac{1}{4}$-in. nozzle into molasses ($\rho_f \approx 1.2$ g/cm^3, $\mu_f = 1000$ poise).

9.9. Solve Example 9.1 for $R_b = 0.005$, 0.05, 0.1, 0.3, and 1.0 cm.

9.10. Solve Example 9.1 and Prob. 9.9 if the fluid is aniline for which $\rho_f = 63.7$ lb/ft^3, $\mu_f = 2.93$ cp, and $\sigma = 41.7$ dynes/cm.

9.11. A bubble is nucleated in a 100-ft-high column of water which is supersaturated with carbon dioxide and open at the top. According to Calderbank and Moo-Young,[49] the mass-transfer coefficient for rising bubbles is independent of diameter so that the rate of gas evolution per unit surface area is constant. If this rate of evolution is 10^{-4} g/(cm^2)(sec) estimate the bubble volume and rise velocity at different points in the column if it starts from the bottom with a diameter of 10 μ. Ignore evaporation into the bubble, but do not neglect hydrostatic pressure changes or surface-tension effects.

9.12. Nicklin[46] measured the drift velocity, v_{gj}, of air bubbles as a function of gas flux through stagnant water. His results are shown in Fig. 9.14. Derive the relationship between j_{gf} and α, the values of v_∞ and n, and explain the existence of the minimum value of v_{gj} shown in the figure.

9.13. Marrucci and Gioia[47] studied cocurrent air-water flow in a vertical pipe of 5.3-cm diam. Their results are shown in Fig. 9.15a. Evaluate n, v_∞, v_{gj}, and j_{gf} as a func-

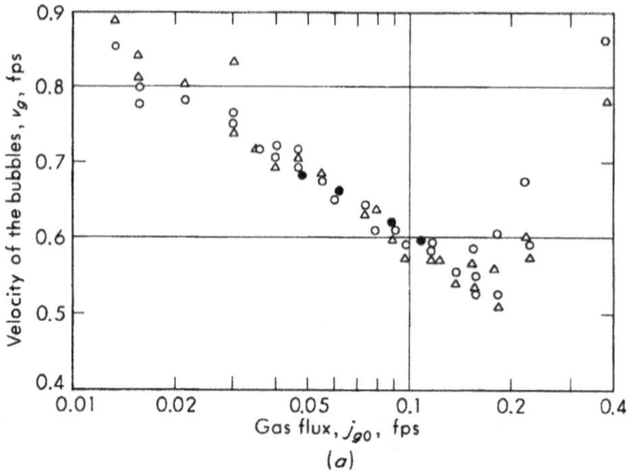

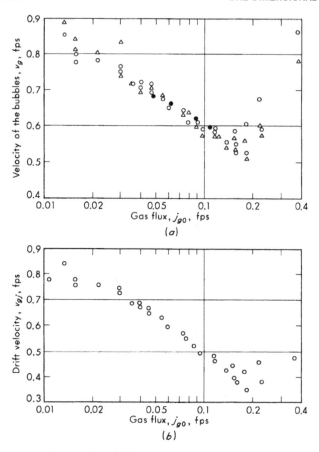

Fig. 9.14 (*a*) Bubble velocity and (*b*) drift velocity for bubbling of gas through stagnant liquid. (*Nicklin.*[46])

tion of α from these data. What regime of bubbly flow do you diagnose? The same authors plotted v_{gj} versus α as shown in Fig. 9.15*b*. What values of v_∞ and n are predicted from this method of plotting? Do the conclusions from the two figures agree?

9.14. Air and benzene (σ = 28.8 dynes/cm, ρ_f = 54.7 lb/ft^3, μ_f = 0.647 cp) flow vertically upward in a 1-in.-diam pipe at flow rates of 1 and 1000 lb/hr, respectively. The bubble diameter is 2 mm at the bottom of the pipe where the pressure is 50 psia and the temperature 15°C. Evaluate the void fraction and pressure as a function of height. How long a pipe is needed to drop the pressure to atmospheric?

9.15. Solve Prob. 9.14 if the flow rates are each increased by a factor of 10.

9.16. A series of air bubbles with R_b = 0.5 in. flow upward in stationary fluid in a vertical pipe of 2 in. diam. What is their velocity, (*a*) in glycerin at 20°C and (*b*) in water at 20°C? What is the gas flow rate in each case if the bubbles are 3 in. apart from nose to nose?

9.17. An air flux of 3 cm/sec is suddenly supplied to the bottom of a 50-cm depth of clear water containing a strong foaming agent in a long vertical tube. The bubbles which are produced all have a volume of 0.002 cm³. Show that Fig. 9.16 describes the qualitative behavior. Identify the various features of the sketches and evaluate the key parameters in detail. Which parts of the process, if any, are likely to be physically impossible or unstable.

9.18. What is the velocity of sound in a hydrogen-water mixture at 1000 psia, 70°F, and with mean density 40 lb/ft³?

9.19. Solve Prob. 9.18 if the pressure is 5 psia and the radius of the bubbles is 1 mm.

9.20. Solve Example 9.4 if flow is horizontal.

9.21. 3 scfm of air are to be bubbled through a tank of 1 ft diam containing 20 gal of water at 20°C. The designer has a choice of the number of nozzles or orifices to use as well as diameters. Show how the following parameters depend on the design of the injection system: (*a*) void fraction, (*b*) bubble residence time in the liquid, (*c*) total interfacial area in the tank, and (*d*) operating limits.

9.22. In a 4-in.-diam countercurrent-flow bubble column, Shulman and Molstad[23] found that flooding occurred for the following mass fluxes of carbon dioxide and water

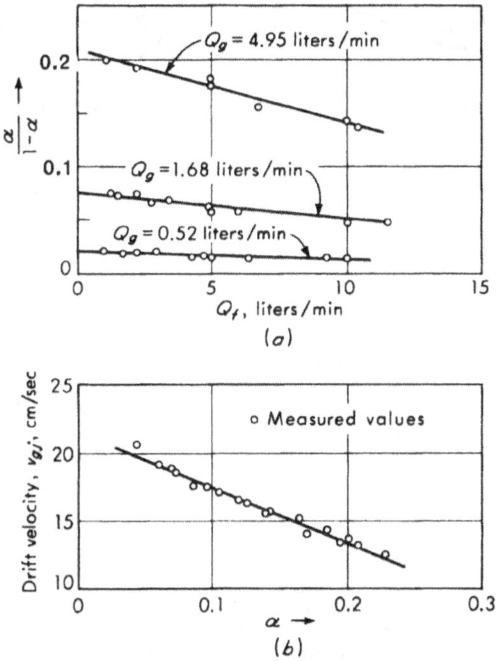

Fig. 9.15 Some results of Marrucci and Gioia for vertical bubbly flow.[47]

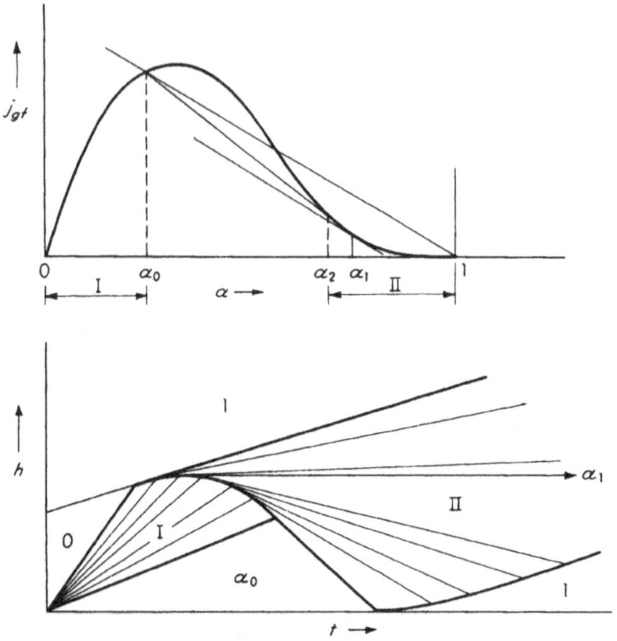

Fig. 9.16 Transient bubbling behavior of the system described in Prob. 9.17.

(in pounds per hour per square foot at 70°F and 15 psi):

$-W_f$	15,000	10,000	5000
W_g	38	42	48

Compare these results with the theoretical values at 15 psia, 70°F.

9.23. Derive Eq. (9.72).

9.24. Show that, in isothermal frictionless homogeneous nozzle flow,

$$-\frac{1}{p}\frac{dp}{dz} = \frac{1}{j}\frac{dj}{dz}\left[M^2\left(1+\frac{1}{\delta}\right)\right] = -\frac{1}{A}\frac{dA}{dz}\left[\frac{M^2}{1-M^2}\left(1+\frac{1}{\delta}\right)\right]$$
$$= \frac{1}{M^2}\frac{dM^2}{dz}\frac{M^2(1+1/\delta)}{2(1+M^2/\delta)}$$

and that these results may be combined with Eq. (9.72) to form a basis for numerical prediction of all the variables in terms of δ^* and M.

9.25. In a choked nozzle of given geometry, with given upstream stagnation pressure, how does the liquid flow rate depend on the value of δ_0? Deduce the relationship beteeen W_f and W_g for the valve described in Prob. 2.21. Design such a valve if it is to control oil flow rates between 1 and 10 tons/hr of oil with density 60 lb/ft^3 at 60°F and 100 psia using compressed air.

9.26. By assuming that the ratio $(1 - \alpha)/C$ is constant through a nozzle, where C is the coefficient of apparent mass, show that the limiting value of the slip ratio in an accelerating bubbly flow is

$$\frac{v_g}{v_f} = \frac{2(1 - \alpha)}{\sqrt{C^2 + 4C(1 - \alpha)} - C}$$

If C has the usual value of $\frac{1}{2}$, show how the velocity ratio depends on the average void fraction.

If the bubbles distort to form ellipsoids with axes in the ratio 1 : 6, the value of C drops to 0.045. What effect does this have on the velocity ratio?

9.27. Show that the velocity ratio across a normal shock wave in a bubbly mixture is given, in terms of the upstream conditions, by

$$\frac{j_2}{j_1} = \frac{M_1^2 + \delta_1}{M_1^2(1 + \delta_1)}$$

9.28. 100 cfm of water and 200 scfm of air at 100 psi and 70°F are supplied to an experimental convergent-divergent nozzle. What is the throat area when the flow is choked? If the nozzle discharges into the atmosphere at 14.7 psia, what values of exit area will cause shocks to form in the nozzle? Plot the pressure variation down the nozzle as a function of area for various values of the exit area. What is the maximum thrust which the nozzle can exert on its mountings?

9.29. Discuss the various forces which act on the bubbles in one-dimensional flow. Under what conditions is the flow governed by a balance between (*a*) gravity and drag, (*b*) inertia and pressure forces, (*c*) pressure and the Basset force, and (*d*) surface tension and buoyancy?

9.30. Show that an error is introduced by assuming ϵ to be small in Eqs. (9.39) and (9.40), which tends to underestimate the duration of stage 1. Show that stage 1 in Example 9.2 actually ends when $t = 30$ sec when an amount 3.87 cm^3 of liquid has drained from the foam.

9.31. A bottle of beer is shaken vigorously and poured into a long straight vertical vessel (a "yard"). If the initial void fraction is 80% and no bubbles form or burst after the first few seconds, describe what happens.

9.32. Solve Prob. 9.31 if the initial void fraction is 30%.

9.33. According to Harrison and Leung,[48] bubbles formed at orifices in fluidized beds obey Eq. (9.5). Estimate the void fraction as a function of the number of nozzles when a bed of glass balls, 0.01 cm diam, is fluidized with air at a volumetric flux of 3 cm/sec, measured at atmospheric pressure and temperature. The bed is 10 ft deep. Assume that all the air flow in excess of that needed to fluidize the bed forms bubbles. State other assumptions made.

REFERENCES

1. Kutateladze, S. S., and M. A. Styrikovich: "Hydraulics of Gas-Liquid Systems," Moscow, Wright Field trans. F-TS-9814/V, 1958.
2. Jackson, R.: *Chem. Eng.*, vol. 42, pp. 107–118, May, 1964.
3. Siemes, W., and J. F. Kaufmann: *Chem. Eng. Sci.*, vol. 5, pp. 127–139, June, 1956.
4. Davidson, J. K., and D. Harrison: "Fluidised Particles," Cambridge University Press, London, 1963.
5. Davidson, J. F., and B. O. G. Schüler: *Trans. Inst. Chem. Engrs.*, vol. 38, pp. 144–154 and 335–342, 1960.

6. Taylor, G. I.: *Proc. Roy. Soc.* (London), vol. A210, p. 192, 1950.
7. Zuber, N.: AEC Rept. U-4439, 1959.
8. Fritz, W.: *Physik. Z.*, vol. 36, p. 623, 1933.
9. Hinze, J. O.: *A. I. Ch. E. J.*, vol. 1, p. 289, 1955.
10. Peebles, F. N., and H. J. Garber: *Chem. Eng. Progr.*, vol. 49, pp. 88–97, 1953.
11. Haberman, W. L., and R. K. Morton: David W. Taylor Model Basin Rept. 802, 1953.
12. Stokes, G. G.: "Mathematical and Physical Papers," vol. 1, Cambridge University Press, London, 1880.
13. Hadamard, J.: *Compt. Rend. Acad. Sci. Paris*, vol. 152, pp. 1735–1738, 1911.
14. Rybczynski, W.: *Bull. Acad. Sci. Cracovie*, vol. A, pp. 40–46, 1911.
15. Davies, R. M., and G. I. Taylor: *Proc. Roy. Soc.* (London), vol. 200, ser. A, pp. 375–390, 1950.
16. Harmathy, T. Z.: *A. I. Ch. E. J.*, vol. 6, p. 281, 1960.
17. Collins, R.: *J. Fluid Mech.*, vol. 28, part 1, pp. 97–112, 1967.
18. Ladenburg, R.: *Ann. Physik*, vol. 23, p. 447, 1907.
19. Edgar, C. B., Jr.: AEC Rept. No. NYO-3114-14 by G. B. Wallis, pp. 19–21, 1966.
20. Jameson, G. J.: *Chem. Eng. Sci.*, vol. 21, pp. 35–48, 1966.
21. Jameson, G. J., and J. F. Davidson: *Chem. Eng. Sci.*, vol. 21, pp. 29–34, 1966.
22. Wallis, G. B.: Paper no. 38, *Intern. Heat Transfer Conf.*, Boulder, Colo., ASME, 1961.
23. Shulman, H. L., and M. C. Molstad: *Ind. Eng. Chem.*, vol. 42, p. 1058, 1950.
24. Kutateladze, S. S., and V. N. Moskvicheva: *Zh. Tech. Fiz.*, vol. 29, no. 9, pp. 1135–1139, 1959.
25. Gaylor, R., N. W. Roberts, and H. R. C. Pratt: *Trans. Inst. Chem. Engrs.*, vol. 31, p. 57, 1953.
26. Miles, G. D., L. Shedlovsky, and J. Ross: *J. Phys. Chem.*, vol. 49, p. 93, 1943.
27. Zuber, H., and J. Hench: Rept. no. 62GL100, General Electric Company, Schenectady, N.Y., 1962.
28. Zuber, N., and J. A. Findlay: *Trans. ASME J. Heat Transfer*, vol. 87, ser. C, p. 453, 1965.
29. Radovcich, N. A., and R. Moissis: Rept. no. 7-7673-22, Department of Mechanical Engineering, Massachusetts Institute of Technology, 1962.
30. Meyer, P. E., and G. B. Wallis: AEC Rept. No. NYO-3114-12 (EURAEC 1530), 1965.
31. Penny, W. G., and M. Blackman: Note 282, Ministry of Home Security, Great Britain, 1943. See also J. Hermans, "Flow Properties of Disperse Systems," North-Holland Publishing Company, Amsterdam, 1953.
32. Sibree, T. O.: *Trans. Faraday Soc.*, vol. 31, p. 325, 1943.
33. Raza, S. H., and S. S. Marsden: *Soc. Pet. Engrs. J.*, pp. 359–368, December, 1967.
34. Rose, S. C., Jr., and P. Griffith: Rept. 5003–30, Massachusetts Institute of Technology, 1964.
35. Wood, A. B.: "A Textbook of Sound," The Macmillan Company, New York, p. 327, 1930.
36. Karplus, H. B.: Rept. C00-248, *Armour Res. Found.*, June, 1958.
37. Gouse, S. W., Jr., and G. A. Brown: E.P.L. Rept. DSR 8040-1, Massachusetts Institute of Technology, April, 1963.
38. Marchal, R.: *Compt. Rend. Acad. Sci.*, vol. 254, pp. 2524–2526, 1962.
39. Silberman, E.: *J. Acoust. Soc. Am.*, vol. 29, no. 8, pp. 925–933, 1957.
40. Huey, C. T., and R. A. A. Bryant: *A. I. Ch. E. J.*, vol. 13, no. 1, pp. 70–76, 1967.
41. Huey, C. T.: *Can. J. Chem. Eng.*, vol. 44, no. 6, pp. 313–321, December, 1966.
42. Tangren, R. F., C. H. Dodge, and H. S. Seifert: *J. Appl. Phys.*, vol. 20, no. 7, pp. 637–645, 1942.

43. Shapiro, A. H.: "Dynamics and Thermodynamics of Compressible Fluid Flow," The Ronald Press Company, New York, 1953.
44. Muir, J. F., and R. Eichhorn: *JSME Semi-intern. Symp.*, Tokyo, pp. 81–92, September, 1967.
45. Eddington, R. B.: AIAA paper 66–87, 1966. Also Rept. 32-1096, Jet Propulsion Laboratories, California Institute of Technology, 1967.
46. Nicklin, D. J.: *Chem. Eng. Sci.*, vol. 17, pp. 693–702, 1962.
47. Marrucci, G., and F. Gioia: *Chim. Ind.*, vol. 45, no. 10, pp. 1205–1211, Milan, Italy, 1963.
48. Harrison, D., and L. S. Leung: *Trans. Inst. Chem. Engrs.*, vol. 39, p. 409, 1961.
49. Calderbank, P. H., and M. B. Moo-Young: *Chem. Eng. Sci.*, vol. 16, pp. 39–54, 1961.
50. Foster, J. M., J. A. Botts, A. R. Barbin, and R. I. Vachon: *Trans. ASME J. Basic Eng.*, ser. D, vol. 90, pp. 125–132, March, 1968.
51. Wallis, G. B.: *Int. J. Multiphase Flow,* Vol. I, pp. 491-511, 1974.

10
Slug Flow

10.1 INTRODUCTION

The *slug-flow* regime is characterized by a series of individual large bubbles which almost fill the available flow cross section. Some familiar examples of slug flow are

1. Flow in a drinking straw when the glass is almost empty
2. Flow in the "riser" section of a coffee percolator
3. Flow in the neck of a bottle which is being emptied too rapidly

A typical example of a slug-flow bubble in a vertical pipe under laboratory conditions is shown in Fig. 10.1.

10.2 GENERAL THEORY

BUBBLE DYNAMICS

Several of the overall properties of the slug-flow regime can be predicted if one can describe the dynamics of a typical individual bubble. This

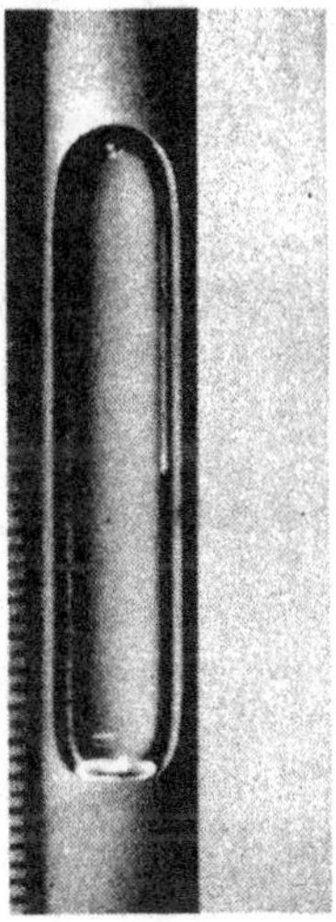

Fig. 10.1 A slug-flow bubble rising in glycerin in a vertical tube.

is perhaps best done by considering a *unit cell* consisting of one bubble and part of the liquid slugs on each side of it as shown in Fig. 10.2. For a specified total volumetric flux j the mean velocity of the liquid in the slug is then simply

$$j = \frac{Q_g + Q_f}{A} \tag{10.1}$$

The bubble dynamics are then determined by this velocity, the corresponding velocity profile, the bubble length, pipe geometry, and fluid properties. Apart from the effects of the wake from the preceding bubble, the velocity profile in the liquid slug is a function of the pipe roughness and the Reynolds number,

$$\mathrm{Re}_j = \frac{jD\rho_f}{\mu_f} \tag{10.2}$$

Therefore the bubble dynamics are dependent on j but not on the individual fluxes j_f and j_g of the liquid and the gas. Furthermore, as long as each unit cell corresponding to Figure 10.2 is independent, the bubble dynamics are not a function of the void fraction α.

BUBBLE VELOCITY

In accordance with the above paragraph the bubble velocity is a function of the average volumetric flux j, the pipe geometry, the fluid properties, and the body force field. In almost every case the bubble length is not

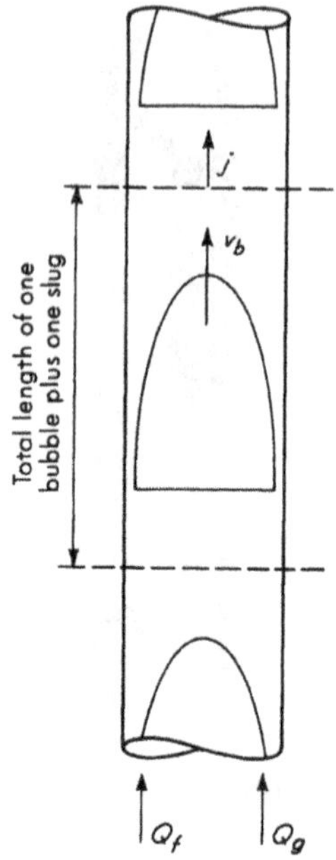

Fig. 10.2 "Unit cell" for slug-flow analysis.

found to be an important variable since the dynamics of the nose and tail of the bubble govern the motion entirely.

The drift velocity of the gas is the bubble velocity minus the overall volumetric flux. Therefore

$$v_{gj} = v_b - j \tag{10.3}$$

Hence the drift velocity is also independent of void fraction and depends on j and not on j_f and j_g independently.

VOID FRACTION

The void fraction—or mean volumetric gas concentration—can be derived from Eq. (1.10) if the bubble velocity and the gas flux are known. Thus

$$\alpha = \frac{j_g}{v_b} \tag{10.4}$$

PRESSURE DROP

The pressure drop is conveniently divided into three parts:

1. Pressure drop in the liquid slug
2. Pressure drop around the ends of the bubble
3. Pressure drop along the body of the bubble

Usually the gas viscosity and density are much lower than the liquid viscosity and density. In this case the gas in the bubble is substantially at constant pressure. The body of the bubble is approximately cylindrical and the gas-liquid interface has a constant curvature. Therefore there is no pressure drop down the body of the bubble and item 3 above is zero.

The pressure drop, item 1, in the liquid slug can be calculated by single-component flow techniques.

The pressure drop, item 2, around the ends of the bubble remains to be predicted in terms of fundamental quantities.

10.3 VERTICAL SLUG FLOW

RISE VELOCITY OF SINGLE BUBBLES IN STAGNANT LIQUID

A bubble rises through a denser liquid because of its buoyancy. The velocity v_∞ with which a single bubble rises through stagnant liquid is governed by the interaction between buoyancy and the other forces acting on the bubble as a result of its shape and motion. If the viscosity of the gas or vapor in the bubble is negligible, the only three forces besides buoyancy which are important are those from liquid inertia, liquid viscosity, and surface tension. The balance between buoyancy and these three forces may be expressed in terms of three dimensionless groups

$$\frac{\rho_f v_\infty^2}{Dg(\rho_f - \rho_g)} \qquad \frac{v_\infty \mu_f}{D^2 g(\rho_f - \rho_g)} \qquad \frac{\sigma}{D^2 g(\rho_f - \rho_g)}$$

where D is a characteristic dimension of the duct cross section. The general solution to the problem is a function of these three parameters which may be combined to generate new dimensionless quantities if this is convenient. As long as the bubble length is greater than the tube diameter it does not influence the rise velocity. Alternatively, from Eqs. (9.26) and (9.29), the bubble equivalent diameter must exceed 60 percent of the tube diameter.

The simplest solutions are obtained when only one dimensionless group governs the motion. These limiting cases are as follows:

Inertia dominant When viscosity and surface tension can be neglected the bubble rise velocity is given solely in terms of the first dimensionless group. Thus

$$v_\infty = k_1 \rho_f^{-1/2}[gD(\rho_f - \rho_g)]^{1/2} \tag{10.5}$$

Approximate analytical solutions to this problem for a cylindrical duct have been obtained by Dumitrescu[1] and by Davies and Taylor.[2] Values of the constant which were obtained for round tubes by these authors were:

Dumitrescu[1] $k_1 = 0.351$ (10.6)
Davies and Taylor[2] $k_1 = 0.328$ (10.7)

Dumitrescu's experiments gave a slightly different result

$$k_1 = 0.346$$

and this is close to the value (0.345) obtained in a series of experiments by White and Beardmore.[3] The preferred value of k_1 is therefore

$$k_1 = 0.345 \tag{10.8}$$

If one measures the rise velocity of a bubble in a tube with an open top there is an apparent dependence on bubble length due to the expansion of the gas as it rises in the hydrostatic pressure gradient. This expansion leads to a value of j ahead of the bubble which is not zero, and the rise velocity is augmented by the velocity of this liquid. By timing bubbles of different lengths in tubes which were closed at the top and did not allow any motion of the liquid ahead of the bubble, Nicklin[4] has shown that the bubble rise velocity relative to stationary liquid is indeed given by Eqs. (10.5) and (10.8) and is independent of the bubble length.

The case of tubes of noncircular cross section was investigated by Griffith.[5] For *rectangular channels* of dimensions D_b by D_s the larger dimension D_b was found to be the more significant one for insertion into Eq. (10.5). The constant k_1 can then be expressed as a function of D_s/D_b and a good approximation to the final result is

$$k_1 = 0.23 + 0.13 \frac{D_s}{D_b} \tag{10.9}$$

The limiting case in which D_s/D_b goes to zero agrees with Birkhoff and Carter's[6] theoretical predictions for plane two-dimensional bubbles.

Griffith's paper also gives results for bubbles rising in annuli and tube bundles. For annuli the outer diameter D_o is the significant dimension for insertion into Eq. (10.5). The variation of k_1 as a function of the ratio of the internal to external diameters (D_i/D_o) is shown in Fig. 10.3. A remarkable conclusion is that the smaller the annular spacing is made, the faster the bubble will rise.

In the limit $D_i = D_o$ the result is approximately

$$k_1 = 0.23\sqrt{\pi} \tag{10.10}$$

which would be expected from Eq. (10.9) for a plane bubble which is wrapped around the annulus so that the liquid streams coalesce on one side. Symmetrical bubbles are usually not observed in annuli.

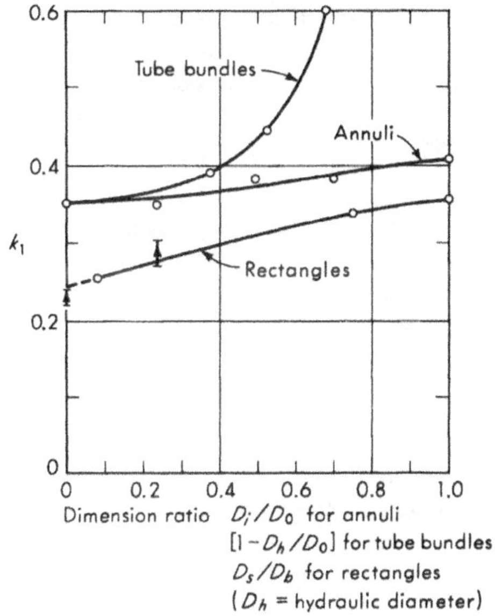

Fig. 10.3 Dependence of the factor k_1 on duct geometry. (*Griffith.*[5])

In tube bundles the characteristic dimension is more likely to be the overall housing diameter D_o rather than the dimension of individual tubes. Experimental evidence for slug-flow bubbles in tube bundles is contained in Griffith's paper.[5]

The reader is reminded that the above equations are valid only when surface tension and viscous forces are negligible. The limits of validity of this regime will be discussed later.

Viscosity dominant When viscosity dominates, the form of the equation for rise velocity is readily obtained from the relevant dimensionless group. Thus

$$v_\infty = k_2 \frac{gD^2(\rho_f - \rho_g)}{\mu_f} \tag{10.11}$$

Experimental observations confirm this equation for vertical round tubes with the following values of k_2

$$k_2 = 0.010 \qquad \text{Wallis}^7 \tag{10.12}$$
$$k_2 = 0.0096 \qquad \text{White and Beardmore}^3 \tag{10.13}$$

Surface tension dominant Surface tension dominates when the bubble does not move at all. The static interface adopts a particular shape so that hydrostatic forces are completely balanced by surface forces. For *vertical round tubes*, it has been shown analytically by Bretherton[8] and by Hattori[9] that this will occur when

$$N_{\text{Eö}} = \frac{gD^2(\rho_f - \rho_g)}{\sigma} < 3.37 \tag{10.14}$$

The above dimensionless number was called the *Eötvös number* by Harmathy.[10]

An alternative variable which is sometimes used is the *Bond number*, defined as follows:

$$N_{\text{Bo}} = \frac{gR^2(\rho_f - \rho_g)}{\sigma} = \frac{N_{\text{Eö}}}{4} \tag{10.15}$$

The general case Since the general solution is governed by three parameters it can be presented as a two-dimensional plot of any two chosen dimensionless groups with a third independent dimensionless group as a parameter. The actual manner in which these groups are chosen is simply a matter of convenience. For example, the *dimensionless bubble velocity* k_1, defined by Eq. (10.5), may be plotted versus the *dimensionless inverse viscosity* N_f given by

$$N_f = \frac{[D^3g(\rho_f - \rho_g)\rho_f]^{1/2}}{\mu_f} \tag{10.16}$$

which is obtained by eliminating v_∞ from the first two dimensionless groups. A convenient third independent parameter is obtained by eliminating both v_∞ and D entirely to obtain the *Archimedes number* which depends only on fluid properties and gravitational acceleration and is a constant for a given fluid at a particular temperature. Thus

$$N_{\text{Ar}} = \frac{\sigma^{3/2}\rho_f}{\mu_f^2 g^{1/2}(\rho_f - \rho_g)^{1/2}} \tag{10.17}$$

The result of plotting experimental data in this way is shown in Fig. 10.4. The three asymptotic solutions equivalent to Eqs. (10.8), (10.12), and (10.14) are clearly satisfied. They are

Inertia dominant	$N_f > 300 \quad N_{\text{Eö}} > 100$ $k_1 = 0.345$	
Viscosity dominant	$N_f < 2 \quad N_{\text{Eö}} > 100$ $k_1 = 0.01N_f$	(10.18)
Surface tension dominant	$N_{\text{Eö}} = 3.37$ $N_f^2 = 6.2N_{\text{Ar}}$	(10.19)

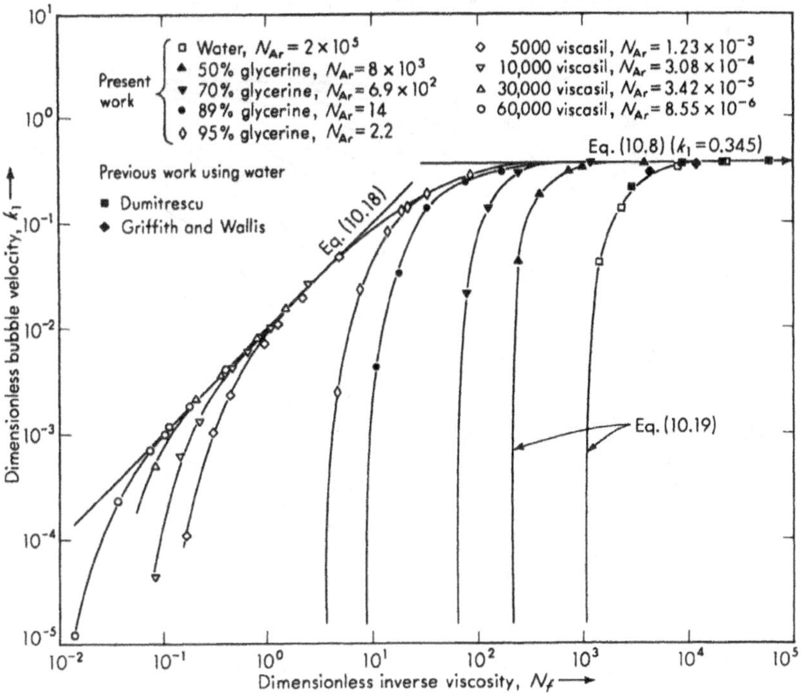

Fig. 10.4 General dimensionless representation of bubble rise velocity in slug flow. (*Wallis.*[7])

Alternative methods of plotting the result have also been used by combining the dimensionless quantities in various ways. A parameter used by White and Beardmore[3] is the *property group*

$$Y = \frac{g\mu_f^4}{\sigma^3\rho_f} \tag{10.20}$$

Y is simply equal to $1/N_{Ar}^2$ when the gas density is low compared with the liquid density. A plot of k_1 versus $N_{Eö}$ as a function of Y is shown in Fig. 10.5.

An equation which has the property that it reduces to Eqs. (10.8) and (10.11) in their appropriate range of applicability is

$$k_1 = 0.345(1 - e^{-0.01N_f/0.345}) \tag{10.21}$$

Equation (10.21) also gives a good approximation in the intermediate range of N_f shown in Fig. 10.4, when surface-tension effects are negligible.

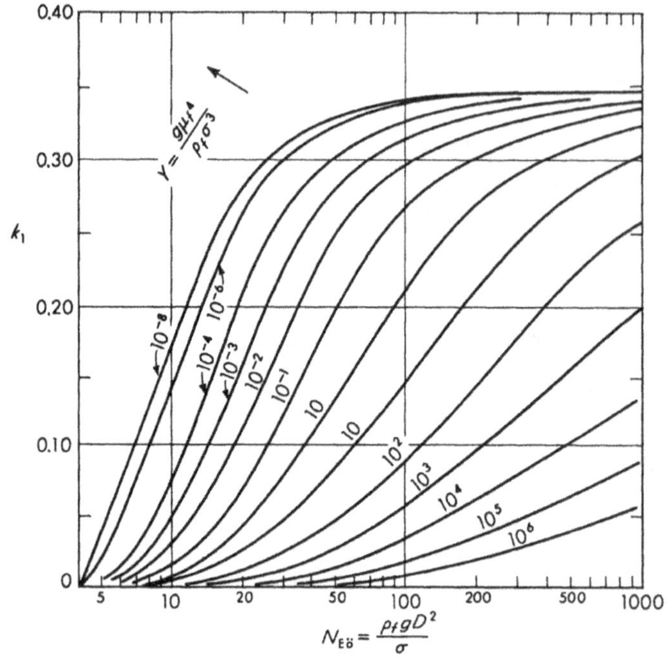

Fig. 10.5 An alternative representation of Fig. 10.4. (*White and Beardmore.*[3])

Surface-tension effects can be approximated algebraically by a further modification

$$k_1 = 0.345(1 - e^{-0.01N_f/0.345})(1 - e^{(3.37 - N_{Eö})/m}) \tag{10.22}$$

where m is a function of N_f and takes on the following values:

$$\begin{aligned} N_f > 250 \qquad & m = 10 \\ 18 < N_f < 250 \qquad & m = 69N_f^{-0.35} \\ N_f < 18 \qquad & m = 25 \end{aligned} \tag{10.23}$$

Equation (10.22) is a general correlation for bubble rise velocity in terms of all the relevant variables.

In the inviscid region when N_f is large, Eqs. (10.23) and (10.22) yield

$$k_1 = 0.345(1 - e^{(3.37 - N_{Eö})/10}) \tag{10.24}$$

This result correlates the data which were used by Masica and Petrash[11] equally as well as their equation, which was written in terms of the Bond number,

$$k_1 = 0.34 \left[1 - \left(\frac{0.84}{N_{Bo}}\right)^{N_{Bo}/4.7}\right] \tag{10.25}$$

The bubble shape is different for the various regimes. In a highly viscous fluid ($N_f < 2$) both the bubble nose and tail are rounded and the wake is laminar, whereas in a fluid of low viscosity ($N_f > 300$) the bubble tail is flat and the wake is turbulent.

Example 10.1 What is the rise velocity for a slug-flow bubble composed of gas with density 10^{-3} g/cm³ in a vertical round tube 2 cm in diameter? The liquid properties are $\rho_f = 1$ g/cm³, $\sigma = 100$ dynes/cm, $\mu_f = 1$ poise.

Solution First, evaluate the following dimensionless groups, neglecting the gas density in comparison with the liquid density:

$$N_{E\ddot{o}} = \frac{gD^2\rho_f}{\sigma} = \frac{(981)(2)^2(1)}{100} = 39.2$$

$$N_f = \frac{g^{1/2}D^{3/2}\rho_f}{\mu_f} = \frac{(981^{1/2})(2^{3/2})(1)}{1} = 88$$

$$N_{Ar} = \frac{\sigma^{3/2}\rho_f^{1/2}}{\mu_f^2 g^{1/2}} = \frac{(100^{3/2})(1^{1/2})}{(1^2)(981^{1/2})} = 32$$

$$Y = \frac{1}{N_{Ar}^2} = 10^{-3}$$

Using Eq. (10.22) we have

$$m = \frac{69}{N_f^{0.35}} = 15$$

$$k_1 = 0.345(1 - e^{-0.88/0.345})(1 - e^{(3.37-39.2)/15}) = 0.29$$

Therefore, from Eq. (10.5),

$$v_\infty = 0.29(1^{-1/2})[(981)(2)(1)]^{1/2}$$
$$= 12.7 \text{ cm/sec}$$

Alternatively, using Fig. 10.5 we find $k_1 = 0.26$, $v_\infty = 11.4$ cm/sec.

USE OF THE BUBBLE VELOCITY IN THE DRIFT-FLUX MODEL

The techniques of Chap. 4 can be applied to the slug-flow regime as long as the effects of wall shear stress on the bubble dynamics are small. This is approximately true for values of N_f greater than 300 and when the frictional pressure drop, estimated from homogeneous theory, is small compared with the gravitational pressure drop.

In Sec. 10.2 it was found that the drift velocity v_{gj} was independent of α but was a function of j. However, for one-dimensional flow with no wall shear effects the drift velocity is independent of j but is a function of α. The only way in which these conditions can both be satisfied is for v_{gj} to be a constant. This constant can be evaluated by choosing the special case of a single bubble in stagnant liquid for which

$$v_{gj} = v_{\infty} \tag{10.26}$$

Therefore, for all values of α,

$$j_{gf} = \alpha v_{\infty} \tag{10.27}$$

The bubble velocity when there is net flow is, from Eq. (10.3),

$$v_b = j + v_{\infty} \tag{10.28}$$

Therefore the mean void fraction, from Eq. (10.4), is

$$\langle \alpha \rangle = \frac{j_g}{j + v_{\infty}} \tag{10.29}$$

In terms of the volumetric flow rates, Eq. (10.29) can be written[12]

$$\langle \alpha \rangle = \frac{Q_g}{Q_f + Q_g + A v_{\infty}} \tag{10.30}$$

In view of the assumption of negligible wall shear stress the pressure gradient in nonaccelerating flow is

$$-\frac{dp}{dz} = g[\rho_f(1 - \alpha) + \rho_g \alpha] \tag{10.31}$$

and this may be calculated by using the value of α from Eq. (10.29).

IMPROVEMENTS TO THE SIMPLIFIED THEORY

In reality the bubble drift velocity is not strictly constant since it is influenced by the velocity profile in the liquid slug. This profile is a function of j—or more strictly a function of the Reynolds number defined by Eq. (10.2)—and is also influenced by the wake of a preceding bubble. These effects may be taken into account by applying correction factors to Eq. (10.28). Thus

$$v_b = C_1 j + C_2 v_{\infty} \tag{10.32}$$

The coefficient C_1 is a measure of the fact that the bubble does not simply move relatively to the average liquid velocity but relative to a weighted average value. C_2 is a measure of the change in relative velocity due to the approaching velocity profile. C_1 has the same overall effect on the equations as the coefficient C_0 in Eq. (4.24) although the physical reasoning behind the derivation of these coefficients is not identical.

A simple result is obtained in the case of fully developed turbulent flow in the liquid slug (Reynolds number greater than approximately 8000). Nicklin[4] has found that in this case the bubble moves with a velocity v_∞ relative to the approximately uniform velocity in the turbulent core. For a circular pipe and fully developed flow the result is

$$\mathrm{Re}_j > 8000 \qquad C_1 = 1.2 \qquad C_2 = 1 \tag{10.33}$$

Griffith[5] has suggested appropriate values of C_1 for rectangular channels and annuli at high Reynolds numbers and these are shown in Fig. 10.6.

Accepted correlations for C_1 and C_2 in the laminar flow region are not yet available. Griffith and Wallis[12] plotted C_2 versus Reynolds number with some success but failed to separate the effects of the two correction factors. When viscosities are high, C_1 probably approaches the limiting value of 2.27 which was predicted by Taylor[13] for horizontal flow.

The effect of the wake behind one bubble on the rise velocity of the next bubble has been rationalized by Moissis[14] who expressed the factor C_2 as a function of the ratio between bubble separation (liquid slug length) L_s and pipe diameter D for circular tubes:

$$C_2 = 1 + 8e^{-1.06L_s/D} \tag{10.34}$$

This means that the following bubble is continually tending to catch up the preceding bubble and coalesce with it. Slug flow is therefore never strictly stable and fully developed, although eventually the slug length becomes sufficiently long for the second term in Eq. (10.34) to be negligible. In the entrance region in a pipe or when slug flow is developing from bubbly flow, or when bubbles are continually being formed as a result of boiling, the interaction between bubbles is significant and should be taken into account. It is rather awkward to use Eq. (10.34) in prac-

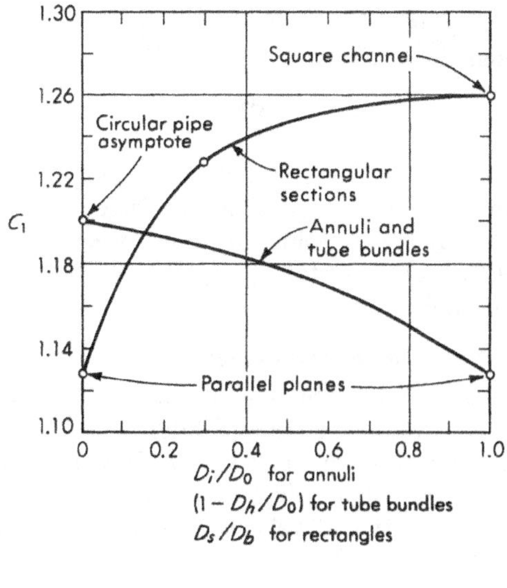

Fig. 10.6 Variation of the factor C_1 with duct geometry. (*Griffith.*[5])

tice since the slug length L_s is indeterminate solely in terms of the overall flow rates of the components. Griffith[5] suggests that in a channel in which boiling is occurring it is sufficiently accurate to assume that

$$C_2 = 1.6 \tag{10.35}$$

The coefficients C_1 and C_2 modify Eq. (10.30) to give an improved equation for the mean void fraction

$$\langle\alpha\rangle = \frac{Q_g}{C_1(Q_f + Q_g) + C_2 A v_\infty} \tag{10.36}$$

The correction to Eq. (10.31), which is necessary when wall shear stress is significant, is difficult to estimate. The average shear stress can be either positive or negative since some liquid is actually running down the wall around the bubble. A possible procedure is to calculate the shear stress in the liquid slug from the single-phase friction factor based on j, i.e.,

$$\tau_w = C_f \tfrac{1}{2} \rho_f j^2 \tag{10.37}$$

where C_f is the usual function of Re_j. Approximately a fraction $(1 - \alpha)$ of the pipe length is occupied by slugs, therefore, if the wall shear stresses around the bubble are neglected, Eq. (10.31) becomes, with the addition of the drag on the liquid slugs,

$$-\frac{dp}{dz} = g[\rho_f(1 - \alpha) + \rho_g\alpha] + (1 - \alpha)C_f \frac{2\rho_f j^2}{D} \tag{10.38}$$

If the gas density is much smaller than the liquid density we have

$$\rho_m \approx (1 - \alpha)\rho_f \tag{10.39}$$

and the last term in Eq. (10.38) can be rewritten as

$$\left(-\frac{dp}{dz}\right)_F = \frac{2C_f \rho_m j^2}{D} \tag{10.40}$$

which could be regarded as a homogeneous flow frictional pressure drop in which the mean density is calculated from the void fraction in Eq. (10.30). The techniques which were used in the second part of Example 9.4 should therefore be approximately applicable to slug flow as well as to bubbly flow.

CORRECTION FOR LONG BUBBLES

If the slug-flow bubbles are long (greater than 15 diameters for example) a substantial amount of liquid is held up in the film surrounding the bubble. For purely potential flow the liquid downward velocity relative to the bubble is $\sqrt{2gh}$ at a distance h below the nose. Eventually, however, a terminal velocity is reached at which the weight of the film is completely balanced by the wall shear stress. The film now falls at a steady speed and has a uniform thickness which can be calculated from falling film theory. Since its weight is completely balanced by the wall shear stress, the liquid in the film does not contribute to the pressure drop. Furthermore the liquid slug length is also decreased by the amount of liquid which is held up in the film. Both of these effects can be accounted for by treating the liquid in the film as if it were gas and modifying the void fraction in Eq. (10.38). . For a long cylindrical bubble surrounded by a film of thickness δ (Fig. 10.7) the ratio of the bubble volume to the tube volume is

$$\left(\frac{D - 2\delta}{D}\right)^2$$

Therefore the effective void fraction for pressure-drop prediction in Eq. (10.38) should be modified to

$$\alpha' = \alpha\left(1 - \frac{2\delta}{D}\right)^{-2} \tag{10.41}$$

The film thickness is calculated as follows. Consider a section across the tube in the cylindrical part of the bubble shown in Fig. 10.7.

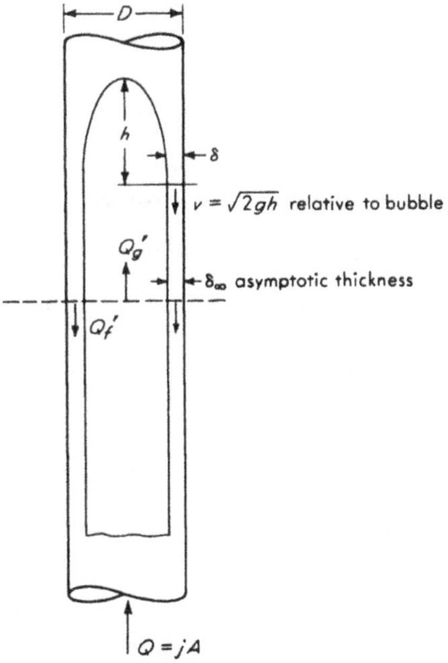

Fig. 10.7 Falling film flow around a long bubble in a vertical tube.

The gas velocity upward in the center is the same as the bubble velocity. Therefore

$$v_g = v_b \tag{10.42}$$

For inviscid bubbles and turbulent flow ($\mathrm{Re}_j > 8000$) Eqs. (10.32) and (10.33) give

$$v_g = 1.2j + v_\infty \tag{10.43}$$

The gas flow rate across the section is

$$Q'_g = \frac{\pi}{4} v_g (D - 2\delta)^2 \tag{10.44}$$

Let the liquid flow *downward* across the section be Q'_f. Since the total volumetric flow across any cross section of the pipe is constant from continuity,

$$Q'_g - Q'_f = Q = \frac{\pi D^2}{4} j \tag{10.45}$$

Combining Eqs. (10.43) to (10.45) we find eventually that the liquid flux downward is

$$j'_f = \frac{4Q'_f}{\pi D^2} = (1.2j + v_\infty)\left(1 - \frac{2\delta}{D}\right)^2 - j \tag{10.46}$$

Now, Q'_f is related to δ by the falling film theory of Chap. 11. Therefore δ can be found by solving Eq. (10.46) and the falling film equations simultaneously.

The theory is only valid if the value of N_f is greater than 300 so that the bubble velocity is given by Eq. (10.5) and is

$$v_\infty = 0.345 \rho_f^{-1/2} [gD(\rho_f - \rho_g)]^{1/2} \tag{10.47}$$

[Note: Similar techniques can be applied to bubbles in viscous fluids as soon as an equation can be developed to replace Eq. (10.43) for $\mathrm{Re}_j < 8000$.]

Using the relationship $j'_f/v_\infty = \dfrac{j_f^{*\prime}}{0.345}$, falling film theory from Eqs. (11.68) and (11.76) is now put into the convenient form

$$\mathrm{Re}_\Gamma = j_f^{*\prime} N_f = \frac{j'_f}{v_\infty}(0.345) N_f \tag{10.48}$$

$$\mathrm{Re}_\Gamma < 3500 \qquad \frac{j'_f}{v_\infty} = 3.85 N_f \left(\frac{\delta}{D}\right)^3 \tag{10.49}$$

$$3500 < \mathrm{Re}_\Gamma < 30{,}000 \qquad \frac{j'_f}{v_\infty} = 190 \left(\frac{\delta}{D}\right)^{3/2} \tag{10.50}$$

Dividing Eq. (10.46) by v_∞ we have

$$\frac{j'_f}{v_\infty} = \left(1.2\frac{j}{v_\infty} + 1\right)\left(1 - \frac{2\delta}{D}\right)^2 - \frac{j}{v_\infty} \tag{10.51}$$

Equations (10.49) to (10.51) are now to be solved simultaneously. Figure 10.8 shows a graphical solution. Equation (10.51) is used to plot j'_f/v_∞ versus δ/D for various values of j/v_∞. The intersections with Eqs. (10.49) and (10.50) give the value of δ/D, which is substituted into Eq. (10.41) to give a modified value of α for use in Eq. (10.38). In general, Eq. (10.38) will tend to overestimate the pressure drop, whereas the modification using Eq. (10.41) will underestimate it.

Acceleration pressure drop can be treated as in the bubbly flow regime and as shown in Example 9.4. However, since the slug-flow pattern is not as homogeneous as bubbly flow, the choking condition given by Eq. (r) of Example 9.4 is likely to be in error. Furthermore, since some of the liquid is moving with a velocity j in the slugs, whereas the rest is moving downward in the falling film, the assumption of uniform liquid velocity is incorrect. Further studies are worthwhile in this area.

Example 10.2 Calculate the mean pressure gradient in a vertical oil well for the following conditions, $Q_f = Q_g = 35.4$ liters/sec, $D = 15$ cm, $\mu_f = 1$ poise, $\sigma = 25$ dynes/cm, $\rho_f = 0.85$ g/cm³, $\rho_g = 0.0025$ g/cm³. The local pressure is 60 psia.

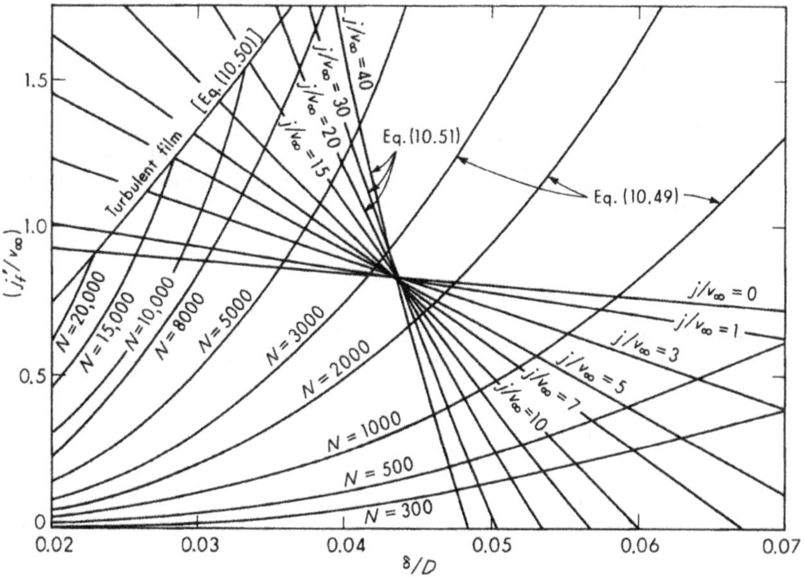

Fig. 10.8 Graphical solution for film thickness around a slug-flow bubble.

Solution The long length of vertical pipe in an oil well is conducive to the formation of slug-flow bubbles. Furthermore, if sufficient length is available for agglomeration to occur, these bubbles may be very long. Depending on the particular details of the well design the pressure gradient should therefore lie between the predictions of Eq. (10.38) which are obtained using Eq. (10.36) or (10.41), as long as the slug flow is fully developed and no correction is introduced by Eq. (10.34).

First we calculate the appropriate dimensionless groups as follows:

$$N_f = \frac{(981^{1/2})(15^{3/2})(0.85)}{1} = 1440$$

$$N_{E\ddot{o}} = \frac{(981)(15^2)(0.85)}{25} = 7250$$

$$N_{Ar} = \frac{(25^{3/2})(0.85^{1/2})}{(1^2)(981^{1/2})} = 3.2$$

From Fig. 10.4 it is found that $k_1 = 0.345$. Therefore

$$v_\infty = 0.345(981 \times 15)^{1/2} = 42 \text{ cm/sec}$$

The cross-sectional area is

$$A = \frac{\pi}{4}(15)^2 = 177 \text{ cm}^2$$

The volumetric fluxes of the oil and gas are

$$j_f = \frac{Q_f}{A} = \frac{35{,}400}{177} = 200 \text{ cm/sec}$$

$$j_g = \frac{Q_g}{A} = \frac{35{,}400}{177} = 200 \text{ cm/sec}$$

The Reynolds number for the liquid slug is

$$\text{Re}_j = \frac{\rho_f j D}{\mu_f} = \frac{(0.85)(400)(15)}{1} = 5100$$

There is now a problem in deciding what to do since the theory is not yet adequate for predicting C_1 and C_2 for Reynolds numbers below 8000. The only procedure which is available is to assume that the error in applying the turbulent flow theory with a Reynolds number of 5100 is not significant. Then the value of $\langle\alpha\rangle$ from Eq. (10.36) with $C_1 = 1.2$ and $C_2 = 1$ is

$$\langle\alpha\rangle = \frac{35{,}400}{(1.2)(70{,}800) + (42)(177)} = 0.384$$

For a smooth pipe with a Reynolds number of 5100 the friction factor is 0.009. Therefore the sum of the gravitational and frictional pressure drops in Eq. (10.38) is

$$-\frac{dp}{dz} = 981[(0.85)(0.616) + 0.0025(0.384)] + \frac{(0.616)(0.009)}{2}(0.15)(400)^2\frac{4}{15}$$
$$= 513 + 100 = 613 \text{ dynes/cm}^3$$

The correction for acceleration pressure drop is found using Eq. (q) of Example 9.4. The denominator of that equation is, with sufficient accuracy,

$$1 - \frac{\rho_f j_f^2 j_g}{(j_f + v_\infty)p} = 1 - \frac{(0.85)(200)^3}{(242)(60)(69{,}000)} = 0.993$$

The pressure gradient is therefore

$$-\frac{dp}{dz} = \frac{613}{0.993} = 618 \text{ dynes/cm}^3$$

To make the correction for long bubbles we use Fig. 10.8 with the values $j/v_\infty = 400/42 = 9.5$ and $N_f = 1440$. The value of δ/D is found to be 0.0485. From Eq. (10.41) the modified void fraction is

$$\alpha' = \frac{0.384}{[1 - 2(0.0485)]^2} = 0.471$$

Substituting this value into Eq. (10.41) the predicted pressure drop is decreased in the ratio $(1 - \alpha')/(1 - \alpha)$ to the value

$$-\frac{dp}{dz} = \frac{(618)(0.529)}{0.616} = 531 \text{ dynes/cm}^3$$

The pressure gradient therefore should lie between 531 and 618 dynes/cm³ and is probably closer to the lower value.

VISCOSITY EFFECTS

At low values of $N_f(<300)$, or for values of Re_j less than 8000, viscous effects become important and the above equations are incorrect. A complete study of these phenomena is not available. However, as long as buoyancy effects are small compared with viscous and surface-tension effects, the results to be described in Sec. 10.4 under the heading "Horizontal Slug Flow" are approximately valid. It is suggested that these results should be used when v_∞, as derived from Eq. (10.22), is much less than j, for example, when

$$\frac{v_\infty}{j} < 0.1 \tag{10.52}$$

10.4 HORIZONTAL SLUG FLOW

BUBBLE VELOCITY

There is no drift flux in horizontal flow due to buoyancy effects. Therefore v_∞ loses its significance. The bubbles, however, do not move with the same average velocity as the liquid. This is apparent from Fig. 10.9.

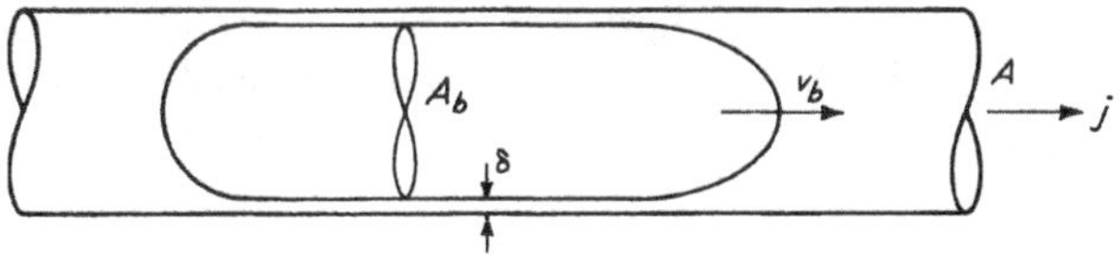

Fig. 10.9 Horizontal slug flow.

Since there is no pressure drop along the length of the bubble, the liquid film on the wall is substantially stationary. If this film has a mean thickness δ, the available flow area for gas in the bubble is

$$A_b = \pi \left(\frac{D}{2} - \delta\right)^2 \tag{10.53}$$

Therefore, for continuity of volumetric flux at this cross section

$$C_1 = \frac{v_b}{j} = \frac{A}{A_b} \tag{10.54}$$

or, if the film is thin

$$v_b \approx \left(1 + \frac{4\delta}{D}\right) j \tag{10.55}$$

Hence, v_b is greater than j.

A useful parameter which represents the fraction of the bubble cross section which is occupied by the liquid is

$$m = 1 - \frac{A_b}{A} = 1 - \frac{j}{v_b} \tag{10.56}$$

In the absence of effects due to the gas viscosity and inertia and for bubbles which move independently, the factors influencing bubble velocity can be combined into the following dimensionless groups:

$$\frac{j}{v_b} \qquad \frac{jD}{\nu_f} \qquad \frac{j\mu_f}{\sigma} \qquad \frac{(\rho_f - \rho_g)gD^2}{\sigma}$$

The first group represents the ratio between the liquid velocity in the slug and the bubble velocity; the second group is the Reynolds number for the liquid in the slug; the third group represents the relative importance of viscous and surface-tension effects; and the final group is the ratio between buoyancy and surface-tension forces. In fact the last three groups are directly analogous to the groups which describe the balance between inertia, viscosity, surface tension, and buoyancy in the case of vertical flow. The first group is the inverse of C_1.

The second and third groups can be combined to yield a parameter which is independent of velocity and is a constant for a particular fluid in a particular pipe, i.e.,

$$\lambda = \frac{\mu_f^2}{D\rho_f\sigma} \tag{10.57}$$

Suo[15] performed experiments in the range

$$\frac{(\rho_f - \rho_g)gD^2}{\sigma} < 0.88 \tag{10.58}$$

for which stratification effects were small and the buoyancy forces could be neglected. His results are shown in Fig. 10.10 plotted in terms of the dimensionless quantities mentioned above. It can be seen that at very low velocities the bubbles move at the slug velocity and the liquid film around the bubbles is therefore very thin in view of Eq. (10.55). At high velocities and high Reynolds numbers the ratio j/v_b tends to about 0.84, that is,

$$v_b \approx 1.19j \tag{10.59}$$

The value of the coefficient in Eq. (10.59) is very close to the equivalent value ($C_1 = 1.2$) for vertical flow at high Reynolds numbers. The Reynolds numbers Re_j corresponding to the approach to the asymptotic value in Fig. 10.8 are approximately 3000. Therefore a simple expression for bubble velocity in terms of the overall flow rates for $\mathrm{Re}_j > 3000$ is

$$v_b = 1.2\,\frac{Q_f + Q_g}{A} \tag{10.60}$$

The line corresponding to large values of λ in Fig. 10.10 represents the regime in which inertia effects are negligible and the motion is governed by the balance between viscosity and surface tension. In this case the results of Taylor[13] can be represented by the empirical equation

$$m = 0.56(1 - e^{-2.64(\mu v_b/\sigma)^{0.567}}) \tag{10.61}$$

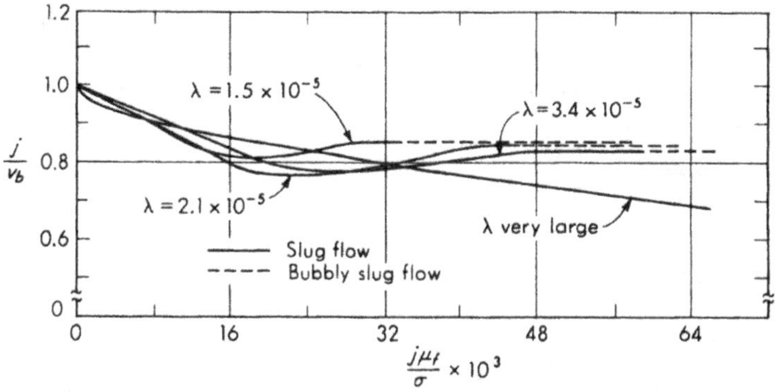

Fig. 10.10 Velocity ratio as a function of $j\mu_f/\sigma$. (*Suo and Griffith.*[15])

which reduces to

$$m \approx 1.48 \left(\frac{\mu v_b}{\sigma}\right)^{0.567} \tag{10.62}$$

at low values of $\mu v_b/\sigma$. Equation (10.62) is intermediate between Bretherton's[8] equation

$$m = 2.68 \left(\frac{\mu v_b}{\sigma}\right)^{2/3} \tag{10.63}$$

and one derived by Fairbrother and Stubbs[16]

$$m = 1.0 \left(\frac{\mu v_b}{\sigma}\right)^{1/2} \tag{10.64}$$

and represents Bretherton's data better than either Eq. (10.63) or (10.64).

The parameter m is very useful if one is interested in predicting the film thickness around a bubble, but the form of Eqs. (10.61) to (10.64) forces an iteration procedure since v_b is not usually known. An alternative expression which gives C_1 to an accuracy of 2 percent is

$$C_1 = 1 + 1.27(1 - e^{-3.8(\mu_f j/\sigma)^{0.8}}) \tag{10.65}$$

If an accurate value of m is required, Eq. (10.61) should be used after employing Eq. (10.65) to estimate v_b.

When C_1 exceeds 2, the bubble actually moves faster than the center streamline in the liquid slug. The character of the liquid flow relative to the bubble is therefore quite different[21,26] for the two cases $C_1 < 2$ and $C_1 > 2$.

VOID FRACTION

The void fraction α is calculated directly from Eq. (10.4) by using the bubble velocity which was found as described above. For the asymptotic limit at Reynolds numbers above 3000, Eqs. (10.60) and (10.4) give the simple result

$$\alpha = 0.84 \frac{Q_g}{Q_f + Q_g} \tag{10.66}$$

This equation can be compared with Eq. (4.31) and agrees well with the experimental results of Armand.[17]

PRESSURE DROP

The pressure drop in the liquid slug can be calculated by single-component flow techniques. The pressure drop along the cylindrical part of the bubble is zero. Therefore one is left only with an additional pressure drop *per bubble* due to effects at the nose and tail. Since different numbers of bubbles can make up the same overall flow rates in the same pipe, the pressure drop is not determinate unless the bubble length is specified separately.

The pressure drop per bubble was found by Suo[15] to be correlated by the following equations for low values of Re_j:

$$\mathrm{Re}_j < 270 \qquad \Delta p_b = \frac{90\mu_f v_b}{D} \tag{10.67}$$

$$270 < \mathrm{Re}_j < 630 \qquad \Delta p_b = 0.163\rho_f v_b^2 \tag{10.68}$$

The sudden change at Re = 270 from purely viscous terms in Eq. (10.67) to purely inertial terms in Eq. (10.68) is unusual, since no sudden change in flow dynamics occurs at that point, and it is likely that equations of the form of Eqs. (10.67) and (10.68) are valid at low and high Reynolds numbers, respectively. The pressure drop per bubble is then approximately equal to the pressure drop in a length of about four pipe diameters. An approximate correlation is then for all Reynolds numbers for the typical unit cell comprising one bubble and one slug,

$$\Delta p = 4C_f \tfrac{1}{2}\rho_f j^2 \, \frac{L_s + 4D}{D} \tag{10.69}$$

where L_s is the slug length.

The mean pressure gradient is then

$$-\frac{dp}{dz} = \frac{2C_f\rho_f j^2}{D} \, \frac{L_s + 4D}{L_s + L_b} \tag{10.70}$$

The evaluation of the last fraction in Eq. (10.70) depends on the conditions of the problem. Perhaps the volume $\mathcal{V}_b$ of each bubble is known. Then the length of a unit cell follows from a knowledge of the void fraction, fraction,

$$L_s + L_b = \frac{\mathcal{V}_b}{A\alpha} \tag{10.71}$$

Since $1/C_1$ represents the fraction of the cross section which is occupied by the bubble, in view of Eq. (10.54), we have a good approximation for long bubbles

$$L_b = \frac{C_1\mathcal{V}_b}{A} \tag{10.72}$$

Substituting Eqs. (10.71) and (10.72) into Eq. (10.70) gives

$$-\frac{dp}{dz} = \frac{2C_f\rho_f j^2}{D}\left(\frac{1}{\alpha} - C_1 + \frac{4DA}{\mathcal{V}_b}\right)\alpha \tag{10.73}$$

Since

$$\alpha = \frac{j_g}{C_1 j} \tag{10.74}$$

Eq. (10.73) can be expressed entirely in terms of the fluxes and C_1 as follows:

$$-\frac{dp}{dz} = \frac{2C_f\rho_f j}{D}\left(j_f + \frac{4DA}{\mathcal{V}_b C_1}j_g\right) \tag{10.75}$$

10.5 SLUG FLOW IN INCLINED PIPES

Slug flow in inclined pipes was studied by Runge.[18] The parameters which describe this regime are precisely those which were used to describe vertical flow together with the angle θ made between the axis of the tube and the vertical. For the bubble rise velocity one obtains curves similar to those in Fig. 10.4 at each value of θ, and a book of graphs would be necessary to represent such a four-dimensional surface.

Suppose, for example, that the bubble velocity is rendered dimensionless by the use of the parameter k_1. In the general case we should have

$$k_1 = k_1[N_f, N_{\text{Eö}}, \theta] \tag{10.76}$$

A simpler and more convenient method of presentation is to divide Eq. (10.76) by the value of k_1 which would occur for the same fluid in a vertical pipe and which can be calculated from the results of Sec. 10.3. Canceling the common parameters we then find that

$$\frac{v_\theta}{v_\infty} = \frac{k_1(N_f, N_{\text{Eö}}, \theta)}{k_1(N_f, N_{\text{Eö}}, 0)} \tag{10.77}$$

v_θ is the bubble velocity in the inclined tube. Equation (10.77) then states that the ratio of the bubble velocity in the inclined tube to the bubble velocity in a vertical tube can be expressed as a function of N_f, $N_{\text{Eö}}$, and θ.

In Refs. 18 and 19 the four-dimensional surface represented by Eq. (10.77) was approximated over selected ranges of the Eötvös number by the curves shown in Fig. 10.11*a* through *d*. These figures can be used in conjunction with the results of Sec. 10.3 to predict the rise velocity of slug-flow bubbles in stationary liquid in inclined tubes to within about 10 percent for a wide range of practical conditions.

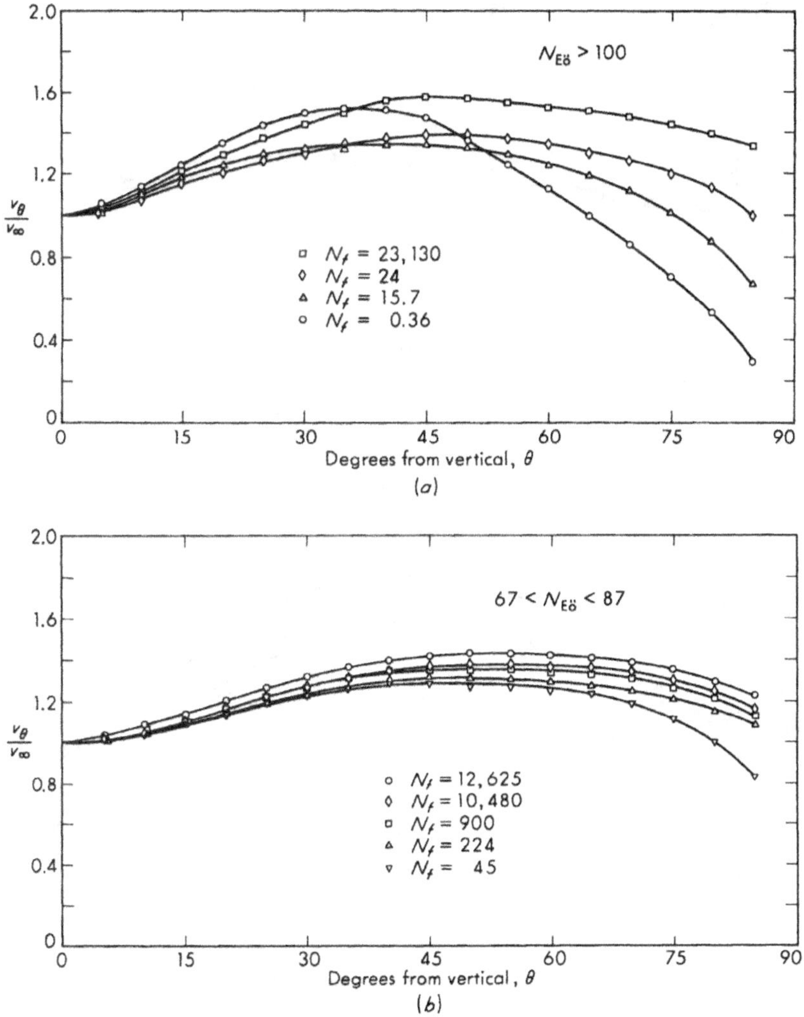

Fig. 10.11 Rise velocity of slug-flow bubbles in inclined pipes. (*Runge and Wallis.*[18])

A remarkable feature of these results is that the bubble velocity in inclined pipes generally exceeds the value in the same pipe placed vertically. In an inviscid liquid the bubble rises faster than in a vertical pipe even when the axis is as low as 2 or 3° from the horizontal. Care should therefore be taken in certain design situations to ensure that "horizontal" pipes are accurately oriented.

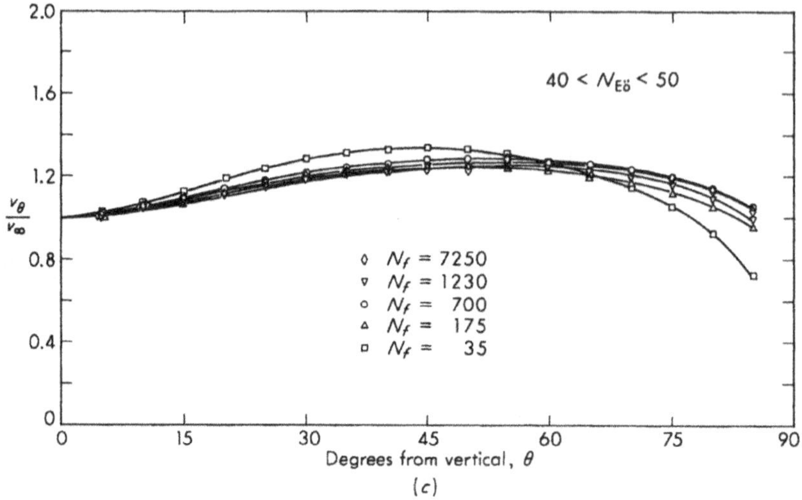

(c)

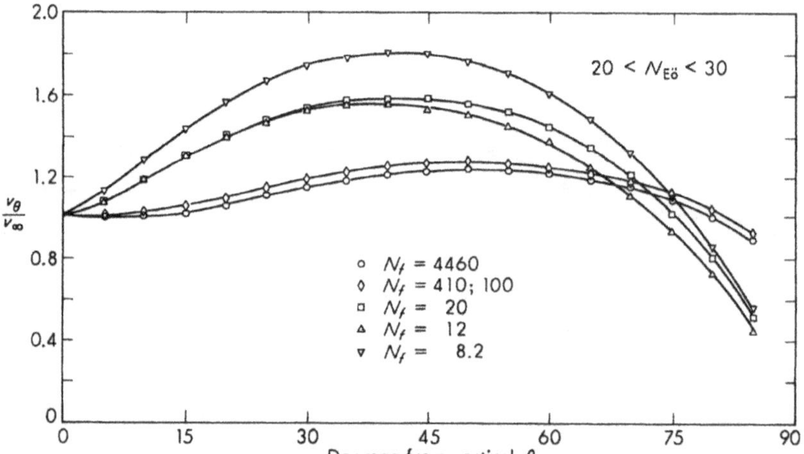

(d)

Fig. 10.11 *(Continued)*

Figure 10.12 shows how the shape of air bubbles rising in glycerin changes as the tube inclination is varied. When the tube is nearly horizontal the bubble appears to slide along the underside of the upper part of the tube. The liquid film along the top of the bubble can be very thin and dry spots will occur if the contact angle is large. This phenomenon is of particular importance in boiler design since evaporation of the thin

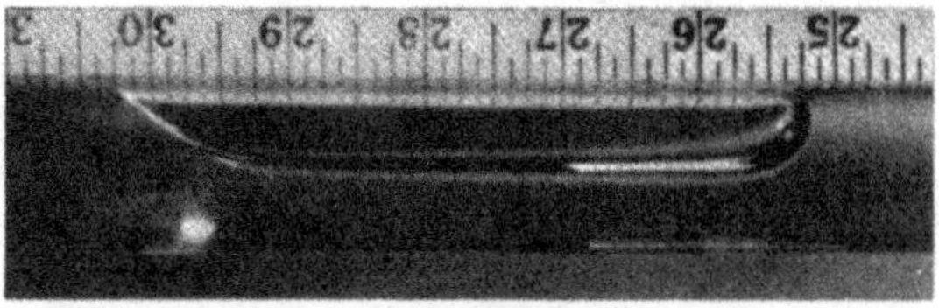

$\theta = 60°$
(a)

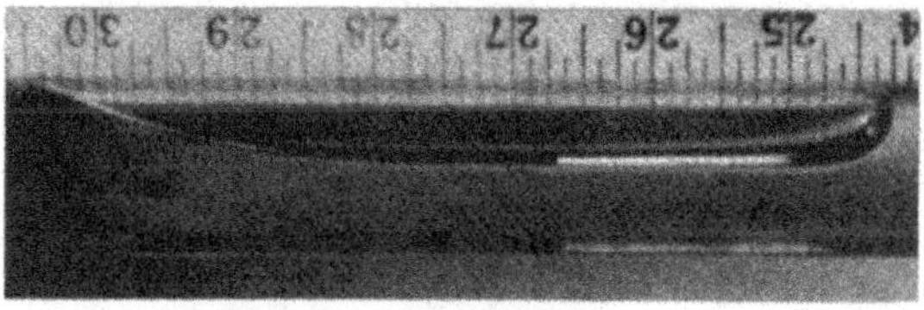

$\theta = 75°$
(b)

Fig. 10.12 Slug-flow bubbles in inclined tubes. (a) $\theta = 60°$; (b) $\theta = 75°$.

film on the top of a bubble can lead to local overheating and blistering of the material of the tube.

Example 10.3 Solve Example 10.1 for values of θ of 30, 60, and 80°.

Solution The value of $N_{Eö}$ is closest to the range of Fig. 10.11c. For $N_f = 88$ the approximate values of v_θ/v_∞ for the given inclinations are

θ	30°	60°	80°
v_θ/v_∞	1.2	1.2	1.0

Therefore, taking a mean of the predictions of Example 10.3, the required velocities are 14.4, 14.4, and 12.0 cm/sec.

10.6 THE LIMITS OF THE SLUG-FLOW REGIME

It is practically certain that there is no lower limit to mean void fraction, quality, or flow rates at which slug flow can be obtained. There is always a tendency for the bubbly flow regime to change to the slug-flow regime at low velocities as a result of agglomeration. Below approximately 10 percent void fraction this process is relatively slow and definite slugs may be observed only in a particularly long pipe. If impurities are present the bubbly regime may be prolonged.

If long slug-flow bubbles are brought to rest by stopping the flow in a small horizontal pipe—or in a zero g environment—they will break up into smaller bubbles about two or three pipe diameters in length.[20,21]

Presumably this instability can also occur at extremely low values of the volumetric flux j. However, this mechanism does not lead to a change in flow regime but merely imposes an upper limit on the length of a stable bubble. Slug-flow bubbles can also break up when they flow through constrictions.[21]

The upper limit of slug flow occurs as a result of shear stresses and drag forces between the liquid and the gas in the slug. One regime boundary is determined by the condition that the slug-flow bubble shall be broken up into smaller bubbles as a result of its motion relative to the surrounding liquid. In a very viscous fluid the bubble breaks up by the formation of long filaments which stream from the bubble tail and eventually split up to form small bubbles. At high Reynolds numbers, on the other hand, the liquid flow separates at the bubble tail to form a very agitated wake in which small bubbles are entrained. A condition for this latter phenomenon in horizontal flow is suggested by Suo[15]; it is

$$\frac{v_b^3 \rho_f^2 D^2}{\mu_f \sigma} > 1.1 \times 10^6 \tag{10.78}$$

This flow regime transition is not an abrupt one and slug-flow bubbles with groups of small bubbles following them can be observed over a range of flow rates.

In vertical flow, similar conditions are obtained. At high Reynolds numbers the relative velocity between the gas and the liquid steadily increases down the bubble length until eventually the liquid film reaches a terminal velocity due to shear stresses at the wall. The condition of bubble destruction due to the effects of relative velocity between the liquid film and the gas in the bubble therefore imposes a condition of maximum stable bubble length and not a condition for the complete destruction of slug flow. For practical purposes a slug-flow bubble with many small bubbles in its wake behaves in much the same way as the equivalent large bubble since the dynamics of the bubble nose determine the drift velocity and the large bubble and its wake move together.

The countercurrent flow of the gas core of the bubble and the liquid film have been recognized by Nicklin and Davidson[22] as a potential cause of flooding and the consequent development of large unstable waves at the interface. Until flooding occurs the drag forces between the gas and the liquid film are essentially zero. Let the liquid film thickness be δ; then the available flow area for the gas is

$$A_b = \pi \left(\frac{D}{2} - \delta\right)^2 \tag{10.79}$$

The gas flow rate in the core is then

$$Q_g' = A_b v_b \tag{10.80}$$

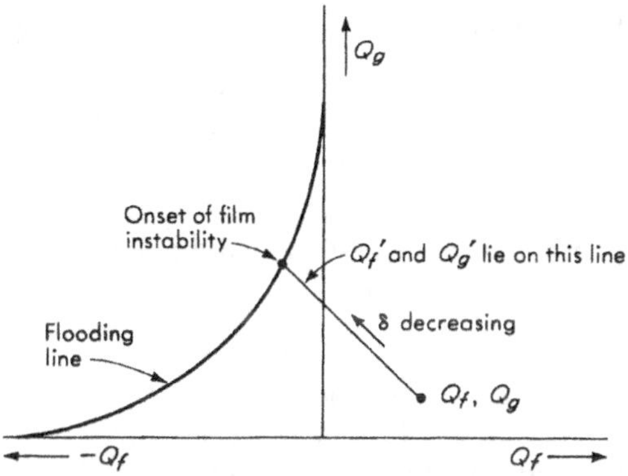

Fig. 10.13 Determination of maximum stable bubble length from Nicklin and Davidson's theory.

where v_b is calculated as in Eq. (10.3) and is a function of j but not of j_f and j_g. The downward liquid flow rate in the film is, from continuity

$$Q_f' = A_b v_b - Aj \tag{10.81}$$

The relationship between Q_f' and Q_g' at which flooding occurs can be determined by a separate experiment or from the equations which will be given in Sec. 11.4. Suppose this flooding locus is plotted as in Fig. 10.13. For a given value of j, Eqs. (10.80) and (10.81) with δ as parameter represent lines of slope -1 through the point Q_f, Q_g which is determined by the imposed flow rates.

The value of the asymptotic film thickness δ_∞ is found, as before, from Eqs. (10.41), (10.50), and (10.51). If the value of δ at which the line cuts the flooding curve is less than δ_∞, then no instability is reached. On the other hand, if the line cuts the flooding curve at a value of δ greater than δ_∞, flooding should occur when the film thickness reaches this value.

For example, consider the case $N > 300$, $\mathrm{Re}_j > 8000$. The upward gas flux in the bubble for this limiting film thickness is, from continuity,

$$j_g' = j + j_f' \tag{10.82}$$

since j_f' is measured downward (we have dispensed with the sign convention for convenience). These values of j_f' and j_g' can now be tested against the appropriate falling film flooding correlation. For $N > 300$ and a condition in which the limiting process occurs in the film alone, the suggested equation is Eq. (11.159), i.e.,

$$j_g^{*1/2} + j_f^{*1/2} = 1 \tag{10.83}$$

Using the definitions of the dimensionless groups we find that

$$j_f^* = \frac{0.35 j_f'}{v_\infty} \tag{10.84}$$

$$j_g^* = 0.345 \sqrt{\frac{\rho_g}{\rho_f}} \frac{j_g'}{v_\infty} \tag{10.85}$$

Furthermore, Eq. (10.82) can be put into the dimensionless form

$$\frac{j_g'}{v_\infty} = \frac{j}{v_\infty} + \frac{j_f'}{v_\infty} \tag{10.86}$$

Now suppose that one starts with the parameters j/v_∞, ρ_f/ρ_g, and N_f. First, Eqs. (10.49) to (10.51) are solved for j_f'/v_∞. Using this result, j_g'/v_∞ is found from Eq. (10.86). Then j_f^* and j_g^* are calculated from Eqs. (10.84) and (10.85). The left-hand side of Eq. (10.83) is evaluated: if it is larger than unity, flooding occurs; otherwise the falling film is stable. The condition of the onset of flooding at the limiting film thickness is reached when Eq. (10.83) is satisfied. This then specifies a relationship between the originally chosen parameters j/v_∞, ρ_f/ρ_g and N_f, which can be determined.

The calculations indicated above have been carried out by Porteous,[23] and the result is shown in Fig. 10.14. This graph has been arranged to show the maximum allowable value of j/v_∞ for specified values of $\sqrt{\rho_f/\rho_g}$ and N_f.

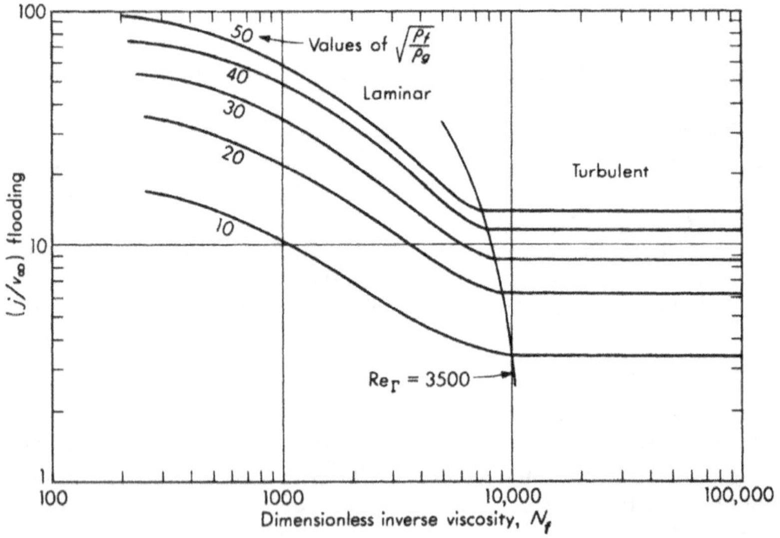

Fig. 10.14 Prediction of the critical value of j/v_∞ for flooding in slug flow. (*Porteous*.[23])

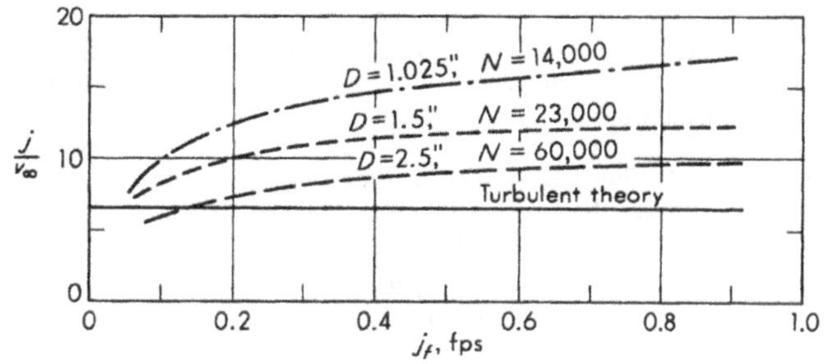

Fig. 10.15 Comparison between the predictions of Fig. 10.14 and Govier and Short's[24] observed slug-froth transition for low liquid rates.

The above theory predicts when flooding will occur once the limiting film thickness δ^* is reached. If the bubbles are not very long then a greater value of j/v_∞ will be necessary before noticeable flooding can occur at values of δ greater than the asymptotic value.

Figure 10.15 shows a comparison between these predictions and the regime boundary between slug flow and "froth flow" observed by Govier and Short[24] for liquid rates below $j_f = 1$ fps. The substantial agreement suggests that the froth was due to the break up of the lower end of the long bubbles as a result of flooding. It is unlikely that this froth will change the void fraction or pressure-drop equations for slug flow. However, there may be a significant effect on mass-transfer phenomena due to the increase of interfacial area which is obtained.

If the total volumetric flux is increased beyond the limiting value given by Fig. 10.14, the effect is to move the onset of flooding up the bubble until eventually no downward velocity is allowed in the liquid film at all.

This limiting value of j equals the volumetric flux of gas at the point $Q'_f = 0$ on the flooding line. This is the basis of Wallis' criterion[25] for the transition to pure annular flow with no downward flow of the liquid, namely,

$$j_g \rho_g^{1/2} = 0.9\,[gD(\rho_f - \rho_g)]^{1/2} \tag{10.87}$$

Below the velocity predicted by Eq. (10.87) the flooding criterion predicts a continual bridging and unbridging of the gas core as flooding and bubble agglomeration occur in sequence. There is thus a region of "slug-annular flow" between the regimes of simple slug flow and continuous film flow. This topic will be discussed further in the following chapter.

Example 10.4 Check whether flooding occurs in the falling film around the bubbles for the conditions of Example 10.2.

Solution From Fig. 10.14 for $\sqrt{\rho_f/\rho_g} = 18.5$ and $N_f = 1440$ we find that the limiting value of j/v_∞ is approximately 16. For the conditions of Example 10.2, j/v_∞ is $400/42 = 9.5$. Since this is less than the limiting value, flooding does not occur.

Example 10.5 Solve Example 2.5 using slug-flow theory.

Solution From the given data we obtain $v_\infty = 0.56$ fps, $C_1 = 1.2$, and $C_2 = 1$. Therefore, from Eq. (10.36),

$$\alpha = \frac{j_g}{1.2j + v_\infty} = \frac{47.2}{(1.2)(50.3) + 0.56} = 0.773$$

The liquid fraction is

$$1 - \alpha = 0.227$$

which agrees well with the observed value of 0.23.

The gravitational and frictional pressure drops are found from Eq. (10.38) to be 8.2×10^{-3} and 8.07×10^{-3} psi/in., respectively. Adding these together and multiplying by 18 in. gives $\Delta p = 0.293$ psi. If the accelerational effects are overestimated by using homogeneous theory (as in Example 2.5), we have $\Delta p = 0.293/0.864 = 0.34$ psi. These two estimates bracket the observed value.

If the correction for long bubbles is applied, the predicted pressure drop is decreased to about 40 percent of the above value. This is plainly in error and the reason is clear from Fig. 10.14. Since $j/v_\infty = 90$, flooding is prevalent and only short bubbles survive.

10.7 PRESSURE OSCILLATIONS IN SLUG FLOW

Equations (10.38) and (10.75) predict the average pressure gradient. However, the bubble is essentially at a constant pressure along its length. Therefore the actual pressure variation over several unit cells appears as

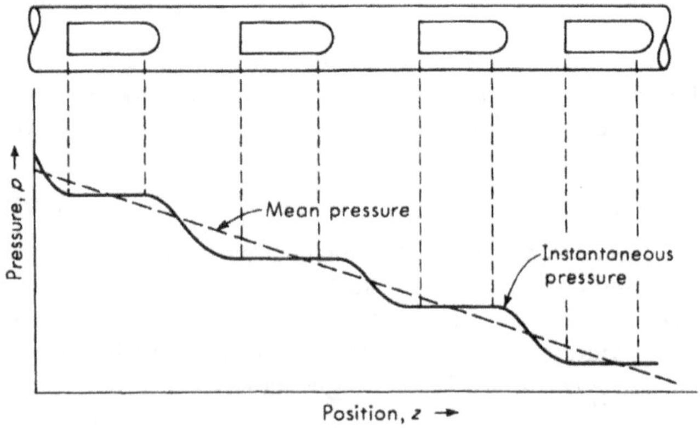

Fig. 10.16 Mean and alternating components of pressure drop in slug flow.

in Fig. 10.16. An approximately triangular or sawtooth alternating component of pressure drop is superimposed on the average pressure gradient and can excite oscillations in two-phase circuits in which slug flow occurs.

PROBLEMS

10.1. A certain silicone fluid has a viscosity of 5000 cp, a surface tension of 21 dynes/cm, and a density of 1 g/cm³. What is the rise velocity of a slug-flow bubble in stationary liquid in vertical pipes with diameters of 0.1, 0.5, 5, and 24 in.?

10.2. When a long bubble rises in a tube closed at the bottom, the value of j ahead of the bubble is not zero because of expansion of the gas in the hydrostatic pressure gradient. A bubble 10 in.³ in volume is injected into a column of water 100 ft high in a 1-in. pipe. If the temperature is 70°F and the pipe is closed at the bottom and open to the atmosphere at the top, how long does it take after release before the bubble breaks the surface?

10.3. Use Eq. (5.32) to show that the film thickness around a slug-flow bubble rising in stagnant viscous fluid is about 34% of the radius. What error would have been introduced by using Eq. (11.68) which does not allow for the curvature of the tube?

10.4. What is the minimum tube size in which large bubbles of air will rise in stationary water at 70°F (*a*) on earth, (*b*) in a spaceship for which "g" = 10^{-4} ft/sec²?

10.5. Water at 100 psia is evaporated slowly in a long vertical ½-in.-diam tube. If $G = 10^5$ lb/(hr)(ft²), the heat flux is 1000 Btu/(hr)(ft²), and the water is saturated at inlet, estimate the void fraction as a function of height. How far up the tube does slug flow persist?

10.6. Solve Probs. 2.30 and 2.31, assuming that the prevailing regime is slug flow.

10.7. Solve Prob. 4.2 if the flow regime is slug flow.

10.8. Estimate the rms value of pressure fluctuations when air and water flow in a vertical pipe of 1-in. diam at 30 psia and 50°F if the flow rates are $W_f = 10^4$ lb/hr, $W_g = 20$ lb/hr, and the volume of each bubble is 10 in.³.

10.9. A large bubble is injected into the lower end of a hydraulic pipeline with 3 ft diam, 2-mi length, and a vertical drop of 1000 ft. How long does the bubble take to reach a closed valve at the top of the pipeline?

10.10. A liquid metal (σ = 300 dynes/cm, μ_f = 0.2 poise, ρ_f = 5 g/cm³) fills a ⅜-in.-diam horizontal pipe. It is desired to blow gas through the pipe to cool the metal and solidify it as a uniform film 0.005 in. thick on the walls. What gas flow rate should be used?

10.11. A series of bubbles, each 1 cm³ in volume, is injected at the rate of 5 per sec into a flow of 10 cm³/sec of glycerol (ρ_f = 1.26 g/cm³, μ_f = 700 cp, and σ = 63 dynes/cm) in a 5-mm-diam horizontal pipe. What are the bubble length and velocity, the film thickness around the bubbles, and the pressure drop?

10.12. Estimate the rate of flow of Vermont maple sirup (ρ_f = 1.42 g/cm³, μ_f = 10,000 cp, and σ = 77 dynes/cm) from a gallon can through a 1-in.-diam pouring spout tilted at 45° to the horizontal.

10.13. Predict the circumferential film thickness variation along a long slug-flow bubble in a horizontal tube as a function of pipe diameter, distance from the bubble nose, j, and fluid properties. Use falling film theory.

10.14. Use Eq. (10.51) to explain why the lines of constant j/v_∞ in Fig. 10.8 are approximately concurrent. Show that the intersections are near the point (δ/D = 1/64.

$j_f'/v_\infty = 5/6$). For what value of N_f will the film thickness around a bubble be independent of j?

10.15. Draw the streamlines in the liquid slug for the cases $C_1 < 2$ and $C_1 > 2$ in laminar flow. Discuss the qualitative effects which can be observed in the two cases.

10.16. A long vertical pipe of diameter 2 inches contains water at 50°F. The air pressure at the top of the pipe is equivalent to a water depth of L_a. The nose of a single slug flow bubble with volume equivalent to a full pipe of length l_0 is at a depth L at time zero. As the bubble rises it expands isothermally. Use (10.33) to derive its velocity as a function of distance traveled and the time taken to reach the top of the pipe. For what conditions is explosive growth predicted?

REFERENCES

1. Dumitrescu, D. T.: *Z. Angew Math. Mech.*, vol. 23, no. 3, p. 139, 1943.
2. Davies, R. M., and G. I. Taylor: *Proc. Roy. Soc.* (London), vol. 200, ser. A, pp. 375–390, 1950.
3. White, E. T., and R. H. Beardmore: *Chem. Eng. Sci.*, vol. 17, pp. 351–361, 1962.
4. Nicklin, D. J., J. O. Wilkes, and J. F. Davidson: *Trans. Inst. Chem. Engrs.*, vol. 40, pp. 61–68, 1962.
5. Griffith, P.: ASME paper no. 63-HT-20, *Nat. Heat Transfer Conf.*, Boston, Mass., 1963.
6. Birkhoff, G., and D. Carter: *J. Rat. Mech. Anal.*, vol. 6, no. 6, pp. 769–780, 1957.
7. Wallis, G. B.: General Electric Company, Schenectady, N.Y., Rept. 62GL130, 1962.
8. Bretherton, F. P.: *J. Fluid Mech.*, vol. 10, p. 166, 1961.
9. Hattori, S.: *Rept. Aeronaut. Res. Inst. Tokyo Imp. Univ.*, no. 115, 1935.
10. Harmathy, T. Z.: *A. I. Ch. E. J.*, vol. 6, p. 281, 1960.
11. Masica, W. J., and D. A. Petrash: NASA Rept. TN D-3005, 1965.
12. Griffith, P., and G. B. Wallis: *Trans. ASME J. Heat Transfer*, vol. 83, Series C, no. 3, pp. 307–320, 1961.
13. Taylor, G. I.: *J. Fluid Mech.*, vol. 10, pp. 161–165, 1961.
14. Moissis, R., and P. Griffith: *Trans. ASME J. Heat Transfer*, Ser. C, vol. 84, p. 29, 1962.
15. Suo, M., and P. Griffith: paper no. 63-WA-96, ASME, 1963.
16. Fairbrother, F., and A. E. Stubbs: *J. Chem. Soc.*, vol. 1, pp. 527–529, 1935.
17. Armand, A.: AERE trans. 828, Harwell, England, 1959.
18. Runge, D. E., and G. B. Wallis: AEC Rept. NYO-3114-8 (EURAEC-1416), 1965.
19. Wallis, G. B.: AEC Rept. NYO-3114-14 (EURAEC), 1966.
20. Griffith, P., and K. S. Lee: ASME paper no. 63-WA-97, 1963.
21. Goldsmith, H. L., and S. G. Mason: *J. Colloid. Sci.*, vol. 18, pp. 237–261, 1963.
22. Nicklin, D. J., and J. F. Davidson: paper no. 4, *Two-phase Flow Symp.*, Institution of Mechanical Engineers, London, 1962.
23. Porteous, A.: Dartmouth College, Hanover, N.H., unpublished work, 1966.
24. Govier, G. W., and W. L. Short: *Can. J. Chem. Eng.*, vol. 36, no. 1, p. 195, 1958.
25. Wallis, G. B.: Rept. AEEW-R 142, Atomic Energy Establishment, Winfrith Heath, England, 1962.
26. Cox, B. G.: *J. Fluid Mech.*, vol. 20, part 2, pp. 193–200, 1964.

11
Annular Flow

11.1 INTRODUCTION

In the *annular* flow pattern a continuous liquid film flows along the wall of a pipe while the gas flows in a central "core." If the core contains a significant number of entrained droplets, the flow is described as *annular mist*, which could be regarded as a transition between ideal annular flow and a fully dispersed drop flow pattern.

Stratified flow occurs in horizontal or inclined pipes and is topologically similar to annular flow because the two components flow side by side without mixing. The lack of symmetry in stratified flow makes analysis difficult although some of the same basic techniques can be applied to either annular or stratified flow.

Annular flow is the predominant flow pattern in evaporators, natural gas pipelines, and steam heating systems.

Theories of annular flow provide an excellent example of the pyramid of analytical techniques that was presented in Chap. 1. Correlations, simple models, and integral and differential methods can all be developed in a hierarchy of complexity. Since each method of analysis can be applied to either the gas or the liquid, numerous combinations are possible.

In the following developments we shall be particularly concerned with showing how the successive levels of sophistication are interrelated. It will usually be found that the more complex analysis provides a method for predicting empirical factors in the simpler models, or for evaluating correction factors which increase the level of accuracy.

Horizontal and vertical flow will be treated separately. However, general techniques which apply to any flow orientation will be derived wherever possible.

11.2 HORIZONTAL FLOW

THE BOUNDARIES OF THE ANNULAR AND STRATIFIED REGIMES IN HORIZONTAL FLOW

The limits of the various horizontal flow regimes are not yet well understood. A simple plot by Baker[1] (Fig. 11.1) may give an indication of the general trend but certainly does not contain all the relevant parameters. Qualitatively, however, high gas flow rates tend to cause entrainment of droplets and high liquid rates to cause the formation of bubbles or slugs as indicated in the figure.

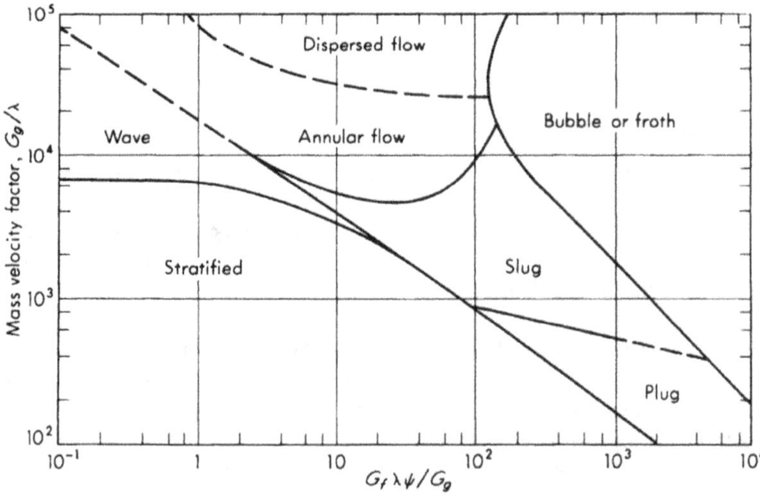

Fig. 11.1 Flow-regime correlations for adiabatic horizontal two-phase two-component flow by Baker[1] (ρ_g and ρ_f in pounds per cubic ft, G_f and G_g in pounds per hour per square foot, σ in dynes/cm, μ_f in cp). $\lambda = [(\rho_g/0.075)(\rho_f/62.3)]^{1/2}$; $\psi = [(73/\sigma)\mu_f(62.3/\rho_f)^2]^{1/3}$.

The ordinate in Baker's plot is not dimensionless and cannot be interpreted physically in its present form. However, it does contain the group $(\rho_g^{1/2} j_g)$ which represents the inertia of the gas. One would expect the balance between this force and gravity to scale stratification effects, probably in the form of the dimensionless group

$$j_g^* = j_g \rho_g^{1/2} [gD(\rho_f - \rho_g)]^{-1/2} \tag{11.1}$$

For the pipe sizes considered by Baker the values of j_g^* at the boundaries between stratified and wavy flow and between slug and annular flow are in the range 0.25 to 1.0.

The dispersed flow boundary is governed by the mechanism of droplet entrainment. This has been found to be virtually insensitive to pipe orientation and is governed primarily by the drag forces that the gas exerts on irregularities at the interface. Several regimes have been identified depending on the relative importance of surface-tension, viscous, and inertia effects.

For an air-water system at atmospheric pressure the critical gas velocity for the onset of entrainment is about 70 fps except at very low liquid rates when the viscous forces in the thin liquid film inhibit the formation of large waves. Steen[2] suggests that as long as viscous forces in the liquid can be ignored the critical gas velocity is given by the equation

$$\frac{j_g \mu_g}{\sigma} \left(\frac{\rho_g}{\rho_f} \right)^{1/2} = 2.5 \times 10^{-4} \tag{11.2}$$

which is compatible with Baker's plot over the range of variables which are represented. Further discussion of entrainment will be given in Chap. 12.

CORRELATIONS

The correlations for separated flow which were introduced in Chap. 3 apply to horizontal annular and stratified flows. In particular, Martinelli's correlation provides a very rapid method for estimating frictional pressure drop and void fraction by using Eq. (3.30) with $n = 3.5$, and Eq. (3.32). Alternatively Figs. 3.4 to 3.8 may be used. The Martinelli parameters ϕ_f, ϕ_g, and X are also useful for expressing the results of more complex methods of analysis.

SEPARATED FLOW, ANNULAR GEOMETRY MODEL

As in Example 3.4, the one-dimensional equations of motion in annular flow can be developed by using average values of the interfacial and wall shear stresses. In steady horizontal flow, with no acceleration, force balances for the gas core and the combined flow relate these shear stresses to the pressure gradient. Thus

$$-\frac{dp}{dz} = \frac{4\tau_i}{D\sqrt{\alpha}} \tag{11.3}$$

$$-\frac{dp}{dz} = \frac{4\tau_w}{D} \tag{11.4}$$

THE INTERFACIAL SHEAR STRESS

The interfacial shear stress will depend, presumably, on the difference between the gas velocity and some characteristic interface velocity. If the gas velocity is much larger than the liquid velocity, then one may neglect the liquid velocity and assume in the usual way that

$$\tau_i = (C_f)_i \tfrac{1}{2}\rho_g v_g^2 = \frac{(C_f)_i \tfrac{1}{2}\rho_g j_g^2}{\alpha^2} \tag{11.5}$$

If the same gas flow filled the pipe, the wall shear stress would be given, in terms of the friction factor $(C_f)_g$, by

$$\tau_{wg} = (C_f)_g \tfrac{1}{2}\rho_g j_g^2 \tag{11.6}$$

and the pressure drop would be

$$-\left(\frac{dp}{dz}\right)_g = \frac{4\tau_{wg}}{D} = \frac{2(C_f)_g \rho_g j_g^2}{D} \tag{11.7}$$

Combining Eqs. (11.3), (11.5), and (11.7) with the definition of ϕ_g^2 in Eq. (3.24) we have

$$(C_f)_i = \alpha^{5/2}\phi_g^2(C_f)_g \tag{11.8}$$

which provides a physical interpretation for the Martinelli parameter ϕ_g^2.

An alternative parameter which is sometimes used is the *superficial gas friction factor*. It is defined as the friction factor which would give the observed pressure drop if the same gas flow rate filled the whole pipe. Denoting this parameter by $(C_f)_{sg}$ we have

$$-\frac{dp}{dz} = (C_f)_{sg}\frac{2}{D}\rho_g j_g^2 \tag{11.9}$$

The relationship between $(C_f)_{sg}$ and $(C_f)_i$ is simply

$$(C_f)_{sg} = \frac{(C_f)_i}{\alpha^{5/2}} = \phi_g^2(C_f)_g \tag{11.10}$$

An explicit expression for $(C_f)_i$ in terms of measurable quantities follows from Eqs. (11.3) and (11.5). Thus

$$(C_f)_i = -\frac{dp}{dz}\frac{\alpha^{5/2}D}{2\rho_g j_g^2} \tag{11.11}$$

Figure 11.2 shows a qualitative plot of the interfacial friction factor versus liquid fraction for both upflow, horizontal flow, and downflow and

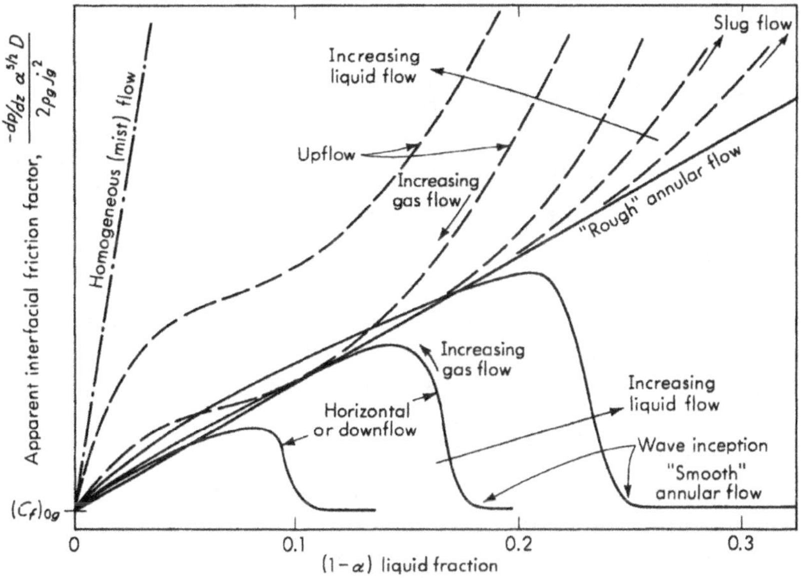

Fig. 11.2 Qualitative aspects of the pressure-drop characteristics of various flow regimes represented in terms of the apparent interfacial friction factor.[3]

serves to indicate the various important regimes of operation.[3] In horizontal or downflow at low gas velocities the liquid film is smooth and the friction factor is approximately the smooth pipe value. Above a critical gas velocity for wave inception the friction factor rises rapidly to a maximum after which it tends to follow the line marked "rough annular flow." At higher gas velocities the tops of the waves are sheared off to form entrained droplets and the density of the core increases. This leads to an increase in the friction factor, in about the ratio ρ_c/ρ_g where ρ_c is now the average homogeneous density of the core.

At very high flow rates of both phases almost all of the liquid flow is entrained and the pressure drop and void fraction are given by homogeneous flow theory as

$$-\frac{dp}{dz} = \frac{2C_f G j}{D} \tag{11.12}$$

In homogeneous flow the liquid fraction is

$$1 - \alpha = \frac{j_f}{j_f + j_g} \tag{11.13}$$

In annular flow the volumetric flow rate of the gas is usually much greater than the liquid flow rate and these equations can be approximated by

$$-\frac{dp}{dz} = C_f \tfrac{1}{2}\rho_g j_g^2 \left(1 + \frac{G_f}{G_g}\right)\frac{4}{D} \tag{11.14}$$

$$1 - \alpha \approx \frac{j_f}{j_g} = \frac{\rho_g}{\rho_f}\frac{G_f}{G_g} \tag{11.15}$$

Using Eq. (11.15) in Eq. (11.14) we find that

$$-\frac{dp}{dz} = C_f \tfrac{1}{2}\rho_g j_g^2 \left[1 + \frac{\rho_f}{\rho_g}(1 - \alpha)\right]\frac{4}{D} \tag{11.16}$$

Since the friction factor in homogeneous flow is relatively unchanged from the value for gas alone, $(C_f)_g$, we can use Eq. (11.16) to show that the apparent value of the interfacial friction factor in homogeneous flow is

$$-\frac{dp}{dz}\frac{D\alpha^{5/4}}{2\rho_g j_g^2} = (C_f)_g \left[1 + \frac{\rho_f}{\rho_g}(1 - \alpha)\right]\alpha^{5/4} \tag{11.17}$$

For air and water at atmospheric pressure $\rho_f/\rho_g \approx 800$, and for homogeneous flow the value of Eq. (11.17) increases very rapidly as a function of $1 - \alpha$, as shown in Fig. 11.2. The curves tend to move toward the homogeneous flow line as the liquid rate is increased and the percent entrained goes up.

In vertical flow the interfacial shear has to support the film against gravity. Furthermore any slugs of liquid or entrained droplets add a gravitational component to the core pressure drop. Thus, at low gas rates in the slug-flow region the friction factor defined by Eq. (11.11) increases well above the annular flow value. At lower liquid rates this is usually at values of liquid fraction above about 0.2. At high liquid rates the transition between slug flow and annular mist flow becomes obscure and there is a general motion toward the homogeneous flow line.

The key to a simple analysis of "rough" or "wavy" annular flow is a plot of the interfacial friction factor versus the dimensionless film thickness as is shown in Fig. 11.3. The points cluster pretty well around a line with the equation

$$(C_f)_i = 0.005\left(1 + 300\frac{\delta}{D}\right) \tag{11.18}$$

For thin films in pipes this is approximately the same as

$$(C_f)_i = 0.005\,[1 + 75(1 - \alpha)] \tag{11.19}$$

We note at this point that the rough pipe correlations of Nikuradse[4] and Moody[5] can be approximated by the equation

$$C_f \approx 0.005\left(1 + 75\frac{k_s}{D}\right) \tag{11.20}$$

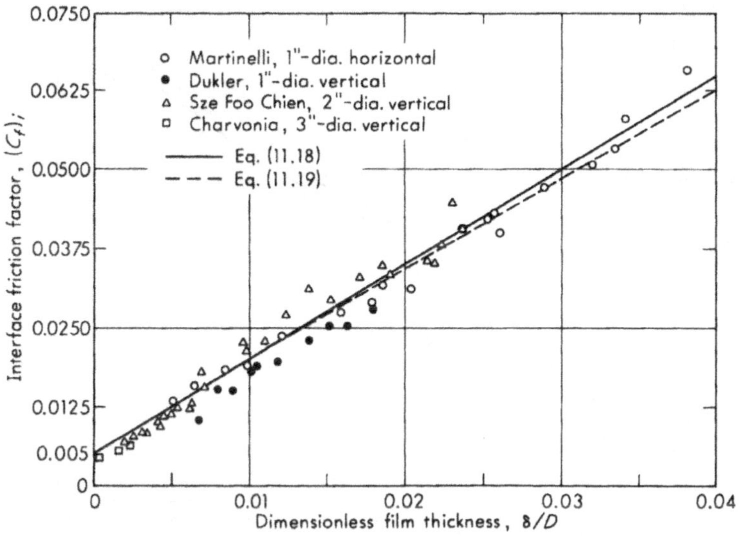

Fig. 11.3 Comparison between Eqs. (11.18) and (11.19) and various air-water data.[3]

over the range $0.001 < k_s/D < 0.03$, where k_s is the grain size of a "sand roughness." Equation (11.18) therefore shows that a wavy annular film is about equivalent to a sand roughness of four times the film thickness.

Using Eqs. (11.19) and (11.11) we obtain for horizontal flow

$$\left(-\frac{dp}{dz}\right)_F = 10^{-2}\frac{\rho_g j_g^2}{D}\frac{1 + 75(1-\alpha)}{\alpha^{5/2}} \tag{11.21}$$

Equation (11.19) may also be used to give an analytical expression for the Martinelli parameter ϕ_g^2. Assuming that $(C_f)_g \approx 0.005$, we have from Eqs. (11.8) and (11.19)

$$\phi_g = \left[\frac{1 + 75(1-\alpha)}{\alpha^{5/2}}\right]^{1/2} \tag{11.22}$$

The alternative representation in terms of the superficial gas friction factor is

$$(C_f)_{sg} = 0.005\,\frac{1 + 75(1-\alpha)}{\alpha^{5/2}} \tag{11.23}$$

A simpler equation which gives much the same result up to $1 - \alpha = 0.1$ is

$$(C_f)_{sg} = 0.005[1 + 90(1-\alpha)] \tag{11.24}$$

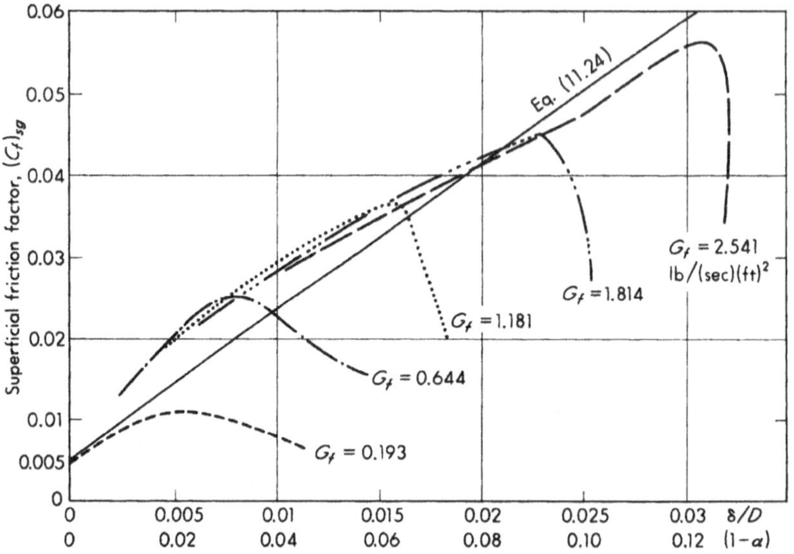

Fig. 11.4 Comparison between Eq. (11.24) and results of Chien and Ibele[6] for vertical downflow.

and is easy to remember by the rule of thumb that a liquid fraction of $\frac{1}{10}$ increases the pressure drop by a factor of 10. The above equation is compared in Fig. 11.4 with experimental curves of Chien and Ibele.[6]

Example 11.1 Air and water flow in a 1-in.-diameter horizontal pipe. The gas mass flow rate is 300 lb/hr and $\rho_g = 0.1$ lb/ft^3. If the void fraction is 90 percent, estimate the pressure gradient using (*a*) Eq. (11.21), (*b*) Eqs. (11.23) and (11.9).

Solution The area of the cross section is

$$\frac{\pi}{(4)(144)} = 5.45 \times 10^{-3} \text{ ft}^2$$

Therefore

$$j_g = \frac{300}{3600} \frac{1}{0.1} \frac{1}{5.45 \times 10^{-3}} = 153 \text{ fps}$$

(*a*) Using Eq. (11.21) we have

$$\left(-\frac{dp}{dz}\right)_F = 10^{-2} \frac{(0.1)(153^2)}{(32.2)(144)(\frac{1}{12})} \frac{8.5}{0.768} = 0.67 \text{ psi/ft}$$

(*b*) The superficial friction factor, from Eq. (11.24), is

$$(C_f)_{sg} = 0.05$$

whence, using Eq. (11.9),

$$-\frac{dp}{dz} = 0.05 \frac{2}{\frac{1}{12}} \frac{(0.1)(153^2)}{(144)(32.2)} = 0.61 \text{ psi/ft}$$

These expressions for the interfacial friction factor can also be compared with Levy's[7] correlation which is expressed in terms of a function F' as follows:

$$F' = \left[\frac{\tau_i}{v_g(v_g - v_f)(\rho_f - \rho_g)}\right]^{1/2} R \tag{11.25}$$

R is an empirical function of density ratio which can be represented quite well[3] by the equation

$$R = \frac{1}{2}\sqrt{\frac{\rho_f}{\rho_g}} \tag{11.26}$$

Making the following approximations, which are consistent with the level of sophistication of the present theory,

$$\rho_f \gg \rho_g$$
$$v_g \gg v_f$$

Equations (11.25) and (11.26) can be combined to give

$$F' = \left(\frac{\tau_i}{4\rho_g v_g^2}\right)^{1/2} \tag{11.27}$$

whence, using Eq. (11.5),

$$F' = \left(\frac{C_{fi}}{8}\right)^{1/2} \tag{11.28}$$

Levy correlated F' versus the ratio of the average film thickness to the pipe radius. In view of Eq. (11.28) this is just what was done in Fig. 11.3 apart from a constant factor. The two theories can therefore be compared directly by substituting Eq. (11.18) into Eq. (11.28). Comparison with the results of Wicks and Dukler[8] is shown in Fig. 11.5.

THE WALL SHEAR STRESS

Performing a similar analysis for the liquid we find that the equivalent of Eq. (11.8) is

$$(C_f)_w = (1 - \alpha)^2 \phi_f^2 (C_f)_f \tag{11.29}$$

where $(C_f)_w$ is the wall friction factor for the film and $(C_f)_f$ the friction factor for the liquid alone in the pipe. One might expect that $(C_f)_w \approx (C_f)_f$, since the wall roughness is the same in both cases, and Eq. (11.29) then gives an expression for ϕ_f as follows:

$$\phi_f \approx \frac{1}{1 - \alpha} \tag{11.30}$$

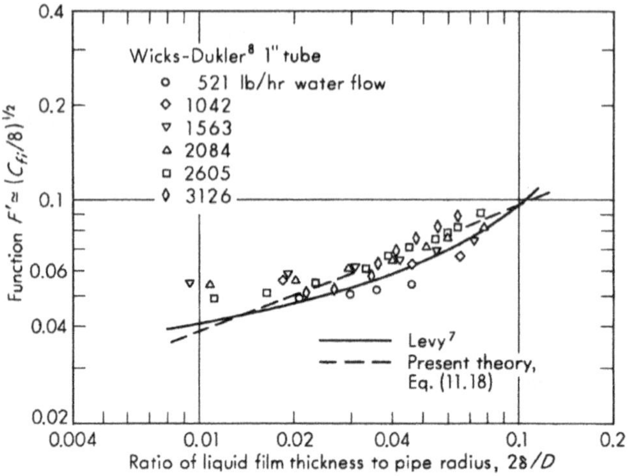

Fig. 11.5 Comparison between Levy's correlation, Eq. (11.18) and data for horizontal flow.

Equation (11.30) has been found to be approximately valid by numerous investigators; it is also in agreement with the results of the integral analysis which will be presented later. Moreover, it will be found that the wall friction factor is almost the same function of the liquid Reynolds number as it is for single-phase flow, apart from the transition region. The liquid Reynolds number is defined as

$$\text{Re}_f = \frac{\rho_f j_f D}{\mu_f} \tag{11.31}$$

In general, from Eq. (11.4)

$$\left(-\frac{dp}{dz}\right)_F = (C_f)_w \frac{2\rho_f j_f^2}{(1-\alpha)^2 D} \tag{11.32}$$

where, for laminar flow,

$$(C_f)_w = \frac{16}{\text{Re}_f} \tag{11.33}$$

and as a simple first approximation in turbulent flow, $(C_f)_w$ can be taken to be 0.005, as usual.

EVALUATION OF PRESSURE DROP AND VOID FRACTION

The above relationships permit the evaluation of the interfacial and wall shear stresses in Eqs. (11.3) and (11.4). These two equations can then be solved simultaneously for the pressure drop and void fraction. For

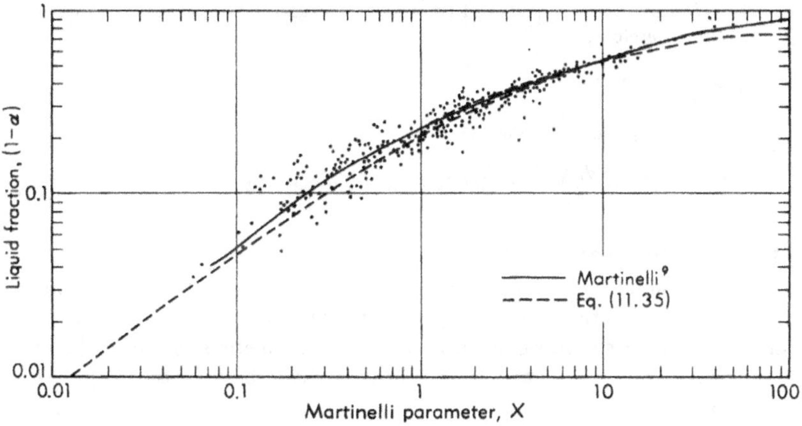

Fig. 11.6 Comparison of liquid fraction predicted by Eq. (11.35) with empirical results of Martinelli.

the void fraction in turbulent-turbulent flow, for example, we eliminate $(dp/dz)_F$ between Eqs. (11.32) and (11.21) to get

$$\frac{(1-\alpha)^2[1+75(1-\alpha)]}{\alpha^{5/2}} = \frac{\rho_f j_f^2}{\rho_g j_g^2} \tag{11.34}$$

Alternatively, in terms of the Martinelli parameter,

$$\frac{(1-\alpha)^2[1+75(1-\alpha)]}{\alpha^{5/2}} = X^2 \tag{11.35}$$

for either laminar or turbulent flow of the liquid. Equation (11.35) is plotted in Fig. 11.6 and compares well with Martinelli's empirical curve.

In practice, X^2 is readily calculated from known quantities; α can then be determined using Eq. (11.35). The pressure drop follows most rapidly by using this value of α in Eq. (11.30).

Example 11.2 Solve Example 3.3 using Eqs. (11.35) and (11.30) if flow is horizontal and the liquid flow rate is reduced by a factor of 3.

Solution The liquid Reynolds number is 23,000/3 = 7700; therefore flow is turbulent and the friction factor from single-phase flow charts is 0.009. The single-phase liquid frictional pressure drop is

$$\left(-\frac{dp}{dz}\right)_f = \frac{(2)(0.009)(31.5^2)(.1)}{2.48} = 7.2 \text{ dynes/cm}^3$$

The gas phase pressure drop, as before, is 9.4 dynes/cm³

Therefore

$$X^2 = \frac{7.2}{9.4} = 0.766.$$

Equation (11.35) is now solved by iteration to give $1 - \alpha = 0.18$. From Eq. (11.30), therefore,

$$\phi_f = \frac{1}{0.18} = 5.55$$

whence the pressure gradient is

$$-\frac{dp}{dz} = \phi_f^2 \left(-\frac{dp}{dz}\right)_f = (5.55^2)(7.2) = 222 \text{ dynes/cm}^3$$

EXTENSION TO THE GENERAL CASE

These expressions for the wall and interfacial shear stresses can be used in the general equations of Example 3.4 to form the basis for an analysis of annular flow in the presence of body forces, area change, phase change, and compressibility effects. Some special cases of this technique will be developed later.

IMPROVEMENTS TO THE THEORY

Up to now we have considered only the effect of the major variables on the dynamics of annular flow. Balances have been made between the dominant forces, whereas secondary effects have been neglected. In general, however, there are at least seven different interacting forces which are due to the pressure gradient, buoyancy, surface tension, and the inertia and viscosity of each phase. A complete representation of the interplay between all of these forces is not available; however, the developments which will be considered in the next few pages provide an example of successive methods for modifying the theory in order to develop a more accurate, but more complex, theoretical structure. Further modifications will be considered in the discussion of vertical annular flow.

INTEGRAL ANALYSIS

The basis of an integral analysis of annular flow is the assumption of a velocity profile. This profile is integrated over the portion of the pipe which is occupied by each phase in order to give the flow rates. The wall and interfacial shear stresses are related to the profile by means of single-phase flow correlations which are already well established. Because of the complications which are introduced by the boundary conditions for the gas at the wavy interface, this analysis has been most successfully developed for the liquid film in symmetrical annular flow.

THE LIQUID FILM

Suppose that there were a single-phase liquid flow with mean velocity $\bar{v}$ which occupied the whole pipe. The frictional pressure drop would be

$$\left(-\frac{dp}{dz}\right)_F = 2C_f \frac{\rho_f \bar{v}^2}{D} \tag{11.36}$$

where the friction factor C_f is a function of the Reynolds number

$$\mathrm{Re}_m = \frac{\bar{v} D \rho_f}{\mu_f} \tag{11.37}$$

Now suppose that only a part of this single-phase flow is present, namely, from the wall inward to a radius $\sqrt{\alpha}\, r_0$, and that the rest of the pipe is occupied by gas, a droplet-laden core, or even a plug of froth or foam. If the velocity profile is $u(r)$, then the liquid flux in this annulus is

$$j_f = \frac{1}{\pi r_0^2} \int_{\sqrt{\alpha}\, r_0}^{r_0} 2\pi r u \, dr = \int_{\sqrt{\alpha}}^{1} 2r^* u \, dr^* \tag{11.38}$$

For laminar flow we have

$$u = 2\bar{v}(1 - r^{*2}) \tag{11.39}$$

and Eq. (11.38) gives eventually

$$j_f = (1 - \alpha)^2 \bar{v} \tag{11.40}$$

On the other hand, for turbulent flow with a one-seventh power velocity profile we have

$$u = {}^{60}\!/_{49} \bar{v}(1 - r^*)^{1/7} \tag{11.41}$$

and, from Eq. (11.38),

$$j_f = (1 - \sqrt{\alpha})^{8/7}(1 + {}^{8}\!/_{7} \sqrt{\alpha})\bar{v} \tag{11.42}$$

Both Eqs. (11.40) and (11.42) are of the general form

$$j_f = J(\alpha)\bar{v} \tag{11.43}$$

where $J(\alpha)$ is some function of α which could be obtained by any assumption about the velocity profile.

In a two-phase flow, if j_f and J are known, then $\bar{v}$ can be calculated and the result used in Eqs. (11.36) and (11.37) to calculate the pressure drop.

These results are readily related to the Martinelli parameter ϕ_f for the liquid. For laminar flow we have

$$\left(-\frac{dp}{dz}\right)_F = \frac{32\bar{v}\mu_f}{D^2} \tag{11.44}$$

and

$$\left(-\frac{dp}{dz}\right)_f = \frac{32 j_f \mu_f}{D^2} \tag{11.45}$$

Comparing Eqs. (11.40), (11.44), and (11.45) we find

$$\phi_f^2 = \frac{\left(-\frac{dp}{dz}\right)_F}{\left(-\frac{dp}{dz}\right)_f} = \frac{1}{(1-\alpha)^2} \tag{11.46}$$

whence

$$\phi_f = \frac{1}{1-\alpha} \tag{11.47}$$

For turbulent flow, on the other hand, the Blasius equation[10] (which is compatible with the one-seventh power law) gives

$$C_f = 0.079\ \mathrm{Re}^{-0.25} \tag{11.48}$$

and, from Eq. (11.36),

$$\left(-\frac{dp}{dz}\right)_F = \frac{0.158\rho_f\bar{v}^2}{D}\left(\frac{\mu_f}{\bar{v}D\rho_f}\right)^{1/4} \tag{11.49}$$

Using Eq. (11.42) to find $(dp/dz)_f$ by putting j_f for $\bar{v}$ in Eq. (11.49) and again evaluating ϕ_f we find that[11,12]

$$\phi_f = \frac{1}{(1-\sqrt{\alpha})(1+\frac{8}{7}\sqrt{\alpha})^{7/8}} \tag{11.50}$$

Table 11.1 shows that the difference between Eqs. (11.47) and (11.50) is negligible for most practical purposes and that one might as well use Eq.

Table 11.1 Comparison between the predictions of Eqs. (11.47) and (11.50) and Martinelli's relation ϕ_{ftt} for turbulent-turbulent flow[11]

α	ϕ_{ftt}	$\phi_f = \frac{1}{1-\alpha}$	% error	$\phi_f = \frac{1}{(1-\sqrt{\alpha})(1+\frac{8}{7}\sqrt{\alpha})^{7/8}}$	% error
0.96	24.4	25.0	2.5	25.7	5.1
0.95	18.5	20.0	8.1	20.5	9.8
0.91	11.2	11.1	−0.9	11.4	1.8
0.86	7.05	7.14	1.0	7.32	3.7
0.81	5.04	5.26	4.4	5.39	6.5
0.77	4.20	4.35	3.6	4.45	5.6
0.69	3.10	3.23	4.2	3.29	5.8
0.60	2.38	2.5	5.0	2.55	6.7
0.52	1.96	2.08	6.1	2.12	7.6
0.47	1.75	1.89	8.0	1.92	8.9
0.34	1.48	1.52	2.7	1.53	3.3
0.24	1.29	1.32	2.3	1.33	3.0
0.16	1.17	1.19	1.7	1.20	2.5
0.10	1.11	1.11	0	1.12	0.9

(11.47) for turbulent as well as for laminar flows. Also shown for comparison are Martinelli's correlated values for ϕ_f in the turbulent-turbulent regime (ϕ_{ftt}).

The result of this analysis is to confirm Eq. (11.30) for both laminar and turbulent horizontal annular flow. Furthermore, Eq. (11.49) can be rearranged with the help of Eq. (11.43) to give an expression for the wall friction factor, as follows:

$$(C_f)_w = 0.079 \text{ Re}_f^{-1/4} \frac{(1-\alpha)^2}{J^{7/4}} \tag{11.51}$$

Since $J^{7/8}$ is almost identical with $1 - \alpha$ for the one-seventh power profile in view of Table 11.1, this equation reduces to

$$(C_f)_w = 0.079 \text{ Re}_f^{-1/4} \tag{11.52}$$

which shows that the wall friction factor can be correlated by the Blasius equation in terms of the Reynolds number based on the overall liquid flux j_f [Eq. (11.31)]. This represents an improvement over the simple assumption that $(C_f)_w$ is always equal to 0.005.

THE GAS CORE

The case of a viscous gas core is of little practical interest. In fact the integral analysis yields exactly the same result as the more fundamental differential analysis which was presented in Example 5.1.

A solution for a turbulent core which is not far from the predictions of Eq. (11.22) at a higher value of film thickness was derived by Turner[11] who assumed a one-seventh power law based on the ratio of the radius to the overall pipe radius. In addition, he required that the gas velocity should equal the liquid velocity at the interface. The liquid interface velocity was, in turn, derived from a further one-seventh power law in the film.

The result equivalent to Eq. (11.43) was

$$j_g = \bar{v}_g \left\{1 - (1 - \sqrt{\alpha})^{8/7}(1 + \tfrac{8}{7}\sqrt{\alpha}) - \frac{\alpha(1-\sqrt{\alpha})^{1/7}}{0.817}\left[1 - \left(\frac{\rho_g}{\rho_f}\right)^{3/7}\left(\frac{\mu_f}{\mu_g}\right)^{1/7}\right]\right\} \tag{11.53}$$

giving a value of ϕ_g as follows:

$$\phi_g = \left\{1 - (1 - \sqrt{\alpha})^{8/7}(1 + \tfrac{8}{7}\sqrt{\alpha}) - \frac{\alpha(1-\sqrt{\alpha})^{1/7}}{0.817}\left[1 - \left(\frac{\rho_g}{\rho_f}\right)^{3/7}\left(\frac{\mu_f}{\mu_g}\right)^{1/7}\right]\right\}^{-7/8} \tag{11.54}$$

DIFFERENTIAL ANALYSIS

A differential analysis for horizontal laminar symmetrical annular flow proceeds as in Example 5.1 except that $\cos\theta = 0$. After some algebra the Martinelli parameters are found to be

$$\phi_f = \frac{1}{1-\alpha} \tag{11.55}$$

$$\phi_g^2 = \frac{1}{2\alpha\mu_g((1-\alpha)/\mu_f + \alpha/2\mu_g)} \tag{11.56}$$

Simple relationships between ϕ_f and ϕ_g result in the limiting cases where $\mu_g \ll \mu_f$ and $\mu_g = \mu_f$. In the former case Eq. (11.56) reduces to

$$\phi_g^2 = \frac{1}{\alpha^2} \tag{11.57}$$

and this coupled with Eq. (11.55) gives

$$\frac{1}{\phi_f} + \frac{1}{\phi_g} = 1 \tag{11.58}$$

In the latter case Eq. (11.56) becomes

$$\phi_g^2 = \frac{1}{\alpha(2-\alpha)} \tag{11.59}$$

Taking Eqs. (11.59) and (11.55) together we find that

$$\frac{1}{\phi_g^2} + \frac{1}{\phi_f^2} = 1 \tag{11.60}$$

Equations (11.58) and (11.60) are special cases of Eq. (3.30).

A differential analysis of the liquid film which covers the whole range of Reynolds numbers was performed by Hewitt,[13] using Eq. (5.4) up to $y^+ = 5$, Eq. (5.10) for $5 < y^+ < 20$, and Eq. (11.78) in the turbulent core. The result can be expressed as the relationship between wall friction factor and liquid Reynolds number which is shown in Fig. 11.7. The differential analysis agrees well with the integral analysis in both laminar and turbulent flow and has the additional advantage of including the "transition" region.

11.3 COUNTERCURRENT VERTICAL ANNULAR FLOW

Vertical annular flow is free from the stratification effects which can occur with horizontal flow but is complicated by the action of gravity, which produces a body force in the direction of flow, and also by the possibility of upward or downward flow of either component. Although the techniques which apply to horizontal flow also give reasonably accurate pre-

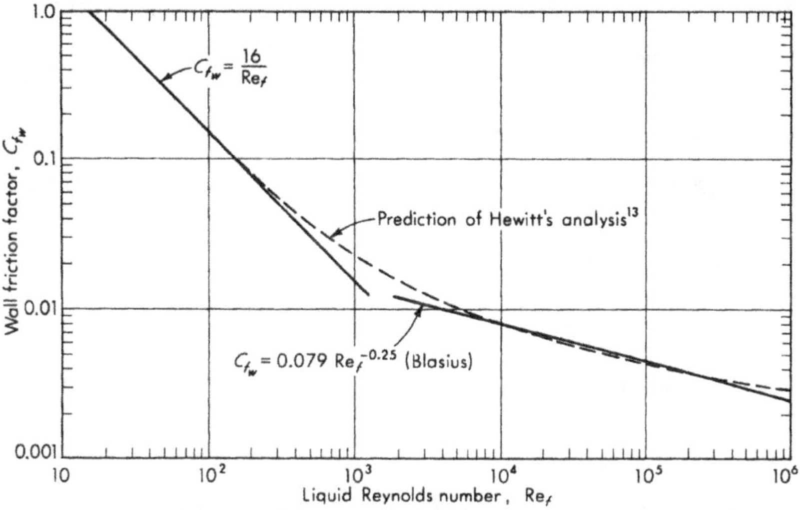

Fig. 11.7 Wall friction factor in horizontal annular flow predicted from integral analysis [Eq. (11.52)] and Hewitt's differential analysis.[13]

dictions in vertical flow at high values of α and sufficiently high velocities, when gravitational effects are relatively small, it is misleading to neglect gravity at lower gas flow rates close to the slug-flow regime boundary and in countercurrent flow.

FALLING FILM FLOW

Falling film flow is a simple case of vertical annular flow. If the gas velocity is sufficiently low, interfacial shear stresses and pressure drop are both negligible and the governing equation for thin films in which wall curvature can be neglected is

$$\tau = g(\rho_f - \rho_g)(\delta - y) \tag{11.61}$$

where δ is the film thickness and y the distance from the wall.

For laminar flow Eq. (11.61) becomes

$$\mu_f \frac{dv}{dy} = g(\rho_f - \rho_g)(\delta - y) \tag{11.62}$$

Integration gives

$$\mu_f v = g(\rho_f - \rho_g)\left(\delta y - \frac{y^2}{2}\right) \tag{11.63}$$

Integrating again over the film, the volumetric flow per unit width[14] is

$$q_f = \frac{\delta^3 g(\rho_f - \rho_g)}{3\mu_f} \tag{11.64}$$

The mass flow per unit width is usually denoted by Γ where

$$\Gamma = \rho_f q_f \tag{11.65}$$

The corresponding total volumetric flow rate of a thin film in a tube of diameter D is

$$Q_f = \frac{\pi}{3} \frac{gD(\rho_f - \rho_g)\delta^3}{\mu_f} \tag{11.66}$$

and the liquid volumetric flux is

$$j_f = \frac{4}{3} \frac{g(\rho_f - \rho_g)\delta^3}{\mu_f D} \tag{11.67}$$

Equation (11.67) can be converted to the equivalent form

$$j_f^* = \frac{4}{3} N_f \left(\frac{\delta}{D}\right)^3 \tag{11.68}$$

Alternatively, a film Reynolds number can be defined as

$$\mathrm{Re}_\Gamma = \frac{4\Gamma}{\mu_f} \tag{11.69}$$

and a dimensionless film thickness as

$$\delta^* = \frac{\delta g^{1/3}(\rho_f - \rho_g)^{1/3}\rho_f^{1/3}}{\mu_f^{2/3}} \tag{11.70}$$

In a circular tube Eqs. (11.69) and (11.70) are equivalent to the following:

$$\mathrm{Re}_\Gamma = \frac{j_f \rho_f D}{\mu_f} = j_f^* N_f \tag{11.71}$$

$$\delta^* = \frac{\delta}{D} N_f^{2/3} \tag{11.72}$$

Equation (11.64) can then be rewritten as

$$\delta^* = 0.909\ \mathrm{Re}_\Gamma^{1/3} \tag{11.73}$$

A plot of experimental data on a basis of δ^* versus Re_Γ shows agreement with Eq. (11.73) up to a value of $\mathrm{Re}_\Gamma = 1000$ (Fig. 11.8). At higher Reynolds numbers the film becomes turbulent and a new correlation is necessary. A line through most of the data in Fig. 11.8 has the equation

$$\delta^* = 0.115\ \mathrm{Re}_\Gamma^{0.6} \tag{11.74}$$

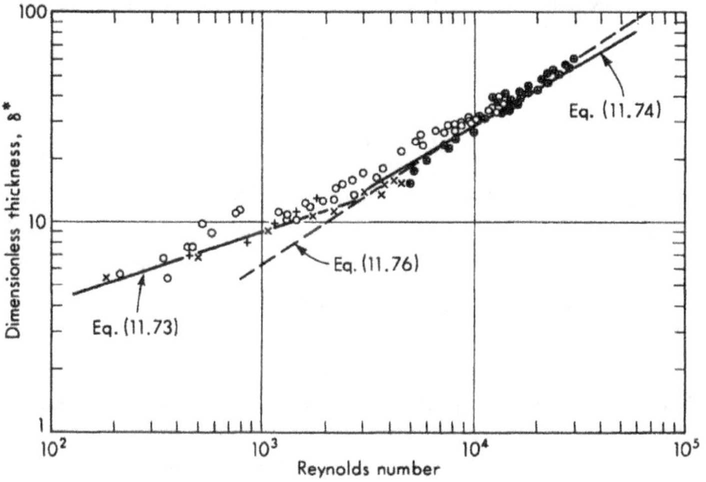

Fig. 11.8 Effect of Reynolds number on falling film thickness with zero shear at the liquid interface. (*Belkin.*[15])

A suggested alternative correlation proposed by Belkin[15] has the form

$$\delta^* = 0.315\,(\mathrm{Re}_\Gamma \sqrt{C_f})^{3/5} \tag{11.75}$$

in which C_f is the appropriate friction factor. For a value of $C_f = 0.008$ this equation becomes

$$\delta^* = 0.063\,\mathrm{Re}_\Gamma^{3/5} \tag{11.76}$$

which is also shown in the figure and gives a good fit to Belkin's own data.

An advantage of Eq. (11.76) is that when it is combined with Eqs. (11.71) and (11.72) the viscosity dependence disappears and we find

$$\frac{\delta}{D} = 0.063 j_f^{*3/5} \tag{11.77}$$

On the other hand, Eq. (11.74) reflects the usual dependence of the friction factor on the -0.2 power of the Reynolds number.

Example 11.3 Water at 70°F flows down a vertical surface at a rate of 10 lb/min per foot width. What is the film thickness?

Solution At 70°F the viscosity of water is 2.5 lb/(hr)(ft). The flow rate per unit width is 600 lb/(hr)(ft). Therefore, from Eq. (11.69),

$$\mathrm{Re}_\Gamma = \frac{(4)(600)}{2.5} = 960$$

Flow is therefore just laminar and Eq. (11.73) applies. The value of δ^* is

$$\delta^* = (0.909)(960)^{1/3} = 8.96$$

Therefore, from Eq. (11.70),

$$\delta = (8.96)\left[\frac{(2.5^2)}{(32.2)(62.5^2)(3600^2)}\right]^{1/3} = 1.4 \times 10^{-3} \text{ ft}$$

A differential analysis of falling films in cocurrent flow was performed by Dukler[16] using Eqs. (5.9) and (5.10) up to $y^+ = 20$ and von Kármán's expression for the eddy viscosity at larger values of y^+. Thus

$$y^+ > 20 \qquad \epsilon = 0.13\left(\frac{dv}{dy}\right)^3\left(\frac{d^2v}{dy^2}\right)^{-2} \tag{11.78}$$

Using these relationships and following the overall plan outlined in Chap. 5, the velocity profile and film thickness were derived. The results shown in Fig. 11.9 give δ^* versus Re_f for various values of a dimensionless interfacial shear parameter β. For downflow in tubes the value of β is

$$-\beta = N_f^{1/3}\frac{\Delta p^*}{4} \tag{11.79}$$

The relative importance of interfacial shear and gravitational effects is scaled by the ratio $4\beta/\delta^*$ which, for thin films, is equivalent to $-\Delta p^*/(1 - \alpha)$ and determines the shear stress variation across the film.

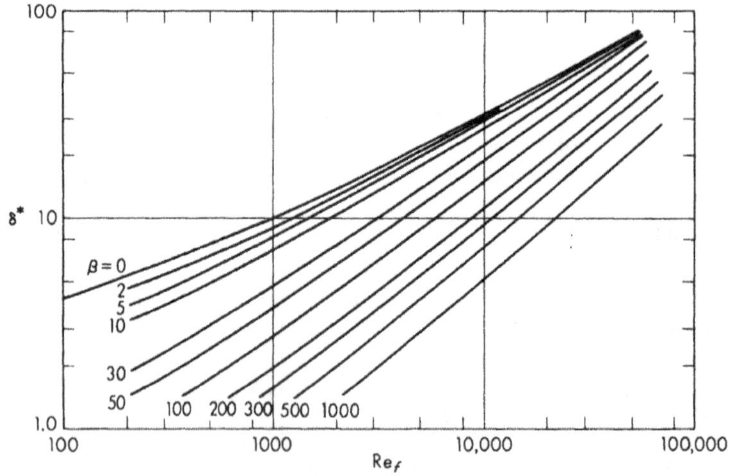

Fig. 11.9 Dimensionless film thickness δ^* versus Reynolds number for various values of dimensionless pressure drop β. (*Dukler.*[16])

STABILITY OF FALLING FILMS

On a vertical surface there is no restoring force for long wavelengths, and the dynamic wave velocity becomes zero. However, the continuity wave velocity is found as in Example 6.1 to be

$$V_w = 3v_f \tag{11.80}$$

a result which was deduced by Benjamin[18] and compares with the value $2.5v_f$ derived by Kapitsa[19] and Portalski.[20]

Since the continuity wave velocity is finite while the dynamic wave velocity is zero, except for surface-tension effects on short wavelengths, a falling film is always unstable in view of Eq. (6.132). The conclusion was reached by Benjamin[18] who also showed that the rate of growth of instabilities was strongly dependent on the Reynolds number. Several authors have claimed that there is a "critical Reynolds Number" for the onset of instability but this is probably due to the fact that the rate of growth of disturbances is so small at low Reynolds numbers that insignificant amplification is noticed in a finite apparatus.

Hewitt and Wallis[21] conducted experiments in which the amplitude of waves on a falling film was measured near the top and bottom of a vertical tube 1¼ in. in diameter and 3 ft long. The wave amplification which occurred in this distance is clearly shown in Fig. 11.10 which shows a trace of film thickness versus time at a given location derived from measurements of the local longitudinal electrical conductivity. The water was made conducting by adding a small amount of salt. Even though the

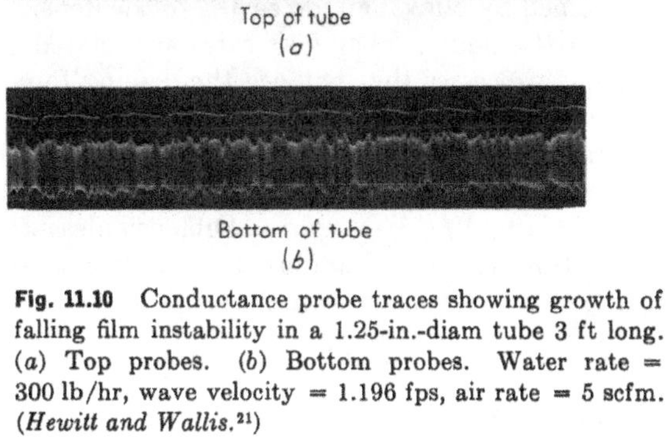

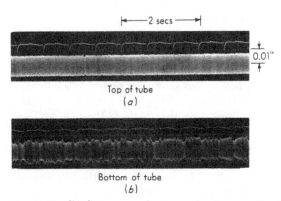

Fig. 11.10 Conductance probe traces showing growth of falling film instability in a 1.25-in.-diam tube 3 ft long. (*a*) Top probes. (*b*) Bottom probes. Water rate = 300 lb/hr, wave velocity = 1.196 fps, air rate = 5 scfm. (*Hewitt and Wallis.*[21])

film was noticeably wavy at the bottom of the tube, Eq. (11.73) was still found to be valid for describing the average film thickness.

In practice, as shown in Prob. 7.4, a whole variety of wavelengths and frequencies are unstable and they all move with velocities less than v_w in view of Eq. (6.126). These waves interact in a rather chaotic way. Hewitt and Wallis[21] found that many wave velocities could be distinguished if individual disturbances were followed on a falling film. The trace shown in Fig. 11.10 is also evidence that a single wavelength does not predominate.

11.4 FLOODING

If gas is blown upward through the center of a vertical tube in which there is a falling film, a shear stress which retards the film is set up at the interface. As long as the film remains fairly smooth and stable, this shear stress is usually small and the film thickness, and consequently the void fraction, is also virtually unchanged from the value which is obtained with no gas flow.[21,28] However, for a given liquid rate there is a certain gas flow at which very large waves appear on the interface, the whole flow becomes chaotic, the gas pressure drop increases markedly (Fig. 11.11), and liquid is expelled from the top of the tube. This condition is known as *flooding*. Unlike the phenomenon which occurs in dispersions of drops or bubbles, the flooding point is not approached as the limit of a continuous process but is the result of a sudden and dramatic instability which increases the pressure gradient by an order of magnitude.

EMPIRICAL FLOODING CORRELATIONS

Turbulent flow in both components Let the gas flux upward and the liquid flux downward be j_g and j_f, respectively. An empirical flooding correlation can be sought in the following way. Countercurrent flow is maintained by buoyancy forces due to the density difference between the gas and the liquid. The flow rates are related to the film thickness by dynamic processes that balance the driving force of buoyancy with dissipative effects in the fluids. By analogy with single-phase flow turbulent systems it can be assumed that the average turbulent stresses are related to the average momentum fluxes of the components, i.e., to the quantities $\rho_g j_g^2/\alpha$ and $\rho_f j_f^2/(1-\alpha)$. Dimensionless groups which relate these momentum fluxes to the hydrostatic forces, apart from any dependence on the dependent variable α, are

$$j_g^* = j_g\rho_g^{1/2}[gD(\rho_f-\rho_g)]^{-1/2} \tag{11.81}$$
$$j_f^* = j_f\rho_f^{1/2}[gD(\rho_f-\rho_g)]^{-1/2} \tag{11.82}$$

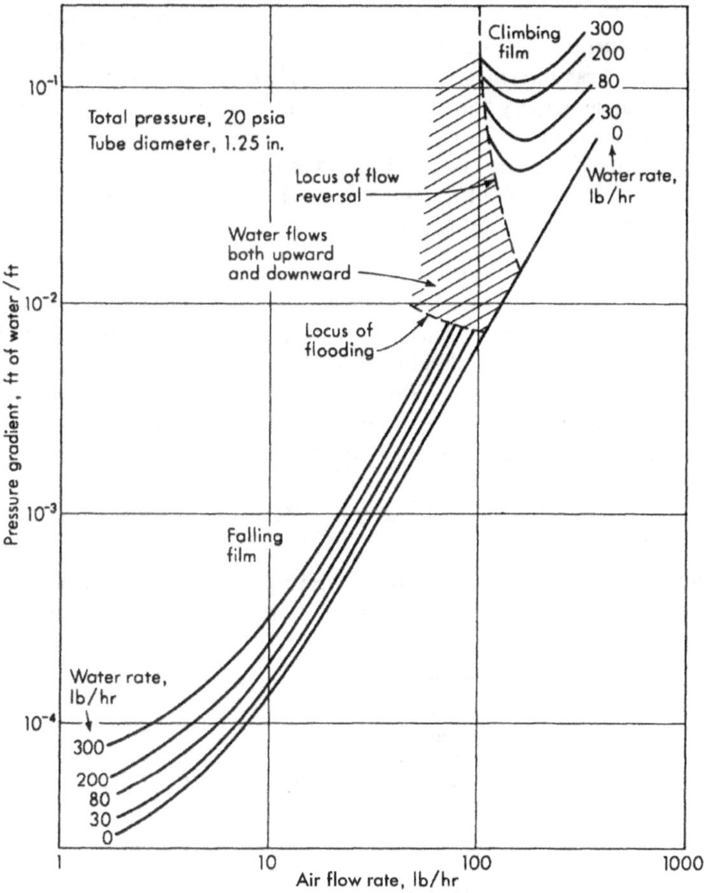

Fig. 11.11 Pressure-drop characteristics near the boundary between countercurrent and cocurrent flow. (*Hewitt, Lacey, and Nicholls.*[22])

A general correlation in terms of these parameters can then be sought in cases where the effects of other forces, such as those due to viscosity or surface tension, can be neglected. This hypothesis receives considerable support from comprehensive studies of flooding in packed towers. For example, Lobo *et al.*[23] and Sherwood *et al.*[24] present their results as a plot of the variables

$$\frac{j_g^2}{g}\frac{a}{F^3}\frac{\rho_g}{\rho_f}\mu_f^{0.2} \qquad \text{versus} \qquad \frac{j_f}{j_g}\left(\frac{\rho_f}{\rho_g}\right)^{1/2} \tag{11.83}$$

where a/F^3 is a characteristic inverse length which is equal to the total packing surface area divided by the total available flow volume.

The first parameter can be identified with j_g^{*2} and the second with j_f^*/j_g^* if, as is usually the case, the assumption $(\rho_f - \rho_g) \approx \rho_f$ is made and a/F^3 is assumed to be the inverse of a characteristic length D. After rearranging the coordinates, the curves of both Lobo and Sherwood can be drawn on a graph of j_f^* versus j_g^* as shown in Fig. 11.12. A good fit to both curves is given by the equation

$$j_f^{*1/2} + j_g^{*1/2} = 0.775 \tag{11.84}$$

The form of this equation is comparable with the results of Dell and Pratt[25] who measured flooding rates for various liquid-liquid combinations in packed columns. Flooding data plotted on a basis of the square roots of the volumetric fluxes were found to lie on straight lines with approximately equal intercepts on the axes. The final correlation was

$$1 + 0.835 \left(\frac{\rho_d}{\rho_c}\right)^{1/4} \left(\frac{j_d}{j_c}\right)^{1/2} = C \left[\frac{\rho_c j_c^2}{g(\rho_c - \rho_d)} \frac{a}{F^3} \sigma^{1/4}\right]^{-1/4} \tag{11.85}$$

where the subscripts referred to continuous and discontinuous components. If the surface-tension factor is absorbed into the coefficient C,

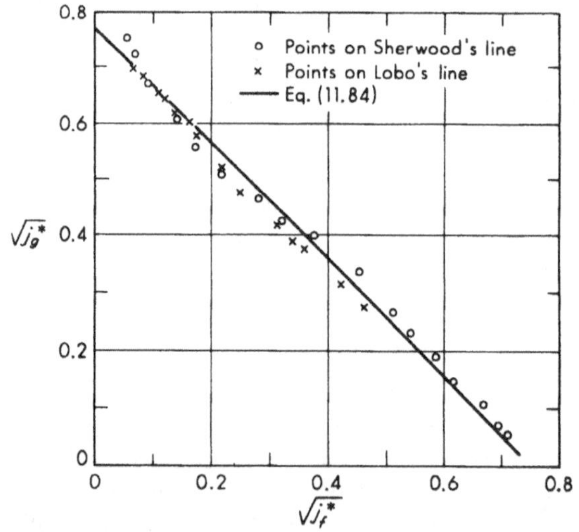

Fig. 11.12 The flooding correlations of Sherwood[24] and Lobo[23] plotted on a basis of the square roots of the parameters j_f^* and j_g^*.

Eq. (11.85) may be multiplied throughout by $j_c^{*1/2}$ to yield

$$j_c^{*1/2} + 0.835\, j_d^{*1/2} = C \tag{11.86}$$

and the similarity with Eq. (11.84) is evident.

The correction factor for viscosity which was used by Sherwood and Lobo is insignificant for most liquids and only amounts to a viscosity ratio raised to the one-tenth power multiplying j_g^*. Similarly, the surface-tension factor in Eq. (11.85) is raised to the one-sixteenth power and has little importance.

Correlations for flooding in vertical tubes resemble Eqs. (11.84) and (11.86) and may be expressed in the general form[26]

$$j_g^{*1/2} + m j_f^{*1/2} = C \tag{11.87}$$

For turbulent flow m is equal to unity. The value of C is found to depend on the design of the ends of the tubes and the way in which the liquid and gas are added and extracted. For tubes with sharp-edged flanges, $C = 0.725$, whereas when end effects are minimized, C lies between 0.88 and 1.[27,20] In inclined tubes the flow rates at the flooding points can be much higher.

Some hysteresis is observed in many flooding experiments. For example, if a smooth film is set up in a vertical tube, then the gas flow rate necessary to cause flooding coincides with the flow rate which will cause the growth of a large wave. Once the wave has formed it disrupts the whole film which becomes very rough and is covered with waves several times its own thickness in amplitude. The flow rate has to be reduced to a considerably lower level before the tube will return to smooth operating conditions. The range of hysteresis seems to lie between $C = 0.88$ and $C = 1$, as shown in Fig. 11.13.

Once the tube is dried out, it is found that the gas flow has to be reduced below the value characteristic of sharp-edged tubes

$$j_g^* \approx 0.5 \tag{11.88}$$

before the liquid will flow back down the tube.[27]

Viscous flow in the liquid If the liquid is very viscous the term due to liquid inertia in Eq. (11.83) should be replaced by a term which is proportional to viscous forces. The ratio between viscous and buoyancy forces is given by the expression

$$\frac{j_f^*}{N_f} = \frac{\mu_f j_f}{g D^2(\rho_f - \rho_g)} \tag{11.89}$$

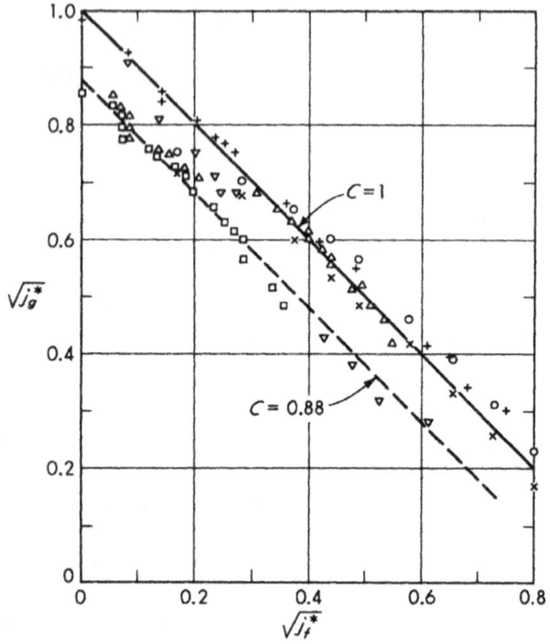

○ 1"-dia. (Nicklin and Davidson[28]) "short column" well-rounded air inlet design.

× 1"-dia. (Nicklin and Davidson[28]) "main column" well-rounded air inlet design.

△ 1¼"-dia. (Hewitt and Wallis[21]) water injection and extraction through porous walls. Points where flooding starts with increasing flow.

□ 1¼"-dia. (Hewitt and Wallis[21]) points where flooding stops with decreasing flow.

\+ ¾"-dia. (Wallis, Steen, and Brenner[29]) flooding starts.

▽ ¾"-dia. (Wallis, Steen, and Brenner[29]) flooding stops.

Fig. 11.13 Flooding velocities for air and water in vertical tubes designed to minimize end effects. All data at atmospheric pressure.

For $N_f < 2$ Wallis[26] suggests that the appropriate flooding relationship for the rough wavy film condition in tubes with smooth flanges is

$$j_g^{*1/2} + 5.6\left(\frac{j_f^*}{N_f}\right)^{1/2} = 0.725 \tag{11.90}$$

In general, flooding for any fluid viscosity is correlated by Eq. (11.87) where m and C are functions of N_f as shown in Figs. 11.14 and 11.15.

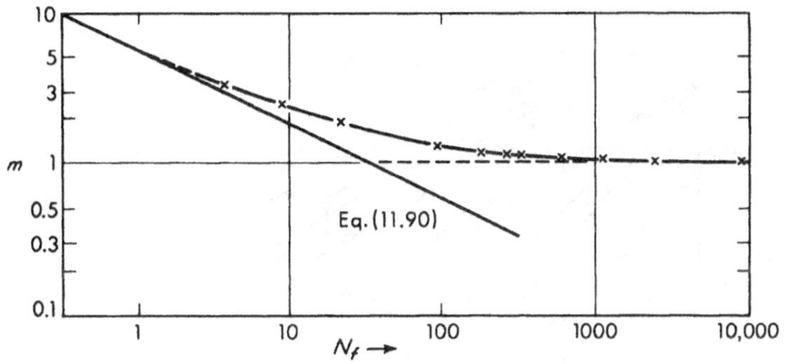

Fig. 11.14 Variation of the coefficient m in Eq. (11.87) as a function of the dimensionless group N_f. (*Wallis.*[26])

The reduction of the coefficient C for viscous fluids to the same value which is characteristic of sharp flanges in the inviscid fluid case is probably due to end effects.[34]

Figure 11.16 compares some flooding data with this correlation scheme.

Example 11.4 Air and water are in countercurrent flow in a vertical pipe of diameter 2 in. If $\rho_g = 0.1$ lb/ft^3 and $\rho_f = 62.5$ lb/ft^3, what is the maximum allowable liquid flow rate when $W_g = 200$ lb/hr? The value of μ_f is 2.5 lb/(hr)(ft).

Solution The cross-sectional area is

$$\frac{\pi(2^2)}{(4)(144)} = 2.18 \times 10^{-2}\ \text{ft}^2$$

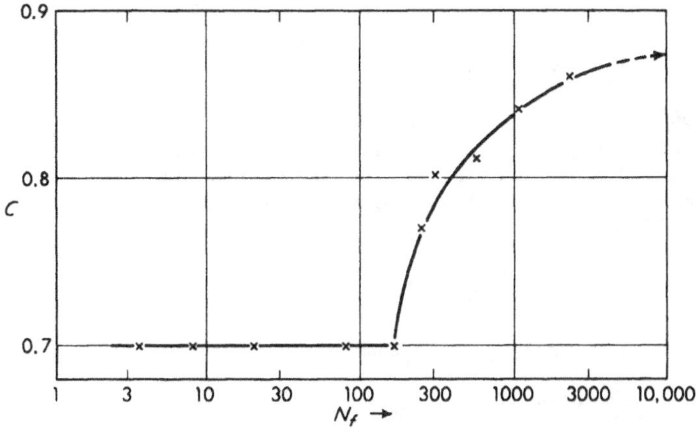

Fig. 11.15 Variation of the coefficient C in Eq. (11.87) as a function of the dimensionless group N_f. (*Wallis.*[26])

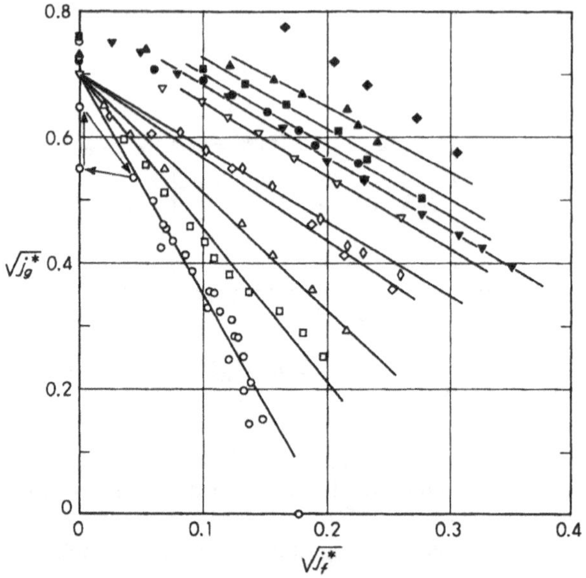

Fluid	Symbol	Viscosity at 10°C, cp	N_f
Glycerol 99% ± 1%	○	3000	3.4
" 95%	□	1270	8.2
" 90%	△	498	21
" 80%	◇	116	90
" 75%	◊	60	160
" 70%	▽	39	250
" 60%	●	17	560
" 50%	■	9	1000
" 33%	▲	4	2200
Water	◆	1	8200
Ethylene glycol	▼	~30	300

Fig. 11.16 Flooding velocities for aqueous glycerol solutions and for ethylene glycol in countercurrent flow with air at atmospheric pressure in a 0.75-in. vertical pipe (*Wallis*[26]) compared with Eq. (11.87) using m and C from Figs. 11.14 and 11.15.

The value of j_g is

$$\frac{200}{3600}\frac{1}{0.1}\frac{1}{2.18 \times 10^{-2}} = 25.5 \text{ fps}$$

From Eq. (11.81),

$$j_g^* = 25.5 \left[\frac{0.1}{(32.2)(2/12)(62.5)}\right]^{1/2} = 0.43$$

Assume that end effects are minimized and that the past history of the flow allows the maximum value of C equal to unity.

The value of $j_g^{*1/2} = 0.655$. Therefore, from Eq. (11.87),

$$m j_f^{*1/2} = C - j_g^{*1/2} = 1 - 0.655 = 0.345$$

N_f is found from Eq. (10.16) as follows:

$$N_f = \frac{[(2/12)^3(32.2)(62.5^2)]^{1/2}}{2.5/3600} = 3.5 \times 10^4$$

Then, from Figure 11.14, $m = 1$ and therefore $j_f^* = (0.345)^2 = 0.119$. The maximum value of j_f is then, from Eq. (11.82),

$$j_f = \left[\frac{(32.2)(2)}{12}\right]^{1/2}(0.119) = 0.276 \text{ fps}$$

The maximum possible liquid flow rate is therefore

$$j_f \rho_f A = (0.276)(62.5)(2.18 \times 10^{-2})(3600) = 1350 \text{ lb/hr}$$

PREDICTION OF FLOODING FROM THE SEPARATE CYLINDERS MODEL[11]

The separate cylinders model was introduced in Chap. 3. It obviously provides a poor representation of the actual geometry in pipe flow, and yet it is the usual theoretical basis for the very successful Martinelli correlation of horizontal annular and stratified flow. Despite its shortcomings it is remarkable as the only analytical derivation of Eq. (11.87).

The model is perhaps more appropriate for the case of packed towers where it represents a possible flow configuration. A similar analysis for this latter case has been presented by Dell and Pratt.[25]

Turbulent flow If the separate cylinders model is applied to vertical flow, the equations have to be adjusted to allow for the effects of gravity. The force per unit volume on the gas changes from $(-dp/dz)$ to $(-dp/dz - \rho_g g)$ and similarly the force per unit volume on the liquid is $(-dp/dz - \rho_f g)$. z is measured upward.

Using Eq. (5.6) and applying the results of Prob. 5.12 with the assumption of a constant mixing length in each cylinder, we have

$$j_g^* = \frac{r_0}{7l_g}\Delta p^{*1/2}\alpha^{3/4} \tag{11.91}$$

$$j_f^* = \frac{r_0}{7l_f}(1 - \Delta p^*)^{1/2}(1 - \alpha)^{3/4} \tag{11.92}$$

where $\alpha = r_g^2/r_0^2$, $1 - \alpha = r_f^2/r_0^2$, and r_g and r_f are the radii of the separate cylinders. We can either assume that l_f and l_g are scaled by the dimensions of each cylinder or by the overall pipe diameter. In the former case we have

$$\frac{l_g}{\alpha^{1/2}} = \frac{l_f}{(1 - \alpha)^{1/2}} = l' \tag{11.93}$$

and in the latter

$$l_g = l_f = l' \tag{11.94}$$

At the inception of flooding the turbulence level in the whole flow field increases immensely. In this case it might be assumed that the mixing length which is usually characteristic of the flow core extends over the whole pipe. This core mixing length is given by Nikuradse[30] as $0.14r_0$ and can be reasonably approximated by

$$l' = \frac{r_0}{7} \tag{11.95}$$

Using Eq. (11.95) in Eqs. (11.91) and (11.92) we obtain

$$j_g^* = \Delta p^{*1/2}\alpha^{n/2} \tag{11.96}$$
$$j_f^* = (1 - \Delta p^*)^{1/2}(1 - \alpha)^{n/2} \tag{11.97}$$

where n has the value 3.5 or 2.5, depending upon whether Eq. (11.93) or (11.94) is valid, and has the same significance as n in Eq. (3.30).

Eliminating Δp^* from Eqs. (11.96) and (11.97) we have

$$\frac{j_g^{*2}}{\alpha^n} + \frac{j_f^{*2}}{(1 - \alpha)^n} = 1 \tag{11.98}$$

which resembles Eq. (6.90). The envelope of Eq. (11.98) as α is varied is

$$j_g^{*2/(n+1)} + j_f^{*2/(n+1)} = 1 \tag{11.99}$$

which is consistent with Eq. (11.87) if $m = 1$ and n takes on the intermediate value of 3.

Viscous flow in the liquid When the liquid flow is viscous, the appropriate equation to replace Eq. (11.97) is

$$j_f^* = \frac{N_f}{32}(1 - \alpha)^2(1 - \Delta p^*) \tag{11.100}$$

Eliminating α from Eqs. (11.96) and (11.100) we obtain

$$\left(\frac{j_g^{*2}}{\Delta p^*}\right)^{1/n} + \left[\frac{32 j_f^*}{N_f(1 - \Delta p^*)}\right]^{1/2} = 1 \tag{11.101}$$

Differentiating to obtain the envelope yields

$$2.8\left(\frac{j_f^*}{N_f}\right)^{1/2} \frac{1}{(1 - \Delta p^*)^{3/2}} = \frac{j_g^{*2/n}}{n(\Delta p^*)^{(1+1/n)}} \tag{11.102}$$

Table 11.2 Values of C predicted by the separate cylinders model for flooding with viscous liquid and turbulent gas flow

Δp^*	0	0.1	0.2	0.3	0.4	0.5	0.6	0.7	0.8	0.9	1
C ($n = 2.5$)	1	0.987	0.956	0.929	0.91	0.899	0.893	0.900	0.920	0.951	1
C ($n = 3.5$)	1	0.876	0.853	0.826	0.816	0.821	0.837	0.863	0.897	0.941	1

Equations (11.101) and (11.102) do not have a simple solution; however, j_f^* and j_g^* can be represented parametrically in terms of Δp^* as follows:

$$5.6\left(\frac{j_f^*}{N_f}\right)^{1/2} = \frac{2(1 - \Delta p^*)^{3/2}}{2 + (n - 2)\,\Delta p^*} \tag{11.103}$$

$$j_g^{*1/2} = \left[\frac{n\,\Delta p^{*(1+1/n)}}{2 + (n - 2)\,\Delta p^*}\right]^{n/4} \tag{11.104}$$

For different values of Δp^* Eqs. (11.103) and (11.104) define values of j_f^* and j_g^* which trace out the flooding curve. These values can then be used to derive the value of C which will agree with the left-hand side of Eq. (11.91). Results shown in Table 11.2 for $n = 2.5$ and 3.5 show that C is remarkably constant.

11.5 VERTICAL UPWARD COCURRENT ANNULAR FLOW

THE BOUNDARIES OF THE VERTICAL ANNULAR FLOW REGIME

There are two mechanisms which determine the boundaries of vertical annular flow,

1. "Bridging" of the gas core by liquid from the film and a consequent transition to slug flow
2. Entrainment of droplets from the film and a consequent transition to annular mist flow

For some combinations of fluid properties and pipe size these two criteria can overlap and no ideal annular flow occurs at all.

The slug-annular transition The various methods for defining the transition between slug flow and annular flow provide an excellent "case study" in flow regime boundary predictions.

The following approaches have been adopted:

1. Definition of criteria for upward or downward flow in a liquid film as a function of gas flow rate
2. Studies of the conditions under which liquid "bridges" across the gas core disappear

3. Measurement of the fractional liquid flow rate at the axis of the tube as a function of flow rates by means of a sampling probe
4. Comparison of void fraction data with theoretical predictions for the slug- and annular-flow regimes
5. Pressure-drop measurements

These techniques will now be discussed in order and related to quantitative theoretical criteria.

Criteria for upward or downward flow in a liquid film If a liquid film is set up on the wall of a vertical pipe, for example, by injecting liquid through a porous sintered metal plug, it can flow upward or downward, or divide and flow partially upward and partially downward, depending on the gas and liquid flow rates involved. The downflow in the film is limited by flooding. For a given upward gas flow rate there is a maximum possible value of downward liquid flow rate and vice versa. The gas flow rate at which no liquid can flow downward gives a possible criterion for the upper limit of slug flow at low liquid rates.

For the observed range of the constant C in Eq. (11.88) the limiting value of gas flow rate at which no liquid will run down the wall is given by the relationships

$$0.5 < j_g^* < 1 \tag{11.105}$$

The upper value in Eq. (11.105) is approached when end effects are minimized.

Figure 11.11 shows typical results which are obtained when water is supplied to a vertical tube through a porous section of the wall and the upward air flow is continuously increased. At low air rates the liquid all flows downward. After flooding occurs some water begins to flow upward and eventually a point is reached at which all of the water flow is upward. The condition that all the water should flow upward gives a way of defining the lower boundary of cocurrent flow.

When there is a very low liquid flow rate the shaded region in Fig. 11.11 becomes narrow, since if one injects liquid at the end of the flooding curve where no liquid will flow downward at all, the only way for it to go is up. Experimental data for both viscous and inviscid liquids[31] correlate with the equation

$$j_g^* \approx 0.9 \tag{11.106}$$

A gas rate lower than Eq. (11.106) allows some downward flow in the film, whereas a flow rate greater than this is needed to cause a finite amount of water to flow upward continuously. If the liquid flows downward in the film and cannot escape at the bottom of the tube, then a liquid slug will eventually be formed.

Equation (11.106) is also approximately consistent with the boundary of the "forbidden region" shown in Fig. 11.24 over a wide range of liquid flow rates.

"Bridging" of the gas core A liquid bridge across the flow cross section can be detected using an electrical probe of the type developed by Griffith[32] and used by Haberstroh and Griffith[33] and Brenner.[35] The probe consists of a conducting tip on the end of an insulated rod which is mounted in the core of a two-phase flow. If the liquid conductivity is very different from the gas conductivity, then a measurement of the electrical resistance between the tip of this probe and the tube wall will indicate whether the tube is bridged by liquid or not.

The method of defining transition is rather arbitrary since there is a rather broad region in which slugs become few and far between but do not disappear altogether. The results of Griffith, Haberstroh, and Brenner all correlate with the equations

$$j_f^* < 1.5 \qquad j_g^* = 0.9 + 0.6 j_f^* \tag{11.107}$$

$$j_f^* > 1.5 \qquad j_g = \left(7 + 0.06\,\frac{\rho_f}{\rho_g}\right) j_f \tag{11.108}$$

"Entrainment" measurements using a sampling probe Another method of detecting a liquid bridge across the gas core is to sample the flow by means of a probe at the tube axis. The probe detects both liquid bridges and entrained droplets but will not collect any liquid in "ideal" annular flow in which all of the liquid flows on the tube wall. The probe measures a liquid flux $(j_f)_{CL}$ near the centerline of the pipe which may be compared with the average liquid flux j_f. In slug flow $(j_f)_{CL}$ can be much larger than j_f because the liquid flow can be up in the slugs and down in the film around the bubbles; the liquid flux profile therefore peaks at the center and is negative at the wall.

Figure 11.17*a* shows entrainment data which were taken in this way in a tube of 1-in. diameter for air and water at atmospheric pressure. Zero entrainment was obtained only over a limited range of gas velocities for low liquid rates. Entrainment at low gas rates is due to liquid bridges, whereas at the higher flows it is due to droplets. It is possible to extrapolate the slug-flow portion of this graph to the abscissa to define a lower critical gas velocity at which the slug-annular transition occurs. The transition line is correlated by the equation

$$j_g^* = 0.4 + 0.6 j_f^* \tag{11.109}$$

which is significantly below the electrical probe transition line.

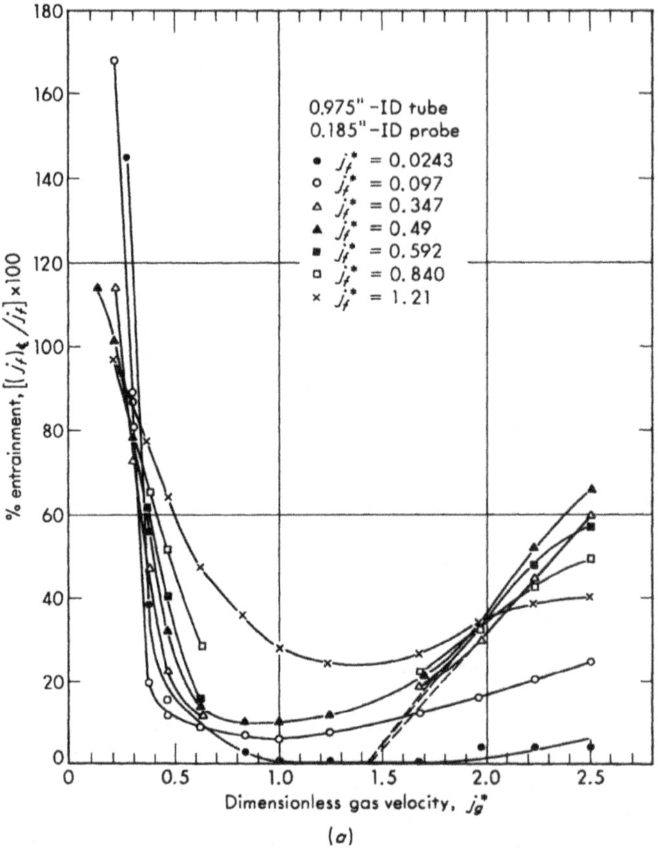

Fig. 11.17a Entrainment as a function of air and water flow rates, upward annular flow.[35]

Comparison of void fraction data with theory A prime purpose of defining the transition between any flow regimes is to predict when one theory should be replaced by another. The slug-annular transition may therefore be defined by seeing where the predictions of slug-flow and annular-flow theories cross. For example, the slug-flow theory for void fraction is, from Eq. (10.36) for turbulent flow,

$$\alpha = \frac{j_g}{1.2(j_f + j_g) + 0.345\sqrt{gD(\rho_f - \rho_g)/\rho_f}} \tag{11.110}$$

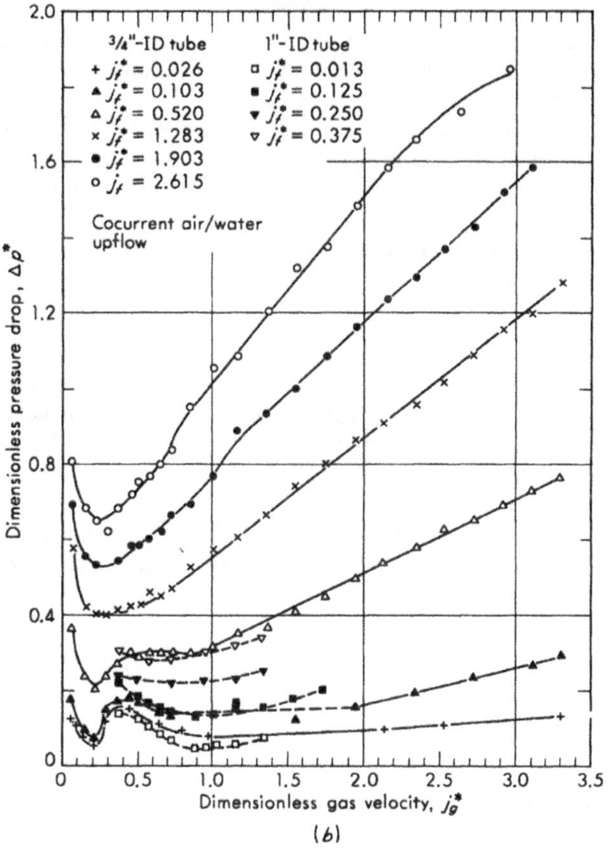

Fig. 11.17b Pressure drop as a function of air velocity at constant water rates, upward annular flow.[35]

This may be rewritten as follows:

$$\alpha = \frac{j_g^* \sqrt{\dfrac{\rho_f}{\rho_g}}}{1.2\left(j_f^* + \sqrt{\dfrac{\rho_f}{\rho_g}}\, j_g^*\right) + 0.345} \tag{11.111}$$

A correlation for void fraction in upwards annular flow that will be introduced later is

$$\frac{j_g^*}{1 - 2.85(1-\alpha)} - \frac{j_f^*}{2.85(1-\alpha)} = 0.775 \tag{11.112}$$

These two equations are compared with data[35] taken in a 1-in. pipe, using air and water, in Fig. 11.18. Data taken in the slug-flow regime contain considerable inherent scatter because the void fraction measurements were taken by using quick-closing valves to isolate a length of tube and the results depended on how many slug-flow bubbles happened to be caught at the instant of closing. The annular flow points cluster around the correlation. The theoretical lines cross at a liquid fraction of about $1 - \alpha = 0.2$ and at values of j_f^* and j_g^* in approximate agreement with Eq. (11.109). In fact the line $1 - \alpha = 0.2$ predicted by Eq. (11.112) has the equation

$$j_g^* = 0.33 + 0.75 j_f^* \tag{11.113}$$

Pressure-drop measurements Several workers (e.g., Govier *et al.*)[36–38] have suggested that regime boundaries could be correlated in terms of maxima or minima in pressure-drop readings. This hypothesis can be compared with other techniques by using data such as are shown in Fig. 11.17*b*.

The predictions using this method are not very accurate since some of the maxima and minima are relatively flat. However, the order of

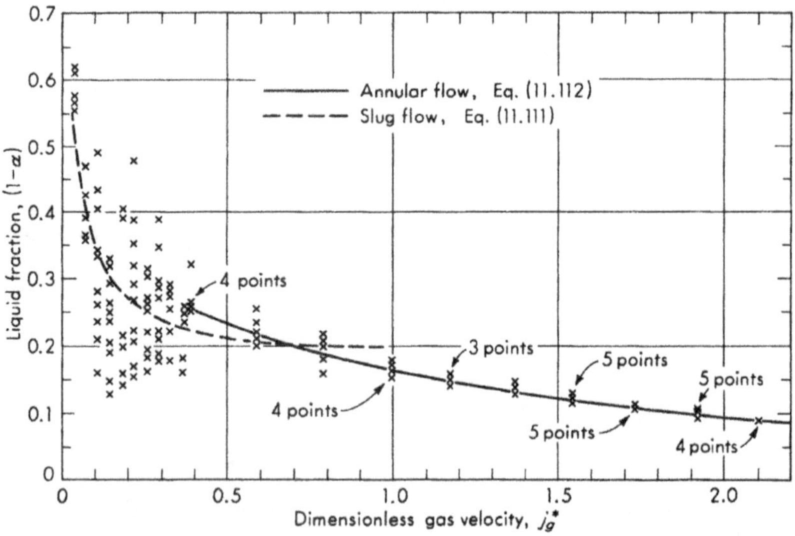

Fig. 11.18 Liquid-fraction data in the vicinity of the slug-annular transition[35] for a dimensionless liquid flux, $j_f^* = 0.53$.

magnitude predicted is correct, showing that pressure-drop behavior is quite closely related to transitions in flow regime.

Discussion Of all the methods for predicting the slug-annular regime boundaries the most meaningful from an engineering standpoint is the one discussed in "Comparison of Void Fraction Data with Theory" and based on the point at which one theory has to be replaced by another. These results are in agreement with the entrainment probe results which are based on an extrapolation of lines on a graph and are hence an idealization. They give a working estimate of when the influence of slugs on the overall fluid mechanics is small.

The electrical probe technique tends to estimate too high a gas velocity by about 50 percent because occasional bridges are still detected long after they have ceased to be the governing mechanism of the flow.

The region between Eqs. (11.109) and (11.107) could be interpreted as a transition region of slug-annular flow in which a rather gradual transition in behavior occurs.

The annular-mist transition Considerable amounts of liquid entrainment usually occur in high-velocity annular flow. If this entrainment is measured by a probe which samples an area a at the center of the tube and measures a liquid flow rate of w_f, a parameter can be defined which gives a rough measure of the fraction of liquid entrained. Thus

$$E = \frac{w_f}{W_f}\frac{A}{a} = \frac{(j_f)_{\mathrm{CL}}}{j_f} \tag{11.114}$$

If the flow were completely homogeneous and one-dimensional, E would be equal to unity, whereas in pure annular flow E would be zero.

Referring to Fig. 11.17*a*, it can be seen that there is a very small range of flow rates over which droplet entrainment is low. In fact most annular flows contain some entrained droplets and can be analyzed by treating the homogeneous mist and the liquid film as the two phases.

The prediction of entrainment will be considered in greater detail in the next chapter.

CORRELATIONS FOR PREDICTING VOID FRACTION AND PRESSURE DROP

The Dartmouth correlation[35] for void fraction The Dartmouth correlation was developed from the hypothesis that there exists a regime of annular flow in which viscous and surface-tension forces are small and the fluid dynamics are governed by a balance between hydrostatic forces and inertia forces in the gas and the liquid. This is partly based on the observation that the film surface is quite markedly wavy and form drag is likely to be far greater than friction drag at the interface. The result-

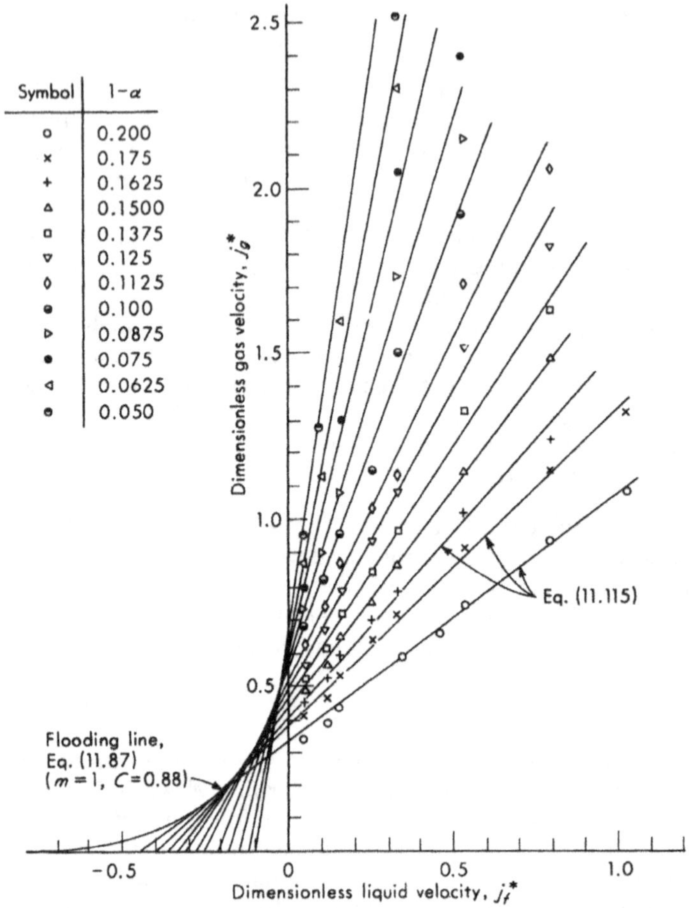

Fig. 11.19 Lines of constant liquid fraction in air-water annular flow versus the dimensionless parameters j_f^* and j_g^*. (*Wallis.*[35])

ing correlation parameters are j_f^* and j_g^* which were used in the previous flooding correlations. Plotting up the data for air and water in a 1-in.-diameter vertical pipe (Fig. 11.19) the constant void fraction lines were observed to be straight and to be tangential to the flooding line. The equation which correlated these lines was

$$\frac{j_g^*}{1 - 2.85(1 - \alpha)} - \frac{j_f^*}{2.85(1 - \alpha)} = 0.775 \tag{11.115}$$

Later tests with greater emphasis on accuracy[11] showed some deviations from linearity in the constant void fraction lines and also indicated that

$$\frac{j_g^*}{1 - 3.1(1-\alpha)} - \frac{j_f^*}{3.1(1-\alpha)} = 1 \tag{11.116}$$

This correlation is valid only over the range of j_f^* and j_g^* shown in the figure and when droplet entrainment is below about 20 percent.

The modified Martinelli correlation[11] The Martinelli correlation was originally derived for horizontal flow and does not apply to vertical flow because it ignores gravity. However, it can be used when velocities are high enough (approximately $j_g^* \gg 2$) for shear stresses to dominate the flow. At lower velocities it is more accurate to follow the procedure outlined in Example 3.7 to express the frictional forces on the two streams. In the absence of acceleration or phase change the momentum equations for the gas core and for the whole flow are then

$$\frac{dp}{dz} + \rho_g g + \phi_g^2\left(-\frac{dp}{dz}\right)_g = 0 \tag{11.117}$$

$$\frac{dp}{dz} + g[\alpha\rho_g + (1-\alpha)\rho_f] + \phi_f^2\left(-\frac{dp}{dz}\right)_f = 0 \tag{11.118}$$

In terms of the dimensionless single-phase frictional pressure drops

$$\Delta p_g^* = \frac{\left(-\frac{dp}{dz}\right)_g}{g(\rho_f - \rho_g)} \tag{11.119}$$

$$\Delta p_f^* = \frac{\left(-\frac{dp}{dz}\right)_f}{g(\rho_f - \rho_g)} \tag{11.120}$$

Eqs. (11.117) and (11.118) can be written as

$$\Delta p^* = \phi_g^2\,\Delta p_g^* \tag{11.121}$$

$$\Delta p^* = \phi_f^2\,\Delta p_f^* + (1-\alpha) \tag{11.122}$$

Assuming that ϕ_f and ϕ_g are known functions of α, Eqs. (11.121) and (11.122) can be solved simultaneously to give Δp^* and α as functions of Δp_g^* and Δp_f^*. This has been done by Turner[11,35] using $\phi_f = 1/(1-\alpha)$ and an empirical expression for ϕ_g. The results are shown in Fig. 11.20 which gives a rapid method for deriving Δp^* and α from known quantities. The lines of constant void fraction are straight and envelop the flooding locus, as usual.

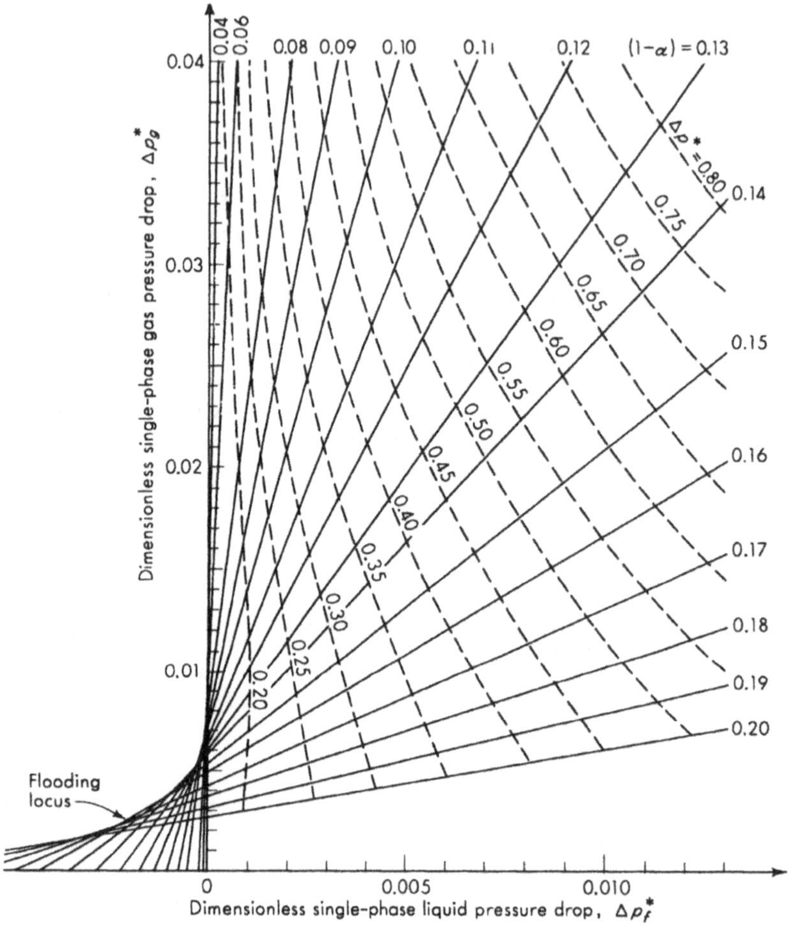

Fig. 11.20 Prediction of void fraction and pressure drop from modified Martinelli correlation.[11,35]

SIMPLE FLOW MODELS

Separate cylinders model The separate cylinders model can be developed for vertical flow in exactly the same way as for horizontal flow. The frictional pressure drop for the gas cylinder is now $-(dp/dz + \rho_g g)$, whereas for the liquid it is $-(dp/dz + \rho_f g)$. This model is quite unrealistic for cocurrent flow unless $-dp/dz > \rho_f g$ and predicts countercurrent flow if $-dp/dz < \rho_f g$. However, it does give a reasonable prediction of some correlations for flooding.

Homogeneous model In Ref. 39 the homogeneous model was tested as a method for correlating the frictional pressure drop in annular-mist flow.

In order to make a fair test of the theory it was first necessary to obtain an estimate of the range of flow rates for which the flow regime was annular-mist rather than some other pattern such as slug flow. When entrainment data were available, the flow regime boundary was determined by requiring that the percent entrainment, measured by the sampling probe technique, should be to the right of the minimum on a graph such as Fig. 11.17 and above 20 percent. When entrainment data were not reported by authors the annular-mist regime was determined by requiring that the gas (or vapor) flow rate should be above both the slug and annular transition line given by Haberstroh and Griffith, Eqs. (11.107) and (11.108), and the critical gas velocity for the onset of entrainment derived by Steen,[2] Eq. (11.2). In all cases it was found that homogeneous theory gave a better estimate of frictional pressure drop than Martinelli's correlation. However, Martinelli's predictions of void fraction were superior to those of homogeneous theory. The homogeneous friction factor was remarkably close to 0.005 for all of the data which were examined.

A more accurate version of the homogeneous model can be obtained by introducing the surface tension as a further correlating factor. This is motivated by the major influence which surface tension exerts on film stability, interfacial roughness, and droplet entrainment. The CISE correlation[40] for frictional pressure drop, for example, is essentially a modified homogeneous flow equation of the form

$$\left(-\frac{dp}{dz}\right)_F = C\,\frac{G^{1.4}\bar{v}^{0.86}\sigma^{0.4}}{D^{1.2}} \tag{11.123}$$

where $\bar{v}$ is the average homogeneous mixture specific volume. The surface tension was found to be a far more significant variable than the viscosity of either component or any form of mixed mean viscosity. Similarly, Bergelin *et al.*[41] found a significant effect of surface tension on the effective friction factor in homogeneous flow.

Separated-flow model The separated-flow model can be developed by using the expressions for wall and interfacial shear stresses which were developed in Sec. 11.2. Figure 11.21 shows that Eq. (11.18) applies to vertical as well as to horizontal flows.

A simple result is obtained by assuming that the single-phase turbulent friction factors are 0.005 for both the gas and liquid.[3] Neglecting the effects of compressibility, relative velocity, and entrainment, the

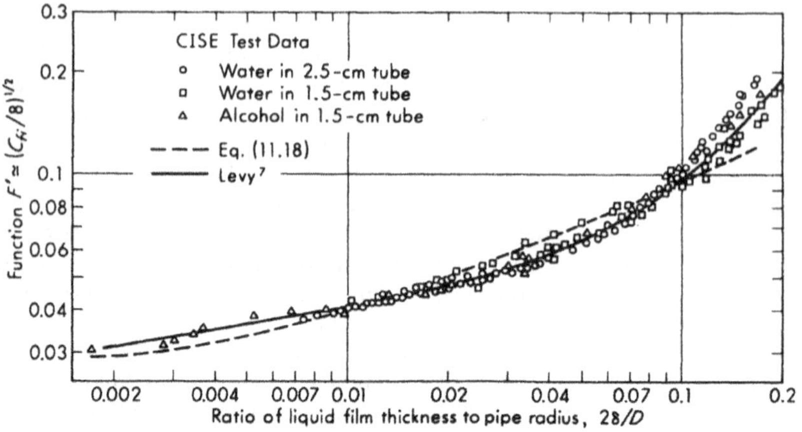

Fig. 11.21 Comparison between Eq. (11.18) and Levy's correlation[7,3] of CISE data for vertical flow.

equivalents of Eqs. (11.121) and (11.122) are

$$\Delta p^* = 10^{-2} j_g^{*2} \frac{1 + 75(1 - \alpha)}{\alpha^{5/2}} \tag{11.124}$$

$$\Delta p^* = \frac{10^{-2} j_f^* |j_f^*|}{(1 - \alpha)^2} + (1 - \alpha) \tag{11.125}$$

For laminar flow, on the other hand,

$$\Delta p^* = \frac{j_f'^*}{(1 - \alpha)^2} + 0.684(1 - \alpha) \tag{11.126}$$

where $j_f'^*$ is defined as

$$j_f'^* = \frac{32 j_f \mu_f}{D^2 g(\rho_f - \rho_g)} = \frac{32 j_f^*}{N_f} \tag{11.127}$$

The factor (0.684) is a correction factor for velocity profile that can be developed from a differential analysis (Eq. (5.32)).

Figures 11.22 to 11.24 show the prediction of Eqs. (11.124) and (11.125) for the turbulent-turbulent flow regime. There are several important qualitative aspects of these graphs.

It is evident from Fig. 11.22 that the character of the flow changes quite dramatically when j_g^* falls below about unity. No downflow is possible unless j_g^* is less than 0.95 and a slight reduction below this value

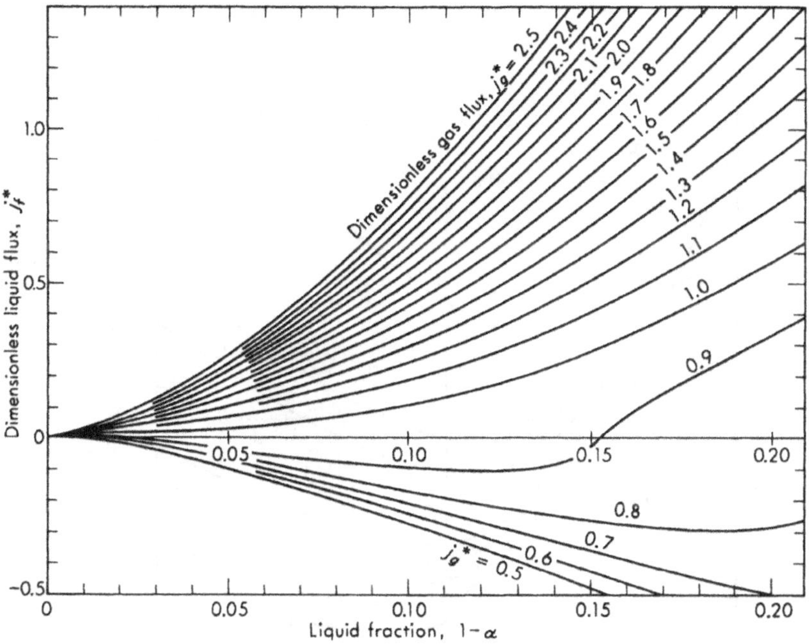

Fig. 11.22 j_f^* versus $(1-\alpha)$ for various values of j_g^* for a turbulent film predicted from Eqs. (11.124) and (11.125). (*Wallis.*[3])

requires high values of $1-\alpha$ before upflow can occur (there is actually a slight region of upflow near the origin which allows j_f^* values only below 0.01). Practically, a value of $j_g^* = 0.9$ corresponds to a situation in which thin liquid films flow downward while thick ones flow upward. A net upflow of liquid then occurs as a result of "waves" of thick film riding over a smoother and thinner falling film. With values of j_g^* below 0.9 these thick films are usually sufficient to bridge the pipe temporarily and bring about a transition to slug flow. This conclusion is consistent with the empirical conclusion reported in Eq. (11.106).

In the case of laminar flow of the liquid the picture is qualitatively the same with the transition occurring at $j_g^* \approx 0.8$.

The pressure-drop graph for low liquid rates (Fig. 11.24) displays the same effects in a different way. Below a gas rate in the range $0.8 < j_g^* < 1$ the pressure drop increases immensely and there is a "forbidden region" in which no solutions can be obtained. Since the pressure drop is often controlled by the external characteristics of a system, this forbidden region can be very significant; an attempt to operate in

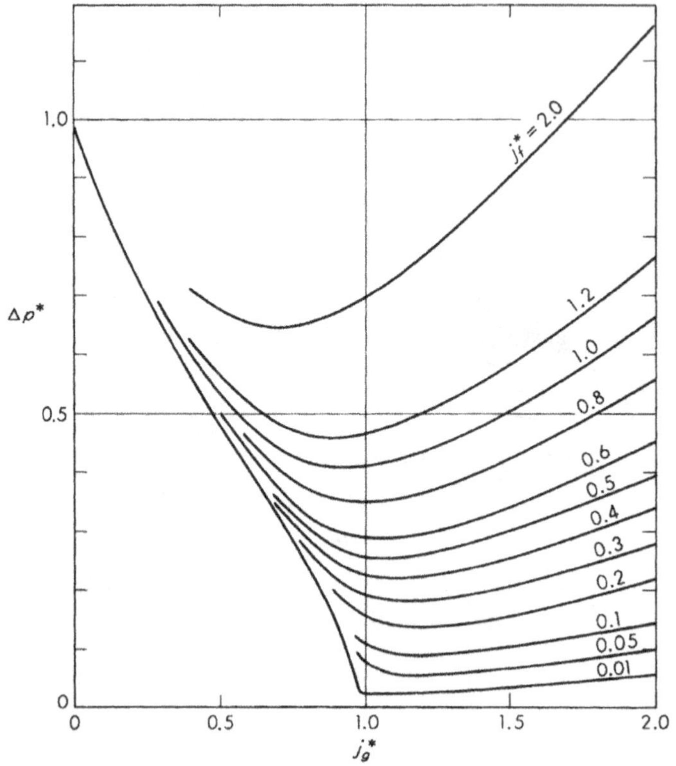

Fig. 11.23 Dimensionless pressure drop Δp^* versus dimensionless gas flux j_g, for various values of j_f^* [from Eqs. (11.124) and (11.125)]. (*Wallis.*[3])

it results in either a new flow regime or an unstable situation in which a large change in the operating conditions occurs. Moreover, at a gas flux which crosses this forbidden region at low liquid flow rates, there are three possible solutions to the pressure drop for a given liquid rate. Rather small fluctuations in any one of the parameters can then lead to a jump from the low-pressure-drop to the high-pressure-drop value.

The pressure-drop minimum at a constant liquid rate is also of significance for stability in a system in which the pressure drop and liquid rate are controlled. If the gas rate is allowed to fall below the value corresponding to the minimum pressure drop, it will continue to fall until the pipe fills with liquid and becomes flooded. Figures 11.23 and 11.24

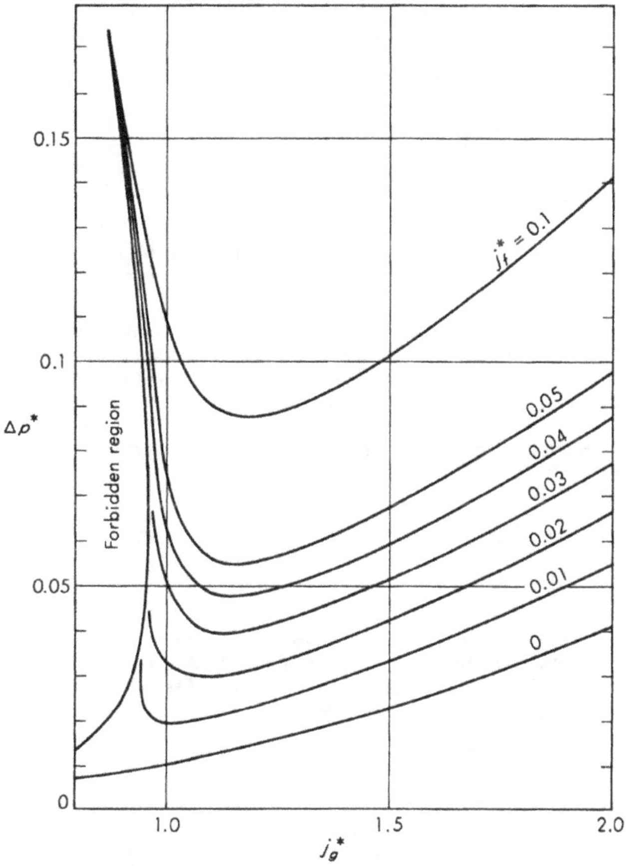

Fig. 11.24 Enlargement of Fig. 11.23 for low values of j_f^*.

show that for Δp^* less than 0.3 the maximum liquid rate at a given pressure drop occurs at a value of j_g^* equal to about 1.1.

Example 11.5 Solve Example 11.2 if flow is vertical. (*a*) Use the modified Martinelli correlation and Fig. 11.20. (*b*) Use Eq. (11.115) and Fig. 11.19. (*c*) Use the separated-flow model.

Solution (*a*) From Example 11.2 we have $(-dp/dz)_f = 7.2$ dynes/cm³ and $(-dp/dz)_g = 9.4$ dynes/cm³. Therefore, from Eqs. (11.119) and (11.120),

$$\Delta p_g^* = \frac{9.4}{980} = 9.6 \times 10^{-3} \qquad \Delta p_f^* = \frac{7.2}{980} = 7.35 \times 10^{-3}$$

From Fig. 11.20, $1 - \alpha = 0.176$, $\Delta p^* = 0.42$.

(*b*) The values of j_g and j_f are 47.2 and 1.03 fps. Therefore the values of j_g^* and j_f^* are

$$j_g^* = 47.2\left[\frac{(0.076)(12)}{(0.98)(32.2)(62.5)}\right]^{1/2} = 1.015$$

$$j_f^* = 1.03\left[\frac{12}{(0.98)(32.2)}\right]^{1/2} = 0.636$$

From Fig. 11.19, therefore, $1 - \alpha = 0.175$.

(*c*) Using the above values of j_g^* and j_f^* in Figs. 11.22 and 11.23 we find $1 - \alpha = 0.205$, $\Delta p^* = 0.3$.

Since the operating point is close to the slug-annular regime boundary and to the minimum pressure gradient in Fig. 11.23, corresponding to zero wall shear, it is likely that the separated-flow model prediction is less accurate than the predictions of methods (*a*) and (*b*).

Example 11.6 Solve Example 2.5 using annular flow theory.

Solution The values of j_g^* and j_f^* are

$$j_g^* = (47.2)\left[\frac{0.076}{(32.2)(0.0817)(62.5)}\right]^{1/2} = 1.015$$

$$j_f^* = (3.1)[(32.2)(0.0817)]^{-1/2} = 1.91$$

These values are outside the range shown in Fig. 11.19, and the use of Eqs. (11.108) and (11.109) shows that the flow pattern is in the transition region between slug and annular flow. Assuming that Eq. (11.116) is reasonably valid in this region we have

$$\frac{1.015}{1 - 3.1(1 - \alpha)} - \frac{1.91}{3.1(1 - \alpha)} = 1$$

whence $1 - \alpha = 0.232$, which is very close to the observed value.

From Fig. 11.23 $\Delta p^* \approx 0.66$. Therefore the pressure drop is $(0.66)(62.5)(^{18}\!/_{1728}) = 0.43$ psi, which is an overestimate. This value is too high because allowance has not been made for the curvature of the velocity profile or for the probable alternating upward and downward flow in the liquid film.

IMPROVEMENTS TO THE SEPARATED-FLOW MODEL[3]

The accuracy of the separated-flow model can be improved in numerous ways by taking many further phenomena into account. In most cases the form of the theory remains unchanged but various correction factors, which depend on additional parameters, are included.

The liquid film A differential analysis of the time-averaged flow in the liquid film can be performed using the methods of Chap. 5. The result for laminar flow was derived in Example 5.1. Over the range $0 < 1 - \alpha$

Table 11.3 Some of the dimensionless parameters used by Hewitt for his differential analysis of the liquid film in annular flow[42]

Hewitt	W^+	β	M	$\frac{1}{\sigma^3}$	Re*
This book	$\frac{Re_f}{4}$	$N_f^{2/3}\frac{\Delta p^*}{4}$	$N_f^{2/3}\frac{\delta}{D}$	$\frac{\Delta p^*}{1-\alpha}-1$	$\frac{N_f}{4}[\Delta p^* - (1-\alpha)]^{1/2}$
Dimensionless form of	Flow rate	Pressure drop	Film thickness	Shear profile curvature	Wall shear

< 0.2, which is usually characteristic of annular flow, the equation for the pressure gradient can be approximated by Eq. (11.126).

The analysis of a film containing both laminar, "buffer," and turbulent layers was performed by Hewitt.[42] Some of the parameters which he used are shown in Table 11.3 together with a translation in terms of the present nomenclature and an indication of physical significance.

The results can be plotted in the form of a wall friction factor versus Reynolds number for various values of the other parameters, as shown in Figs. 11.25 and 11.26.

The lines of constant β correspond to a situation in which the pressure drop is kept constant while the liquid flow rate is varied. At low flow rates the film is thin, the shear profile is approximately linear, and the results agree with the horizontal flow curve. As the liquid rate is increased, the shear profile distorts under the influence of gravity and

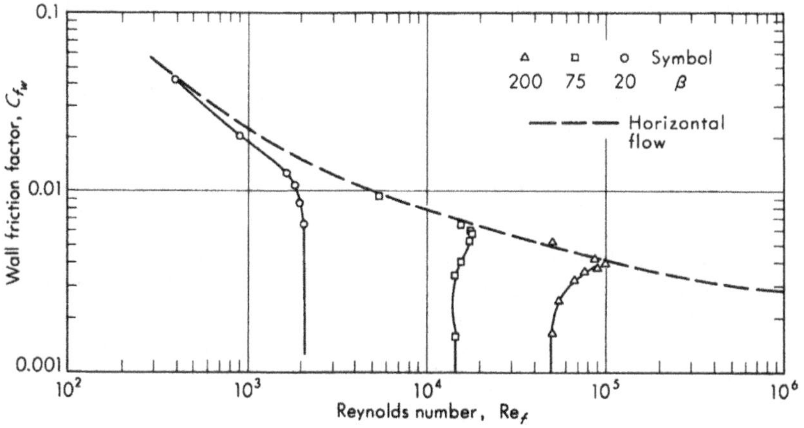

Fig. 11.25 Wall friction factor in vertical flow for Re* = ∞ deduced from Hewitt's differential analysis.[42] (*Wallis.*[3])

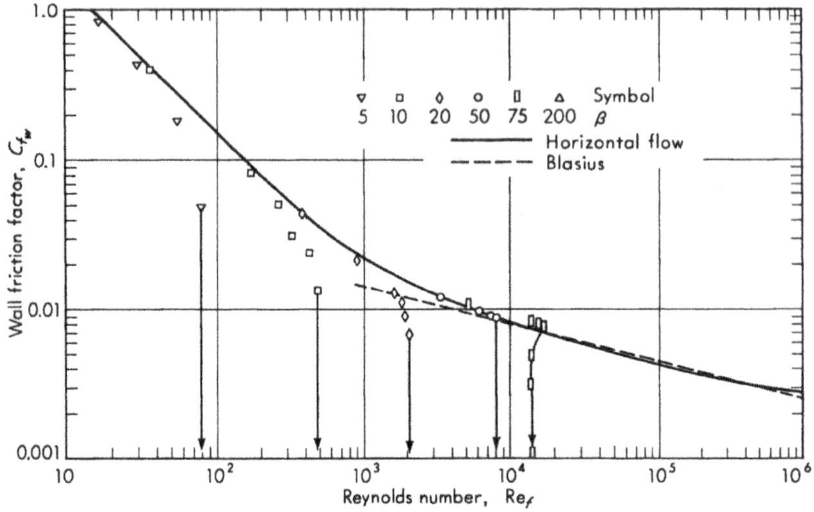

Fig. 11.26 Wall friction factor in vertical flow from Hewitt[42] for Re* = 1000. (*Wallis.*[3])

eventually the wall shear stress, and hence the friction factor, falls dramatically. The liquid Reynolds number at the point of zero wall shear can be correlated approximately with β by the equations[3]

$$\mathrm{Re}_f < 300 \qquad \beta = 1.2\ \mathrm{Re}_f^{1/3} \tag{11.128}$$

$$2000 < \mathrm{Re}_f < 20{,}000 \qquad \beta = 0.13\ \mathrm{Re}_f^{2/3} \tag{11.129}$$

A convenient way of taking account of both viscous effects and the nonlinear shear profile is to introduce two correction factors into Eq. (11.125), which is rewritten as

$$\Delta p^* = \frac{2C'_{fw} j_f^{*2}}{(1-\alpha)^2} + B(1-\alpha) \tag{11.130}$$

C'_{fw} is what the friction factor would be if the shear stress profile were linear, as it is in horizontal flow, and is given by Fig. 11.7. The parameter B takes account of the curvature of the shear stress profile and has been found to give general agreement with Hewitt's results for the following dependence on Reynolds number[3]

$$\begin{aligned} \mathrm{Re}_f &< 1000 & B &= 0.684 \\ 1000 < \mathrm{Re}_f &< 8000 & B &= 0.193\ \mathrm{Re}_f^{0.183} \\ \mathrm{Re}_f &> 8000 & B &\approx 1 \end{aligned} \tag{11.131}$$

These equations are merely one form of approximation to the general function relating Δp^*, $1 - \alpha$, j_f^*, and Re_f, which has the virtue of giving the correct limiting behavior in laminar flow or horizontal flow. Errors tend to be largest near the point of zero wall shear.

The gas core Some of the effects which were neglected in the analysis of the gas core which led to Eq. (11.124) were liquid entrainment, the velocity of the gas-liquid interface, gas viscosity, and compressibility.

The effect of *entrainment* can be taken into account approximately by considering that the core is composed of a homogeneous mixture of gas and droplets. Let the gas mass flow rate be W_g, the film flow rate be W_f, and the entrained mass flow be W_e. The total mass flow rate in the core is

$$W_c = W_g + W_e \tag{11.132}$$

The average density of the core is, approximately (if $\rho_f \gg \rho_g$),

$$\rho_c = \frac{W_c}{W_g}\rho_g \tag{11.133}$$

Therefore the value of j_g^* is increased by the factor $(W_c/W_g)^{1/2}$.

For the *interface velocity* it is not obvious whether one should use the mean liquid velocity, the velocity of waves on the surface, or a time-averaged velocity. Since in most cases the film is thin and the shear stress across it is fairly constant, a reasonable assessment is that the interface velocity is twice the average liquid velocity.

The interfacial shear stress will now be, with these assumptions,

$$\tau_i = 2C_{f_i}\rho_c(v_g - 2v_f)^2 \tag{11.134}$$

The velocity ratio is given in terms of the flow rates by the expression

$$\frac{v_f}{v_g} = \frac{W_f}{\rho_f(1-\alpha)}\frac{\rho_g\alpha}{W_g} \tag{11.135}$$

Using Eq. (11.135) in Eq. (11.134) the interfacial shear stress, and hence Δp^* in Eq. (11.124), is found to be modified by the factor

$$\left(1 - 2\frac{W_f}{W_g}\frac{\rho_g}{\rho_f}\frac{\alpha}{1-\alpha}\right)^2$$

The *viscosity of the gas core* affects the boundary layer near the interface and hence the interfacial shear. A reasonable approximation is to use the value of the friction factor for the gas flowing alone in the pipe instead of the constant value of 0.005 in Eq. (11.19). Up to a Reynolds number of 10^6 the Blasius equation may be used as follows:

$$(C_f)_g = 0.079\,(\mathrm{Re}_g)^{-0.25} \tag{11.136}$$

The gas Reynolds number is calculated from the overall core flow and the gas viscosity. Thus

$$\mathrm{Re}_g = \frac{4(W_g + W_e)}{\pi D \mu_g} \tag{11.137}$$

Combining these correction factors, Eq. (11.124) becomes

$$\Delta p^* = 0.158 j_g^{*2} \left[\frac{1 + 75(1 - \alpha)}{\alpha^{5/2}}\right] \frac{W_c}{W_g} \left[1 - 2\frac{W_f \rho_g \alpha}{W_g \rho_f (1 - \alpha)}\right]^2 \mathrm{Re}_g^{-0.25} \tag{11.138}$$

Typical results of applying these correction factors in order to predict the superficial gas friction factor from the data of Gill and Hewitt[43] are shown in Fig. 11.27. Figure 11.28 shows the application of all of these factors to a wider variety of data taken with two different injection systems. These results are based on the "nonaccelerational" pressure drop which Gill and Hewitt derived by subtracting the acceleration component from the overall pressure drop.

Compressibility effects alter both Eqs. (11.124) and (11.125) because the acceleration of the gas core is felt in both momentum balances. Since $1 - \alpha$ is small and fairly constant, it is usually sufficient to multiply the predicted pressure gradient in both equations by the factor $1/(1 - M_c^2)$ where M_c is the isothermal core Mach number given in the usual way by

$$M_c^2 \approx \frac{G_c j_c}{\alpha^2 p} \approx \frac{(W_e + W_g)Q_g}{\alpha^2 A^2 p} \tag{11.139}$$

It may be convenient to work in terms of a modified core dimensionless flux

$$j_c^* = j_g^* \left(\frac{W_c}{W_g}\right)^{1/2} \tag{11.140}$$

and a dimensionless pressure

$$p^* = \frac{p}{gD(\rho_f - \rho_g)} \tag{11.141}$$

in which case the square of the core Mach number is

$$M_c^2 = \frac{j_c^{*2}}{\alpha^2 p^*} \tag{11.142}$$

Example 11.7 Gill and Hewitt[43] measured film thickness and pressure gradient in upward cocurrent annular flow. A typical data point was $W_g = 300.5$ lb/hr, $W_f = 466$ lb/hr, $\rho_g = 0.0823$ lb/ft³, $\rho_f = 63.3$ lb/ft³, $\mu_g = 0.0423$ lb/(ft)(hr), $\mu_f = 2.615$ lb/(ft)(hr), $W_e = 534$ lb/hr, $\delta = 0.0106$ in., $D = 1.25$ in., $-dp/dz = 23.87$ lb/(ft²)(ft). Compare these results with theoretical predictions.

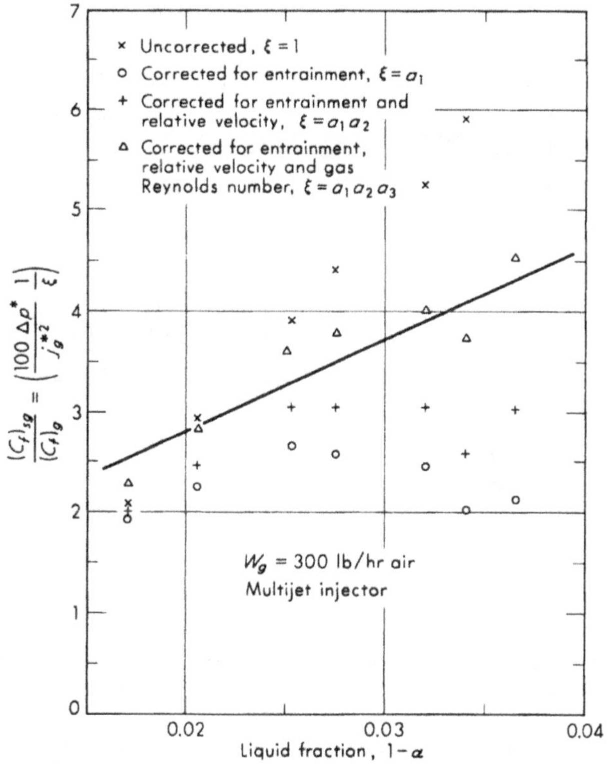

Fig. 11.27 Results of applying successive correction factors to the data of Gill and Hewitt[43] to obtain the ratio $(C_f)_{sg}/(C_f)_g$ as a function of liquid fraction. The three correction factors from Eq. (11.138) are: $a_1 = W_e/W_g$, $a_2 = \left[1 - 2\frac{W_f\rho_g\alpha}{W_g\rho_f(1-\alpha)}\right]^2$ and $a_3 = 0.079\,\mathrm{Re}_g^{-0.25}/0.005$. The line is a modification of Eq. (11.24): $(C_f)_{sg} = (C_f)_g[1 + 90(1-\alpha)]$. (*Wallis.*[3])

Solution The Reynolds number for the liquid film is

$$\mathrm{Re}_f = \frac{4W_f}{\pi D\mu_f} = \frac{(4)(466)}{\pi(1.25/12)(2.165)} = 2180$$

Therefore, from Fig. 11.25 and Eq. (11.131), $(C_{fw})' = 0.015$, $B = 0.79$.

The values of j_f^* and j_c^* are evaluated as 0.133 and 3.94; Re_g is found from Eq. (11.137) to be 2.4×10^5; p^* is 360; $2W_f\rho_g/W_g\rho_f$ is 0.004; M_c^2 from Eq. (11.142) is 0.043. Equations (11.130) and (11.138) with all correction factors included are then

$$\Delta p^*(1 - 0.043) = \frac{5.3 \times 10^{-4}}{(1-\alpha)^2} + 0.79(1-\alpha)$$

$$\Delta p^*(1 - 0.043) = 0.112\,\frac{1 + 75(1-\alpha)}{\alpha^{5/2}}\left(1 - \frac{4 \times 10^{-3}\alpha}{1-\alpha}\right)^2$$

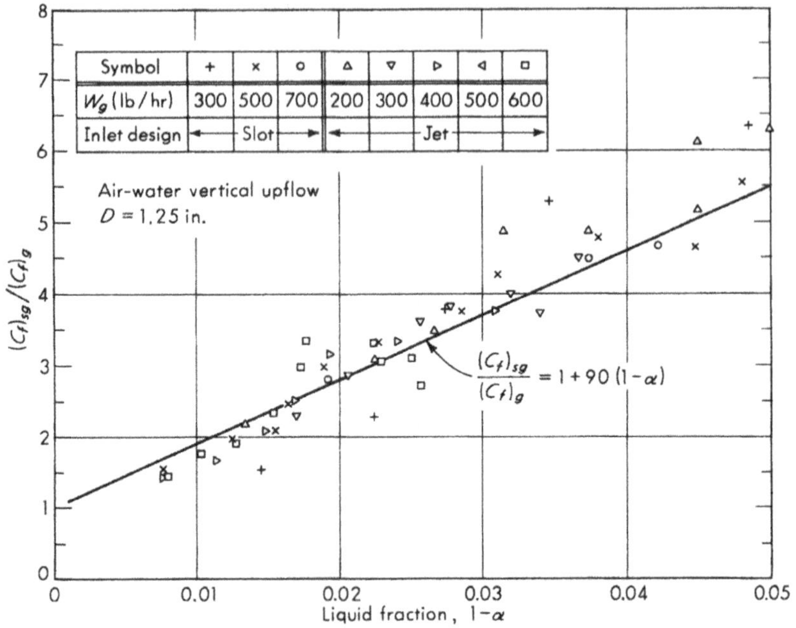

Fig. 11.28 Superficial gas friction factor ratio derived from Gill and Hewitt's data[43] using all correction factors in Eq. (11.138). (*Wallis.*[3])

These equations are solved simultaneously to determine Δp^* and α. The result is $1 - \alpha = 0.038$, $\Delta p^* = 0.40$ compared with the measured values $1 - \alpha = 0.034$ and $\Delta p^* = 0.38$.

The gas-liquid interface Various conditions have been recognized at the interface in annular flow. Very thin films appear smooth, whereas thick ones are covered with large "disturbance waves." In between these is a region of small waves or ripples. Droplet impingement and entrainment modifies the appearance at high gas velocities. These various flow regimes should affect the interfacial friction factor and perhaps also the character of the flow in both phases.

These effects are illustrated in Fig. 11.29 which compares Eq. (11.18) with some results of Shearer and Nedderman[44] taken in upward cocurrent annular flow. Clearly the regime of "rough" annular flow which is described by Eq. (11.18) is characterized by disturbance waves, whereas the friction factor is lower in the "ripple" and "smooth interface" regions. Each series of data points corresponds to a constant gas flow rate.

Figure 11.4 showed the results of Chien and Ibele[6] for downflow plotted on the basis of superficial friction factor versus δ/D for various values of liquid flow rate. The results correlate with each other and approximately with the equation for "rough" annular flow in the regime which the authors describe as "annular mist." Apparently the onset of disturbance waves coincided with the occurrence of entrainment.

Similar results were reported by Dukler[45] and Bergelin and Gazley,[41] who observed a strong influence of surface tension.

It appears, therefore, that the usefulness of Eq. (11.18) is restricted to the large-disturbance wave regime. If greater accuracy is required in the ripple and smooth film regions, new correlations will have to be derived as well as criteria for determining the flow regime of the film. The work of Shearer and Nedderman[44] indicates an effect of gas and liquid flow rates, viscosity, and density, as well as surface tension and pipe diameter. The interaction between all these parameters has yet to be expressed in a form which has any general validity.

ADDITIONAL EFFECTS

Besides the phenomena which have just been described, there are two further effects which make accurate predictions from known quantities

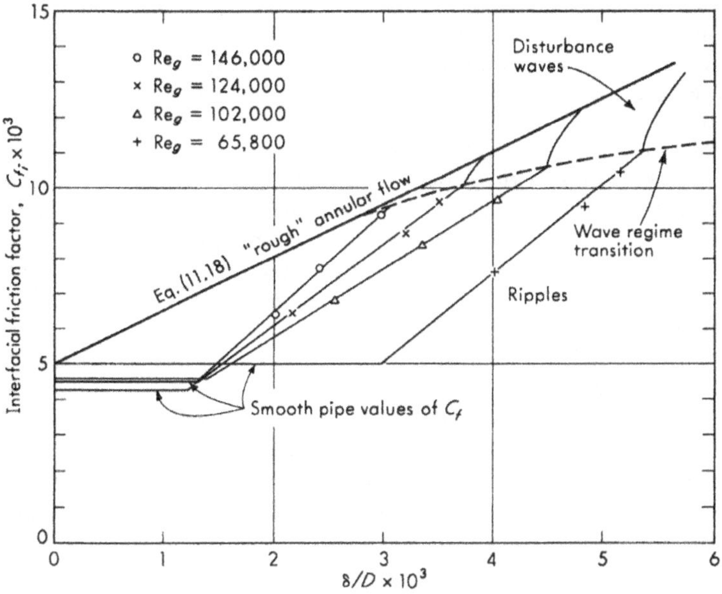

Fig. 11.29 Effect of interfacial wave regimes on friction factor. (*From data of Shearer and Nedderman.*[44]) Pipe diameter = 1.25 in.; air-water upflow. (*Wallis.*[3])

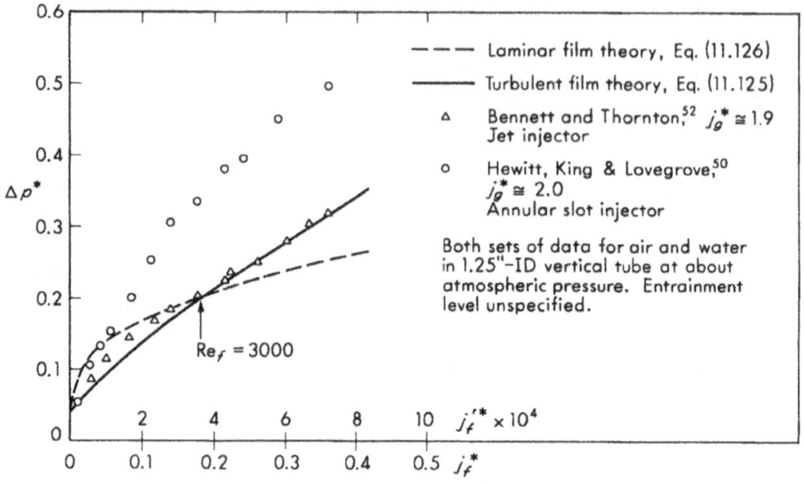

Fig. 11.30 Comparison between theory and data of two different investigations at the same flow rates but using differing injection techniques.

difficult. These are inlet effects and droplet entrainment. Quite different results can be obtained in the same apparatus if the method of introducing the phases is altered. In extreme cases the liquid can be introduced either as a film on the tube wall or as a dispersion of small droplets. For example, pressure-drop results shown in Fig. 11.30 show a difference of a factor of 2 in these two cases. The fraction of the liquid which is entrained may not reach equilibrium some 200 diameters away from the inlet. Even in the equilibrium case, reliable methods for predicting the degree of entrainment are not available. The technique which will be presented in the next chapter (Sec. 12.10) provides only a first approximation.

A changing level of entrainment introduces additional mass-transfer effects between the annular film and the gas core and alters both the accelerational and frictional terms in the momentum equations.

In engineering design it may be worthwhile to allow the degree of entrainment to be an independent variable, to assess its influence on performance, and to seek to control it by suitable inlet design and by introducing devices such as swirl promoters into the flow.

PROBLEMS

11.1. 1000 lb/hr of air and 1000 lb/hr of water flow in a 1-in.-diam horizontal pipe at 70°F and 100 psia. What is the flow pattern? Estimate the pressure drop as a function of the amount of liquid which is entrained and compare the results with Eq. (3.30) with $n = 3.5$.

11.2. Evaluate the ratio $(C_f)_i/(C_f)_g$ from Eq. (11.8) and Martinelli's correlations. Compare the result with the value predicted by Eq. (11.23) with $(C_f)_g$ replacing the coefficient 0.005.

11.3. Compare Eq. (11.20) with single-phase friction factor charts.

11.4. Oil ($\rho_f = 6.5$ lb/gal, $\mu_f = 0.6$ cp) and natural gas ($\rho_g = 3.5$ lb/ft^3, $\mu_g = 0.014$ cp) flow at 1000 psia in a horizontal pipeline of 8 in. diam. The pressure gradient is controlled at 3 psi/1000 ft although the flow rate ratio of the components may vary due to upstream conditions. Estimate the relationship between the flow rates under these conditions.

11.5. For a $(1/n)$th-power velocity profile in turbulent flow the equivalent of Eq. (11.48) is $C_f = A(\mathrm{Re}_m)^{-2/(n+1)}$. What is the value of ϕ_f as a function of n? Compare the result with Table 11.1 for $n = 5$, 9, and 19. What is the limiting expression for ϕ as n tends to infinity?

11.6. Compare Eq. (11.54) with Martinelli's correlation and Eq. (11.22) for typical values of the viscosity and density ratios.

11.7. Deduce Eqs. (11.55) and (11.56) and compare (11.56) with alternative equations relating ϕ_g and α.

11.8. Discuss the various levels of sophistication in the analysis of the liquid film in horizontal annular flow. Under what conditions are the more elaborate theories worthwhile? What is the difference incurred in predicting film thickness by the simplest and by the most complicated theory if $\mathrm{Re}_f = 10^6$, $-(dp/dz)_F = 10^{-1}$ psi/ft, $-(dp/dz)_f = 10^{-3}$ psi/ft?

11.9. What is the film thickness on a vertical wall down which water at 20°C flows at a rate of (*a*) 1 cfm/ft and (*b*) 10^{-3} cfm/ft?

11.10. What are the velocities of continuity waves for the conditions of Prob. 11.9?

11.11. Compare Eq. (11.68) with Eq. (5.32) for $\Delta p^* = 0$. For what value of δ/D is the film thickness prediction in error by 10% if the tube curvature is ignored, for a given value of liquid flow rate?

11.12. Compare the flooding curves predicted by Eqs. (11.116), (11.124), and (11.125) with the modified Martinelli correlation shown in Fig. 11.20.

11.13. 2 cm^3/sec of 99% glycerol at 10°C flow downward inside a vertical pipe of $\frac{3}{4}$ in. diam. What is the maximum allowable countercurrent flow of air at 14.7 psi?

11.14. Compare the limit of the annular flow flooding correlations at very low gas-flow rates with the condition that a slug-flow bubble should be brought to rest in a vertical pipe.

11.15. Shearer and Davidson[46] analyzed the formation of flooding waves on the outside of a tube in terms of the Weber number, $\rho_g j_g^2 \delta/\sigma$, Re_Γ, and Y from Eq. (10.20). Show that this result and the equations presented in Sec. 11.4 are special cases of a function of j_g^*, j_f^*, N_f, and $N_{\mathrm{E\ddot{o}}}$.

11.16. A vertical tube, diameter D, length L, is closed at the bottom and opens at the top into a pool of liquid of low viscosity at saturation temperature. Show that the maximum uniform heat flux which can be supplied to the tube without overheating is

$$\phi_{\max} = C\,\frac{h_{fg} D^{3/2}[\rho_g g(\rho_f - \rho_g)]^{1/2}}{L[1 + (\rho_g/\rho_f)^{1/4}]^2}$$

where $C = \frac{1}{8}$ for a sharp-edged flange at the top and $C = \frac{1}{4}$ for a well-rounded connection.[47] Assume turbulent flow in the film.

11.17. Suppose that a very large wave forms on a falling film in a pipe as a result of a two-phase hydraulic jump. Show that the solution to the countercurrent potential inviscid flow below the wave can be formulated in terms of j_f^* and j_g^*.

11.18. Determine the relationship between mass flux G, quality x, and pressure p at the boundary between slug and annular flow for water in a 1.5-in.-diam vertical pipe.

11.19. Compare the relationships between j_f^* and j_g^* for constant values of void fraction which are predicted for air and water in a 1.25-in.-diam vertical pipe at 20 psia and 70°F by

(*a*) Eq. (11.116);

(*b*) Eqs. (11.121) and (11.122) and Fig. 11.20;

(*c*) Eqs. (11.124) and (11.125) or (11.126), depending on whether the flow is laminar or turbulent;

(*d*) Eqs. (11.124) and (11.125) corrected for relative velocity, Reynolds numbers of both gas and liquid, and compressibility;

(*e*) method (*d*) if there is 20% entrainment of the liquid.

11.20. Show that the minimum value of pressure drop predicted by Eq. (11.126) at a given liquid rate is almost coincident with the point of zero wall shear. Compare with Eq. (11.128).

11.21. Compare the point of minimum pressure drop predicted by Eq. (11.125) with Eq. (11.129) and show that the values will be equal if the wall friction factor is 0.01 rather than 0.005.

11.22. Solov'ev[48] *et al.* found that a plot of $\Delta p^*/(\Delta p^*)_{\min}$ versus $j_g^*/(j_g^*)_{\min}$ gave a unique curve for all liquid rates in upward annular flow with a laminar film. Use Eqs. (11.123) and (11.125) to check this for values of $j_f'^*$ of 10^{-5}, 10^{-4}, 10^{-3}, and 10^{-2}.

11.23. Show that the transition from downflow to upflow in a viscous liquid film occurs at about $j_g^* = 0.8$.

11.24. Hartley and Roberts[49] correlated the interfacial friction factor in annular flow by the equation

$$(C_f)_i - (C_f)_g = 1.5\left[\frac{\delta}{D} - \frac{5}{\mathrm{Re}_g}\sqrt{\frac{2}{(C_f)_g}}\right]$$

Compare this with the correlations in the text.

11.25. Some typical data points obtained by Gill and Hewitt[43] for upward annular-mist flow in a 1.25-in.-diam pipe are given in Table 11.4.

Table 11.4

Δp, $\mathrm{lb}_f/\mathrm{ft}^3$	W_g, lb/hr	W_e, lb/hr	ρ_g, lb/ft³	ρ_f, lb/ft³	μ_f, lb/(ft)(hr)	W_f, lb/hr	δ, in.	μ_g, lb/(ft)(hr)
5.56	306.1	0	0.076	63.3	2.56	20	0.0045	0.043
56.03	269.9	1110	0.086	63.3	2.53	1890	0.0175	0.043
43.7	511.7	379	0.088	63.3	2.68	371	0.0089	0.043
23.43	669.7	1.6	0.088	63.3	2.50	18.4	0.0025	0.043
15.12	300.9	153	0.080	63.3	2.62	197	0.0080	0.042

Compare these results with the predicted values making whichever corrections and assumptions are necessary. W_e is the entrained liquid flow rate and W_f the film flow rate.

11.26. Some air-water data taken by Hewitt, King, and Lovegrove[50] in annular upward flow in a vertical 1.25-in.-diam pipe are shown in Table 11.5. Compare these results with

(*a*) Eqs. (11.115) and (11.116);
(*b*) Fig. 11.20;
(*c*) Eqs. (11.124), (11.125), and (11.126);
(*d*) the separated flow model including all effects;
(*e*) the "unmodified" Martinelli correlation presented in Chap. 3;
(*f*) homogeneous flow theory.

The mean values of the properties which are not given in the table were $\mu_g = 0.0433$ lb/(ft)(hr), $\rho_f = 62.33$ lb/ft³.

11.27. In a vertical air-lift pump the maximum possible liquid flow rate is required for a given pressure gradient. If the fluid is water at 70°F in a 3-in.-diam pipe and the available pressure gradient is 20 psf/ft, what air-flow rate should be used at a pressure of 20 psia? What is the pump "efficiency" in terms of potential energy acquired by the water per unit of energy required to pump the air?

Table 11.5

W_g, lb/hr	W_f, lb/hr	ρ_g, lb/ft³	μ_f, *mean water viscosity*, lb/(ft)(hr)	$1 - \alpha$, %	*Pressure gradient*, lb/ft³
199.7	77.89	0.0774	2.36	2.76	4.189
199.7	95.9	0.0775	2.54	2.91	4.680
200.1	150.5	0.0778	2.58	3.39	5.926
200.6	189.0	0.0794	2.70	3.70	6.281
206.4	295.7	0.0818	2.68	4.21	9.029
206.4	393.4	0.0827	2.60	4.89	10.89
208.3	489.3	0.0844	2.64	5.38	12.36
210.1	718.6	0.0862	2.69	6.66	16.84
211.7	841.6	0.0870	2.69	8.06	18.20
211.6	971.0	0.0878	2.72	7.37	19.78
205.5	639.7	0.0849	2.56	5.95	15.09
207.7	1156.0	0.0891	2.59	8.48	21.80
248.3	39.6	0.0790	2.31	1.73	3.396
254.2	96.1	0.0803	2.43	2.13	6.568
254.4	149.8	0.0803	2.50	2.46	8.255
257.4	196.5	0.0811	2.58	2.78	9.580
255.8	302.2	0.0824	2.55	3.18	12.59
259.8	404.3	0.0846	2.52	3.58	15.74
260.6	499.8	0.0860	2.57	4.09	19.09
262.6	628.3	0.0877	2.58	4.47	20.90
265.7	755.6	0.0901	2.62	4.94	23.62
265.6	855.4	0.0904	2.65	5.57	24.69
263.7	1024.0	0.0920	2.66	6.14	28.07
268.6	1287.0	0.0961	2.68	6.86	30.89

11.28. A complete integral analysis of horizontal annular flow with a smooth gas-liquid interface may be performed[51] by combining Eqs. (11.46) and (11.50) with expressions for ϕ_g which are derived by assuming that the gas flows in a smooth pipe with diameter $D\sqrt{\alpha}$. What are the expressions for ϕ_g in the cases of laminar flow and turbulent flow obeying the Blasius equation? Do not neglect the velocity of the interface. Derive expressions for ϕ_g and α in terms of X for the four combinations of laminar and turbulent flow of each component. Compare with Martinelli's correlations.

This theory is not very accurate because it neglects the roughness of the interface.

11.29. Compare the curve labeled "flow reversal" in Fig. 11.11 with the boundary of the "forbidden region" in Fig. 11.24. What difference does it make if the effects of gas and liquid Reynolds numbers, relative velocity, and shear stress profile are used to give a more accurate estimate?

11.30. In annular flow j_g is usually much larger than j_f. In this case the continuity wave velocity is given approximately by

$$V_w = \left[\frac{\partial j_f}{\partial(1-\alpha)}\right]_{j_g}$$

Show that a dimensionless continuity wave velocity can be deduced as

$$V_w^* = \left[\frac{\partial j_f^*}{\partial(1-\alpha)}\right]_{j_g^*}$$

Using Fig. 11.22 compare V_w^* with the dimensionless liquid velocity $V_f^* = j_f^*/(1-\alpha)$. Under what circumstances will the waves move faster than the liquid velocity?

REFERENCES

1. Baker, O.: *Oil Gas J.*, vol. 53, no. 12, pp. 185–190, 192, 195, July 26, 1954.
2. Steen, D. A.: M.S. thesis, Dartmouth College, Hanover, N.H., 1964; also D. A. Steen and G. B. Wallis, AEC Rept. NYO-3114-2, 1964.
3. Wallis, G. B.: Papers no. 69-FE-45, 69-FE-46, ASME Applied Mech.–Fluids Engg. Conf., Northwestern University, June, 1969.
4. Nikuradse, J.: *Forschungsheft*, p. 301, 1933.
5. Moody, L. F.: *Trans. ASME*, vol. 66, p. 671, 1944.
6. Chien, S. F., and W. Ibele: *Trans. ASME J. Heat Transfer*, ser. C, vol. 86, no. 1, p. 89, 1964.
7. Levy, S.: *Intern. J. Heat Mass Transfer*, vol. 9, pp. 171–188, 1966.
8. Wicks, M., and A. E. Dukler: *A. I. Ch. E. J.*, vol. 6, pp. 463–468, 1960.
9. Lockhart, R. W., and R. C. Martinelli: *Chem. Eng. Progr.*, vol. 45, p. 39, 1949.
10. Blasius, H.: *Forschungsheft*, p. 131, 1913.
11. Turner, J. M.: Ph.D. thesis, Dartmouth College, Hanover, N.H., 1966.
12. Armand, A.: UKAEA, AERE transl. 828, 1959, *Izv. Vses. Teplotekhn. Inst.*, p. 1, 1946.
13. Hewitt, G. F.: unpublished work, 1967.
14. Nusselt, W.: *Z. Ver. Deutsch. Ing.*, vol. 60, p. 541, 1916.
15. Belkin, H. H., A. A. MacLeod, C. C. Monrad, and R. R. Rothfos: *A. I. Ch. E. J.*, vol. 5, pp. 245–248, 1959.

16. Dukler, A. E.: *Chem. Eng. Progr. Symp. Ser.*, vol. 56, no. 30, p. 1–10, 1960.
17. von Kármán, Th.: NACA Rept. TM 611, 1931.
18. Benjamin, T. B.: *J. Fluid Mech.*, vol. 2, p. 554, 1957.
19. Kapitsa, P. L.: *Zh. Eksperm. i Teor. Fiz.*, vol. 18, pp. 2–18 and 19–28, 1948.
20. Tailby, S. R., and S. Portalski: *Trans. Inst. Chem. Engrs.*, vol. 38, pp. 324–330, 1960.
21. Hewitt, G. F., and G. B. Wallis: *Multi-phase Flow Symp.*, ASME, pp. 62–74, November, 1963.
22. Hewitt, G. F., P. M. C. Lacey, and B. Nicholls: *Symp. Two-phase Flow*, Exeter, England, vol. 2, pp. B401–B419, June, 1965.
23. Lobo, W. E., L. Friend, F. Hashmall, and F. Zenz: *Trans. A. I. Ch. E.*, vol. 41, pp. 693–710, 1945.
24. Sherwood, T. K., G. H. Shipley, and F. A. L. Holloway: *Ind. Eng. Chem.*, vol. 30, p. 765, 1938.
25. Dell, F. R., and H. R. C. Pratt: *Trans. Inst. Chem. Engrs.*, vol. 29, pp. 89–109, 1951.
26. Wallis, G. B.: Rept. General Electric Company, 62GL132, Schenectady, N.Y., 1962.
27. Wallis, G. B.: UKAEA Rept. AEEW-R123, 1961.
28. Nicklin, D. J., and J. F. Davidson: paper no. 4, *Symp. Two-phase Flow*, Institution of Mechanical Engineers, London, February, 1962.
29. Wallis, G. B., D. A. Steen, and S. N. Brenner: AEC Rept. NYO-10,487, EURAEC 890, July, 1963.
30. Nikuradse, J.: *Forschungsheft*, p. 356, 1932.
31. Wallis, G. B.: UKAEA Rept. AEEW-R142, 1962.
32. Griffith, P.: Argonne Natl. Lab. Rept. ANL-6796, 1963.
33. Griffith, P., and R. D. Haberstroh: Rept. 5003-28, Mechanical Engineering Department, Massachusetts Institute of Technology, 1964.
34. Clift, R., C. L. Pritchard, and R. M. Nedderman: *Chem. Eng. Sci.*, vol. 21, pp. 87–95, 1966.
35. Wallis, G. B.: AEC Rept. NYO-3114-14, EURAEC 1605, 1966.
36. Govier, G. W., B. A. Radford, and J. S. G. Dunn: *Can. J. Chem. Eng.*, vol. 35, p. 58, 1957.
37. Govier, G. W., and W. L. Short: *Can. J. Chem. Eng.*, vol. 36, p. 195, 1958.
38. Brown, R. A. S., G. A. Sullivan, and G. W. Govier: *Can. J. Chem. Eng.*, vol. 38, p. 62, 1960.
39. Wallis, G. B., and P. E. Meyer: AEC Rept. NYO-3114-10, EURAEC 1480, September, 1965.
40. Casagrande, I., L. Cravarolo, A. Hassid, and E. Pedrocchi: CISE R-73, Milan, 1963.
41. Bergelin, O. P., P. K. Kegel, F. G. Carpenter, and C. Gazley: Heat Transfer and Fluid Mechanics Institute, Berkeley, Calif., 1949.
42. Hewitt, G. F.: UKAEA Rept. AERE-R3680, 1961.
43. Gill, L. E., and G. F. Hewitt: UKAEA Rept. AERE-R3935, 1962.
44. Shearer, C. J., and R. M. Nedderman: *Chem. Eng. Sci.*, vol. 20, pp. 671–683, 1965.
45. Dukler, A. E.: Ph.D. thesis, University of Delaware, 1951.
46. Shearer, C. J., and J. F. Davidson: *J. Fluid Mech.*, vol. 22, part 2, pp. 321–335, 1965.
47. Gambill, W. R.: personal communication, Oak Ridge National Laboratory, Tennessee, 1964.
48. Solov'ev, A. V., E. I. Preobrazhenskii, and P. A. Semenov: *Ind. Chem. Eng.*, vol. 7, no. 1, pp. 59–64, 1967.

49. Hartley, D. E., and D. C. Roberts: Queen Mary College, London, Nuclear Research Memo Q6, May, 1961.
50. Hewitt, G. F., I. King, and P. C. Lovegrove: Holdup and Pressure Drop Measurements in the Two-phase Annular Flow of Air-water Mixtures, UKAEA Rept. AERE-R3764, 1961.
51. Levy, S.: *2nd Midwest Conf. Fluid Mech.*, p. 337, 1952.
52. Bennett, J. A. R., and J. D. Thornton: UKAEA Rept. AERE-R3195, 1965.

12
Drop Flow

12.1 INTRODUCTION

The behavior of droplets suspended in a fluid is similar in many ways to the behavior of bubbles. In fact, many of the equations which will be deduced in this chapter are exactly analogous to the equivalent equations in Chap. 9.

Qualitative differences between the behavior of drops and bubbles are most noticeable when the density difference between the components is high, as in gas-liquid systems at low pressure. In bubbly flow most of the inertia is in the continuous phase and as a result the drag forces on bubbles are large compared with their momentum. Bubbles therefore follow the motion of the surrounding fluid very closely in forced convection. Drops, however, take far longer to adjust to the motions of the surrounding gas. For this reason the homogeneous flow model is usually better for bubbly than for drop flow. Drop-annular flow also has no analog in bubbly flow.

12.2 SINGLE-DROP FORMATION

Quasi-static drop formation at an orifice is an inversion of the problem of bubble formation. The drop radius is therefore given by the analogy to Eq. (9.3), namely,

$$R_d = \left[\frac{\sigma R_0}{g(\rho_2 - \rho_1)}\right]^{1/3} \tag{12.1}$$

where the subscript 2 refers to the discontinuous component. Similarly, when drops form on a vertical surface facing downward, as a result of condensation on a ceiling, for example, the radius is scaled by the wavelength for Taylor instability. Therefore, from Eq. (9.9)

$$R_d \approx \left[\frac{\sigma}{g(\rho_2 - \rho_1)}\right]^{1/2} \tag{12.2}$$

When the liquid velocity through the orifice is increased, the critical velocity given by Eq. (9.8) is soon exceeded, because of the comparatively high density of the fluid, and drops are now formed by the breakup of the resulting jet. This problem is the classical one studied by Rayleigh[1] who showed that such jets are always unstable with the most unstable wavelength being about 4.5 times the jet diameter if the density of the surrounding fluid can be neglected. When a jet breaks up in this way, the radius of the resulting drops is approximately

$$R_d \approx 1.9R_0 \tag{12.3}$$

Further increase of the jet velocity eventually leads to a more severe instability due to the motion relative to the surrounding fluid. As a result the jet becomes violently unstable close to the orifice and breaks up into a shower of very small droplets. This regime of operation is called atomization.

12.3 ATOMIZATION

Most atomization processes produce a large number of small liquid drops by shattering a continuous jet or sheet of liquid. The liquid is usually broken up by aerodynamic forces due to relative motion between the phases. Mechanical, centrifugal, electrical, and ultrasonic force fields can also be used.

The most important dimensionless group for determining the stability of a single droplet is the Weber number based on the relative velocity and the gas density[2–8]

$$\mathrm{We} = \frac{\rho_g(v_g - v_f)^2 d}{\sigma} \tag{12.4}$$

For nonviscous fluids the critical value of the Weber number above which droplets will break up is about 12.

Liquid viscosity apparently has a stabilizing effect which is scaled by the *stability number*, $\mu_f^2/\rho_f\, d\sigma$. The results of Hinze[5,6] and Isshiki[7] can be represented quite well by the equation

$$\mathrm{We} = 12\left[1 + \left(\frac{\mu_f^2}{\rho_f\, d\sigma}\right)^{0.36}\right] \tag{12.5}$$

for values of the stability number less than 5. The presence of the liquid viscosity in the stability criterion implies that instability starts as dynamic oscillation in the droplet shape. In the eventual process of breakup, however, the drop is punched into a baglike shape by the dynamic pressure of the gas acting at the stagnation point. The bag finally bursts to form a ring of smaller droplets.[4,9]

If a drop is introduced into a gas stream at high values of the Weber number, several generations of droplets will be produced by successive shatterings. An expression for the final drop size under these conditions is[7]

$$\left(\frac{d}{d_0}\right)^{0.25} = \frac{1.9}{(\mathrm{We})_0^{0.25}} + 0.315\left(\frac{\rho_g}{\rho_f}\right)^{1.5}(C_D)_0(\mathrm{We})_0^{0.125}\ln\left(\frac{d_0}{d}\right) \tag{12.6}$$

where the subscript 0 refers to the initial conditions.

The Nukiyama-Tanasawa equation,[10] which has been widely used for predicting the mean drop size for atomization with ambient air, is

$$d = \frac{585}{v_0}\sqrt{\frac{\sigma}{\rho_f}} + 597\left(\frac{\mu_f^2}{\sigma\rho_f}\right)^{0.225}\left(\frac{10^3 Q_f}{Q_g}\right)^{1.5} \tag{12.7}$$

v_0 is the initial relative velocity. A disadvantage of this equation is that it is not dimensionless and care must be taken over the units which are ρ_f, g/cm^3; v_0, m/sec; d, microns; σ, dynes/cm; μ_f, (dyne)(sec)/cm^2.

Example 12.1 Compare the predicted drop sizes from Eqs. (12.5) and (12.7) for atomization of benzene (σ = 28.8 dynes/cm, ρ_f = 54.7 lb/ft^3, μ_f = 0.647 cp) in a carburetor using air velocities of 250 fps and an air-fuel mass flow ratio of 18.

Solution A first estimate of d is obtained by neglecting the stability number effect in Eq. (12.5). The result is

$$d = \frac{(12)(28.8)}{(0.0012)[(250)(30.5)]^2} = 5 \times 10^{-3}\ \mathrm{cm}$$

The stability number is found to have a value of 0.012, and the correction for viscosity is therefore negligible.

The values of the terms on the right-hand side of (12.7) are

$$\frac{585}{76.3}\sqrt{\frac{28.8}{0.876}} = 44\ \mu$$

$$597\left[\frac{(0.647 \times 10^{-2})^2}{(28.8)(0.876)}\right]^{0.225}\left[\frac{(10^3)(0.076)}{(18)(54.7)}\right]^{1.5} = 0.63\ \mu$$

Therefore the predicted drop size is 4.5×10^{-3} cm.

In practice drop sizes cannot be estimated very accurately because the atomization process is influenced by many uncontrolled variables such as the degree of turbulence, roughnesses or deposits of dirt in the nozzle.

12.4 DROP SIZE SPECTRA

Equations (12.5) and (12.7) predict only an average drop size. The commonest among the many possible "averages" is the *Sauter mean diameter* which is defined as follows[11]:

$$d_{sm} = \frac{\int_0^{d_{max}} d^3 p(d)\, dd}{\int_0^{d_{max}} d^2 p(d)\, dd} \tag{12.8}$$

$p(d)$ is the probability of a drop having diameter between d and $d + \delta d$. A drop with diameter equal to the Sauter mean diameter has the same surface area to volume ratio as the entire spray.

Numerous statistical drop size distributions have been devised by experimenters in order to correlate data. Many authors have used the *normal* and *log-normal* distributions which are also in common use for describing soils and crushed particles.[12–14]

The Nukiyama-Tanasawa[15] correlation has the form

$$p(d) = A'd^m e^{-bd^n} \tag{12.9}$$

A' and b are normalization factors, whereas m and n are usually integers which can sometimes be given physical significance. Quite often, as reported by Ingebo,[16] the same data correlate equally satisfactorily using any of these methods.

Equation (12.9) can be derived by using the "Jaynes formalism" of statistical mechanics, as described by Tribus.[17] It is assumed that the "states" of the droplets are quantized and can be characterized by indices 1, 2, 3, . . . , etc. Denoting a general one of these states by the index i, and the probability of that state by p_i, we have, since the droplet is in some state,

$$\Sigma p_i = 1 \tag{12.10}$$

Furthermore, we assume that the mean value of the droplet diameter raised to some power n is known from physical considerations. Typically, if the mean volume is known, n is 3, whereas if the mean area is specified, the value is 2. Different methods of atomization may give different values of n. Denoting this average value by $\langle d^n \rangle$ we have

$$\Sigma p_i d_i^n = \langle d^n \rangle \tag{12.11}$$

Applying the "maximum entropy" formalism we find that the probability of a droplet being in the ith state, in the absence of further information, is

$$p_i = A e^{-bd_i^n} \tag{12.12}$$

In order to solve this equation for the diameter probability distribution the quantum states i must be related to the droplet geometry. One approach is to say that i represents the number of molecules in the drop, an obvious quantization, in which case we get

$$i \propto d^3 \tag{12.13}$$

On the other hand, it is possible that the major effect of drop formation is the creation of surface area. If the quantum states represent different amounts of surface energy, we have

$$i \propto d^2 \tag{12.14}$$

In general, to allow for empiricism we let

$$i \propto d^{m+1} \tag{12.15}$$

Since there are so many states that the distribution may be treated as continuous,

$$p_i di = p(d)\, dd \tag{12.16}$$

and substitution in Eq. (12.12), using Eq. (12.15) gives

$$p(d) = (m + 1) d^m A e^{-bd^n} \tag{12.17}$$

which is the generalized Nukiyama-Tanasawa distribution function. We may modify it to the nondimensional form used by Shapiro,[18] by employing the diameter d', at which $p(d)$ is a maximum. Differentiating Eq. (12.17) we find

$$\frac{dp(d)}{dd} = (m + 1) A (md^{m-1} - bnd^{m+n-1}) e^{-bd^n} \tag{12.18}$$

whence

$$b = \frac{m}{nd'^n} \tag{12.19}$$

Defining the dimensionless diameter d^* as

$$d^* = \frac{d}{d'} \tag{12.20}$$

we find, using Eq. (12.19),

$$bd^n = \frac{m}{n} d^{*n} \tag{12.21}$$

We have still to use the constraint which is implied by Eq. (12.10). If there are many quantum states, the sum can be replaced by an integral to give

$$\int_0^\infty p_i \, di = \int_0^\infty p(d) \, dd = 1 \tag{12.22}$$

This result, together with Eq. (12.18), leads to evaluation of the constant A. Thus

$$A = \frac{1}{m+1} \left(\int_0^\infty d^m e^{-bd^n} \, dd \right)^{-1} \tag{12.23}$$

For given values of m and n the values of b and A can now be deduced and the whole distribution generated.

Since $p(d)$ has the dimensions of d^{-1}, in view of Eq. (12.16), a dimensionless probability distribution can be found by defining

$$p^* = d' p(d) \tag{12.24}$$

The commonest exponents in the Nukiyama-Tanasawa equation are $m = 2$ and $n = 1$, corresponding to quantization on a basis of drop volume and a specified value of droplet mean diameter. In this case Eq. (12.17) becomes, in dimensionless form,

$$p^* = 4d^{*2} e^{-2d^*} \tag{12.25}$$

In order to use this equation the value of d' must be known. A simple method is due to MacVean,[19] who found that a great deal of data could be correlated by assuming that

$$d' = \frac{d_0}{2} \tag{12.26}$$

where the characteristic diameter d_0 was given by Eq. (12.5) or (12.7).

An alternative to Eq. (12.25) which is not very different numerically and is more convenient for solving mass transfer and evaporation problems[18] is

$$p^* = d^* e^{-d^{*2}/2} \tag{12.27}$$

12.5 THE TERMINAL VELOCITY OF SINGLE DROPS IN A GRAVITATIONAL FIELD

Small drops with a Reynolds number below unity obey the Hadamard-Rybczynski equation (9.14) with the role of the components reversed. If subscripts 1 and 2 are used to indicate the continuous and discontinuous components, Eq. (9.14) becomes

$$v_\infty = \frac{1}{18}\frac{d^2 g|\rho_2 - \rho_1|}{\mu_1}\frac{3\mu_2 + 3\mu_1}{3\mu_2 + 2\mu_1} \tag{12.28}$$

If surface-active agents are present the surface of the drop is not completely nonrigid and Eq. (12.28) should be modified in the way discussed by Levich.[20]

As long as surface tension is sufficiently dominant in determining the shape, the drops will be approximately spherical and Eq. (8.5) and the methods of Chap. 8 can be used to determine the terminal velocity. For Reynolds numbers above 1000 (e.g., mercury drops in air) the drag coefficient is between 0.4 and 0.5 and the terminal velocity is about

$$v_\infty = 1.7\left(\frac{gd|\rho_2 - \rho_1|}{\rho_1}\right)^{1/2} \tag{12.29}$$

An upper limit to the terminal velocity given by Eq. (12.29) is set by the stability criterion, Eq. (12.5). Comparing these equations for inviscid fluids, it is found that the maximum radius of a stable drop is approximately

$$R_{max} \approx \left(\frac{\sigma}{g|\rho_2 - \rho_1|}\right)^{1/2} \tag{12.30}$$

This criterion does not apply to bubbles which can apparently be indefinitely large, as in the case of bubbles released by underwater explosions.

Even before the condition of Eq. (12.5) is reached, noticeable distortion of the drop shape can occur when the surface tension is insufficiently high. For distorted flat drops the terminal velocity is almost independent of size and is given by an equation similar to Peebles and Garber's for bubbles in region 4 of Table 9.1, namely,

$$v_\infty = \left[\frac{4g(\rho_2 - \rho_1)\sigma}{C_D \rho_1^2}\right]^{1/4} \tag{12.31}$$

Levich[21] recommends a value of $C_D \approx 1$, whence

$$v_\infty = 1.4\rho_1^{-1/2}[g(\rho_2 - \rho_1)\sigma]^{1/4} \tag{12.32}$$

which is intermediate between Harmathy's and Peebles and Garber's equations for distorted bubbles. The range of validity of Eq. (12.32) is about the same as region 4 in Table 9.1.

12.6 ONE-DIMENSIONAL VERTICAL FLOW WITHOUT WALL FRICTION

The techniques of Chap. 4 can be applied to suspensions of drops in fluids if an expression for the drift flux can be found. This problem has not been studied as intensively as the equivalent bubbly flow regime; however, there is considerable evidence in support of a direct analogy between the two cases. For example, the extensive work of Pratt and coworkers[22] on liquid-liquid extraction in spray columns and packed towers showed that the relative velocity of drops in a liquid was given as a function of void fraction by the relation

$$v_{21} = (1 - \alpha)v_\infty \tag{12.33}$$

Substituting this value into Eq. (4.1) we have

$$j_{21} = v_\infty \alpha(1 - \alpha)^2 \tag{12.34}$$

This relationship was found to be very successful in interpreting experimental data as long as the characteristic opening in the tower packing was greater than the value

$$d_{FC} = 2.42 \left[\frac{\sigma}{(\rho_2 - \rho_1)g}\right]^{1/2} \tag{12.35}$$

Comparison between Eqs. (12.35) and (12.30) suggests that Pratt's equation is valid when the drop diameter is determined by stability considerations rather than the method of formation or the apparatus dimensions. In this case, unless the viscous forces are large, Eq. (12.32) is valid and can be substituted into Eq. (12.34) to give

$$j_{21} = 1.4\alpha(1 - \alpha)^2 \rho_1^{-1/2}[g(\rho_2 - \rho_1)\sigma]^{1/4} \tag{12.36}$$

The equation suggested by Wallis[23,24] is the same as Eq. (12.36) except for a different value of the constant (1.18) which was taken from the work of Peebles and Garber.

Gaylor, Roberts, and Pratt[25] also indicate that an equation of the form of Eq. (12.34) should be valid in the viscous region in which the terminal velocity is given by Eq. (12.28).

12.7 FLOODING IN DROP FLOW

The analysis of Chap. 4 is valid whichever component is dispersed; therefore one would expect similar results in the drop or bubbly regimes. This is indicated in Fig. 12.1 which shows the flooding results of Blanding and Elgin[26] for dispersions of naphtha drops in water and vice versa, plotted as in Fig. 4.3.

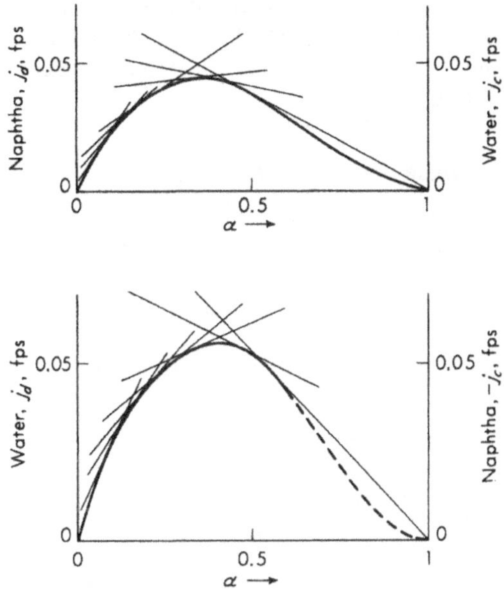

Fig. 12.1 Flooding results of Blanding and Elgin[26] for drops of water in naphtha and vice versa.

The equations for predicting flooding given by Thornton and Pratt[27] are identical with Eqs. (4.16) and (4.17) with $n = 2$.

Other flooding correlations which are given in the literature are sometimes based on a "characteristic velocity" v_0 and have the form[28,29]

$$v_0^{1/2} = |j_1|^{1/2} + |j_2|^{1/2} \tag{12.37}$$

This equation is the consequence of assuming that the relative velocity v_{21} is independent of concentration, i.e.,

$$v_{21} = \frac{j_2}{\alpha} - \frac{j_1}{1 - \alpha} = v_0 \tag{12.38}$$

and may be obtained by forming the envelope of lines of constant α in the second quadrant (Prob. 1.19). This result is intermediate between the results for ideal dispersed flow and churn-turbulent flow discussed in Chap. 9 and may be of use as a general correlation which approximately represents both possibilities. v_0 may be set equal to v_∞ and calculated from the equations in Sec. 12.5.

The comparison between Eq. (12.37) and the flooding equations in annular flow [Eqs. (11.84) and (11.87)] is interesting. In many packed columns, for example, the flooding point is determined by Eq. (11.86) rather than Eq. (12.37) due to coalescence of the drops to form continuous fluid streams. In each case the square roots of the fluxes are linearly related.

12.8 DROP FLUIDIZATION

If a suspension of drops is imagined to be fluidized in exactly the same way as a fluidized bed and one uses Eqs. (12.36) and (4.2) to describe the behavior, the result is

$$j_1 = -1.4(1 - \alpha)^2 \rho_1^{-1/2}[g\sigma(\rho_2 - \rho_1)]^{1/4} \tag{12.39}$$

as long as the drops do not coalesce. Moreover, if the drops are prevented

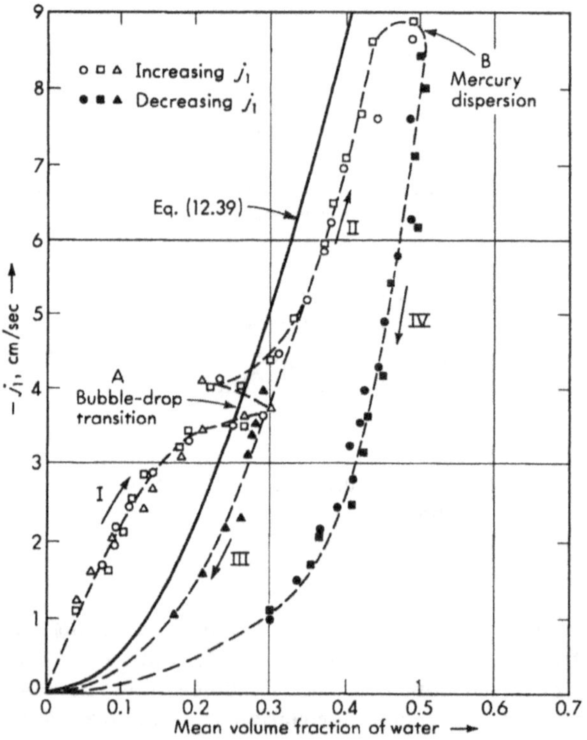

Fig. 12.2 Fluidization of mercury droplets with water (*Kutateladze and Moskvicheva*[30]). I: Bubbles of water in mercury. II, III, IV: Various regimes of mercury droplets in water.

from coalescing by the presence of surface impurities, Eq. (12.39) will describe the behavior over the whole range of α. Because of the difficulties of inhibiting coalescence entirely, few data are available to verify this point; however, the results of Kutateladze,[30] who blew water through mercury, are not far from the theoretical prediction (Fig. 12.2). After fluidization for some time the droplet size apparently decreased and the points followed a lower curve.

A change in character might be expected at the point where the drops become fluidized and no longer rest on one another. For spherical particles this usually occurs at $\alpha \approx 0.6$ or, more accurately, from Fig. 4.2, at $1 - \alpha = 0.38$. Substituting this value into Eq. (12.39) we obtain an expression for the minimum fluidization velocity,

$$(j_1)_{mf} = 0.2\rho_1^{-1/2}[g\sigma(\rho_2 - \rho_1)]^{1/4} \tag{12.40}$$

Wallis[24] tested an equation of this form by blowing air through various liquids from a porous surface and found that the onset of droplet dispersion close to the surface occurred at a value close to the prediction. Above the volumetric gas flux given by Eq. (12.40) a spray of droplets was observed above a shallow liquid pool through which gas was blown.

12.9 PRESSURE DROP IN FORCED CONVECTION

The techniques of Sec. 9.6 are approximately applicable to drop flow in forced convection. However, the analogy should not be stretched too far in the case of gas-liquid flow because of the much greater relaxation times and mean free paths of droplets and their tendency to stick to walls.

For example, the virtual viscosity of an emulsion in laminar flow can be determined, whichever phase is dispersed, in the form[31]

$$\mu = \mu_1\left(1 + 2.5\alpha\,\frac{\mu_2 + \frac{2}{5}\mu_1}{\mu_2 + \mu_1}\right) \tag{12.41}$$

where the subscript 1 refers as usual to the continuous phase. In turbulent flow the homogeneous flow model works reasonably well if the two-phase friction factor is calculated from a Reynolds number based on the overall mass flow rate and the gas viscosity. Thus

$$\mathrm{Re} = \frac{GD}{\mu_g} \tag{12.42}$$

The problem with gas-liquid flow is that the liquid usually has an affinity for the wall so that drops striking the wall stick together to form an annulus. Thus, ideal drop flow usually is never obtained but only the hybrid drop-annular flow. The liquid in the annulus has four effects:

1. It reduces the density of the homogeneous core.
2. It reduces the effective area of flow for the core.
3. It forms waves which increase the effective wall roughness.
4. By means of a continuous exchange of liquid with the core, as a result of drop deposition and entrainment, considerable momentum transfer, which appears as shear stress, can take place.

The above effects are not yet fully understood. However, the methods described in Sec. 11.5 may be used to calculate macroscopic variables such as film thickness and pressure drop.

12.10 ENTRAINMENT

QUALITATIVE OBSERVATIONS

Imagine that a liquid film is caused to flow down the wall of a vertical duct or along the bottom of a horizontal or inclined channel. Then suppose that gas is blown down the same duct over this film. What happens as the gas velocity is increased?

At first, low gas velocities have little effect on the film. However, with increasing relative velocity a destabilizing effect is noted, in accordance with the discussion in Sec. 6.4. Specifically, c^2 becomes more negative with an increase in relative velocity in Eq. (6.83); therefore $v_w^2/u^2 - 1$ is larger in Eq. (6.163) and hence a is larger in Eq. (6.162). In horizontal or inclined channels, gravity acts as a restoring force and may delay the onset of noticeable wave activity. The first waves to appear are small ripples traveling in the direction of the film. Higher velocities of the gas lead to an increase in the amplitude of these ripples and soon three-dimensional disturbances are generated. The interface now has a "pebbled" or "cross-hatched" appearance similar to the waves which are obtained with light squalls on rivers or lakes.[23,33]

At a gas velocity which is about double that which is necessary to produce the cross-hatched wave pattern, the first *roll waves* appear. These have a much larger amplitude and velocity than previous waves and appear to ride over the top of the more uniform small amplitude waves. Roll waves have a steep front and a long region of relatively quiet fluid between crests. They have been the subject of several theoretical and experimental studies.[34–37]

When the gas velocity is sufficiently high, the drag forces on the tops of roll waves are adequate to pull off droplets of liquid that become entrained in the gas stream. The onset of entrainment is usually preceded by a noticeable roughening of the tops of the roll waves which resemble patches of "white water" rushing over the top of the film. Further increase of gas rate increases the entrainment and also has the effect of

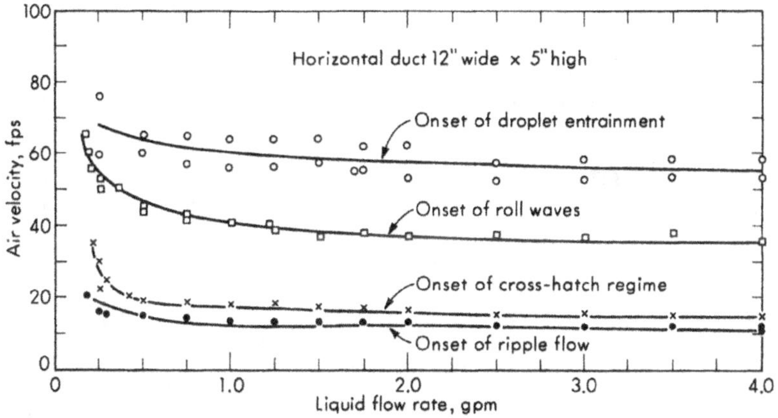

Fig. 12.3 Regions of wave activity in stratified air-water flow. (*Bemberis.*[33])

decreasing the film thickness as a result of both depletion and the higher interfacial shear and consequent augmented liquid velocity.

Figure 12.3 shows a plot of the regions of various wave activity for air flowing over water in a horizontal duct 12 in. wide by 5 in. high. The liquid rate is seen to have little influence on the transition points as long as it is sufficiently high.

The onset of droplet entrainment can be detected by noticing droplets striking the unwetted walls of the duct or by detecting drops in the gas stream. A more precise measurement can be obtained by sampling a given area of the gas stream by means of an extraction probe and measuring the amount of liquid which enters the probe.

For example, Fig. 12.4 shows entrainment data taken by Steen[36] in a 4-in.-diameter vertical tube, and Fig. 12.5 shows Wallis'[38] data for a vertical tube of 0.875 in. in diameter. The ordinate of these graphs is the *percent entrainment* defined by Eq. (11.114). The *air velocity* shown is the volumetric flux j_g, based on the total tube area. The decrease in critical gas velocity at which entrainment starts in Fig. 12.5 at high liquid rates is probably due to the reduction of available area for gas flow because of the thickness of the liquid film.

It is noticeable that the critical gas velocity in both figures displays little sensitivity to duct dimensions, orientation, or liquid flow rate.

Corresponding entrainment results in upward vertical flow have already been shown in Fig. 11.17. Here the persistence of liquid slugs obscures the critical gas flow rate for film instability except at very low liquid rates.

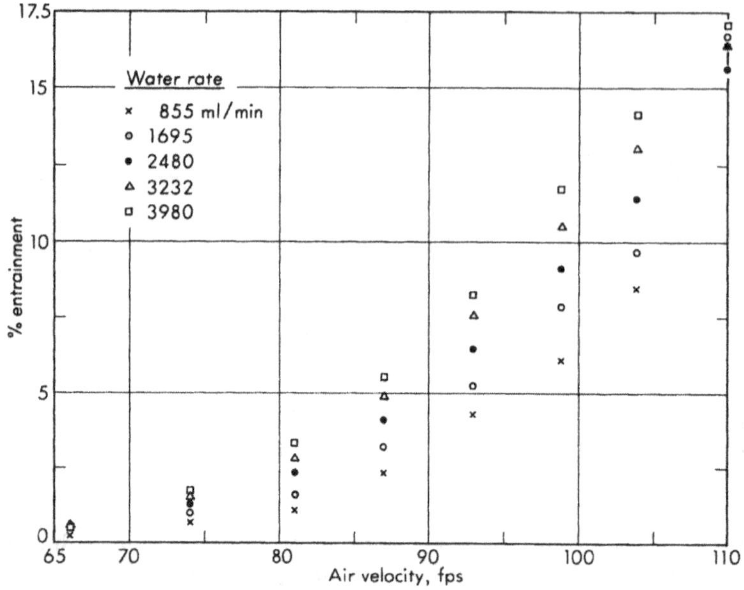

Fig. 12.4 Percent entrainment versus air velocity for 4-in.-ID tube. Sampling at center of tube. (*Steen.*[36]) Cocurrent downflow.

THE EFFECT OF INLET CONDITIONS AND TUBE LENGTH

The results shown in Fig. 12.4 were obtained after considerable care had been taken to prevent entrainment originating where the air and water were introduced into the apparatus. Therefore, the data should give a true measure of film stability. In many practical cases, however, this effect is completely masked by inlet entrainment. Because droplets which are carried along in a gas stream have a high axial but low radial velocity, and because of splashing which occurs when they hit the liquid film, a very long duct is necessary before the effects of inlet conditions can be considered to be negligible.[39] For example, the results of Gill, Hewitt, and Hitchon[40] (Figs. 12.6 and 12.7) show how the total entrained flow rate and mass velocity profiles vary up to 209 in. downstream from the position at which the liquid film was placed on the wall of a $1\frac{1}{4}$-in.-diameter tube.

DEFINITION OF A CRITICAL GAS VELOCITY

A *critical gas velocity* for the onset of entrainment can be specified as long as entrainment results from film stability and from no other causes. Accurate definition of this velocity is not quite as simple as it at first appears. Since zero entrainment cannot be measured, the critical velocity will usually be chosen as the place where the first reasonable amounts of entrainment occur. Referring to Fig. 12.4 we can see that this definition, for the case of 855 ml/min water flow rate, would give values for the critical velocity of 65, 75, or 87 fps depending on whether the measurement sensitivity is 0.1, 1, or 2.5 percent entrainment. Because of the statistical nature of turbulent velocities in the gas and the liquid and also the spectrum of waves on the interface, an occasional combination of circumstances can result in a drop being entrained at quite low gas velocities. Nevertheless, almost all entrainment data display the characteristics shown in Fig. 12.9, that is,

1. A region of negligible entrainment
2. A region of slowly increasing entrainment, with upward curvature

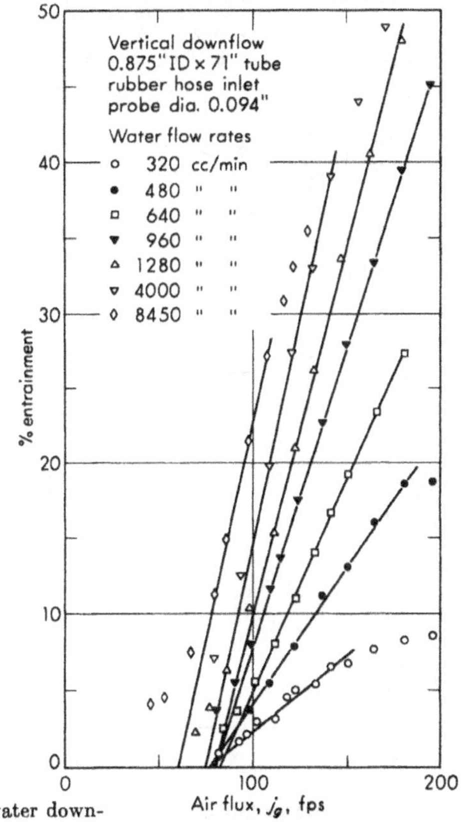

Fig. 12.5 Entrainment data for air-water downflow in a 0.875-in.-diam tube. (*Wallis.*[38])

3. An approximately linear region in which entrainment increases steadily with increasing gas velocity
4. Saturation at high gas velocities where the percent entrainment reaches a limit

The critical velocity is then arbitrarily defined as the point where extrapolation of the linear portion of the curve hits the gas velocity axis.

PREDICTION OF THE CRITICAL GAS VELOCITY

Numerous competing correlations[37,39,41,42,43,44] exist for predicting the onset of droplet entrainment, and a completely general synthesis of these results is yet to be achieved. Steen[36] varied the gas pressure and found that the critical velocity varied approximately inversely as the square root of the gas density (Fig. 12.8). The liquid viscosity was found to be unimportant above a critical liquid flow rate (Fig. 12.9), but the surface tension was very important, being almost proportional to the critical velocity. As a result of these experiments Steen suggested the following criterion for the onset of entrainment:

$$\frac{j_g \mu_g}{\sigma} \left(\frac{\rho_g}{\rho_f}\right)^{1/2} = \pi_2 > 2.46 \times 10^{-4} \tag{12.43}$$

This equation is not universally valid and is incorrect if the liquid occupies a significant proportion of the duct, or if viscous forces in the liquid are significant.

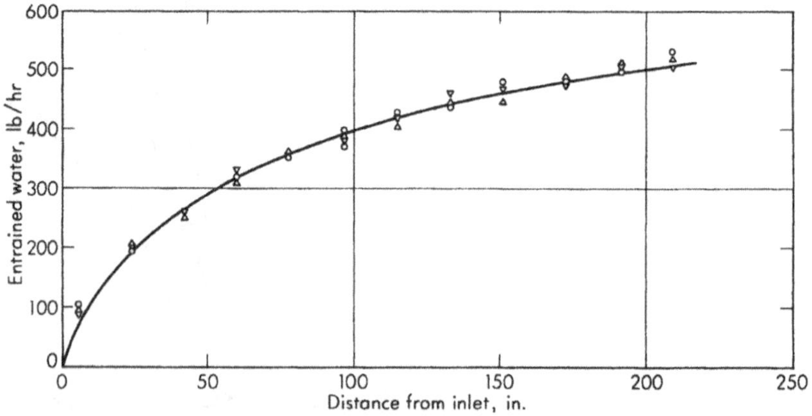

Fig. 12.6 Integrated flux of entrained water through center 1.05 in. of 1.25-in.-diam tube as a function of distance from inlet (AERE-R3954).

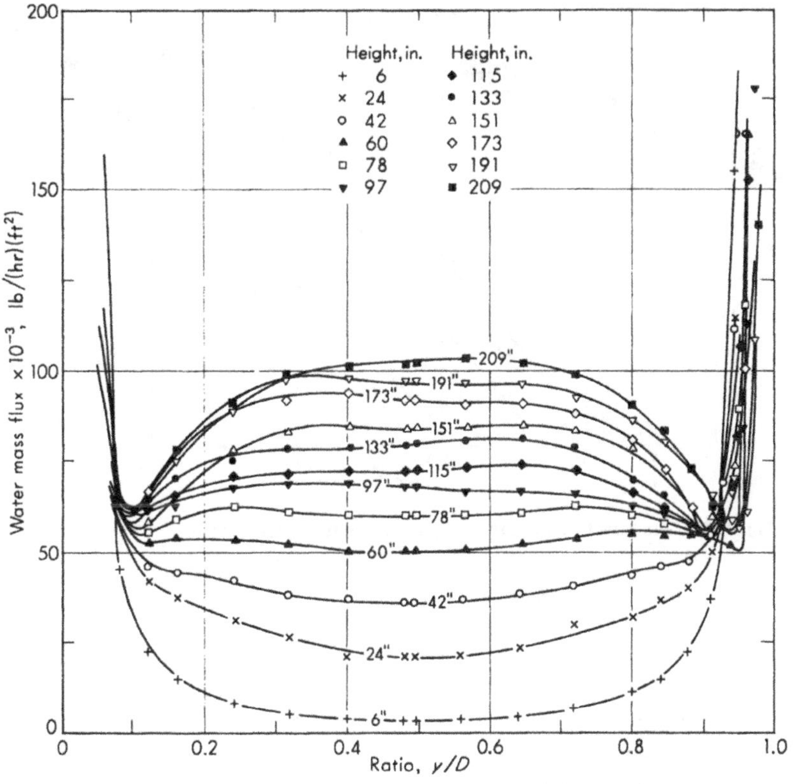

Fig. 12.7 Water mass velocity profiles as a function of distance from inlet (AERE-R3954).

Steen's parameter forms the basis of an approximate correlation of equilibrium percent entrainment in long pipes which was derived by Wallis[43] as a modification of the method of Paleev and Filippovich[44] and is shown in Fig. 12.10. This curve is close to most published data in the thick film regime for entrainment levels below 50 percent. Above this value numerous secondary effects of pipe length, method of experimentation, etc., give rise to considerable scatter. Thin, viscous films are more stable (Fig. 12.9) and give lower amounts of entrainment than are predicted by using Fig. 12.10.

DROPLET CONCENTRATION AND VELOCITY DISTRIBUTIONS

Droplet concentration and velocity profiles have been measured in some detail at Harwell[40,45] and CISE.[46] The droplet concentration measured

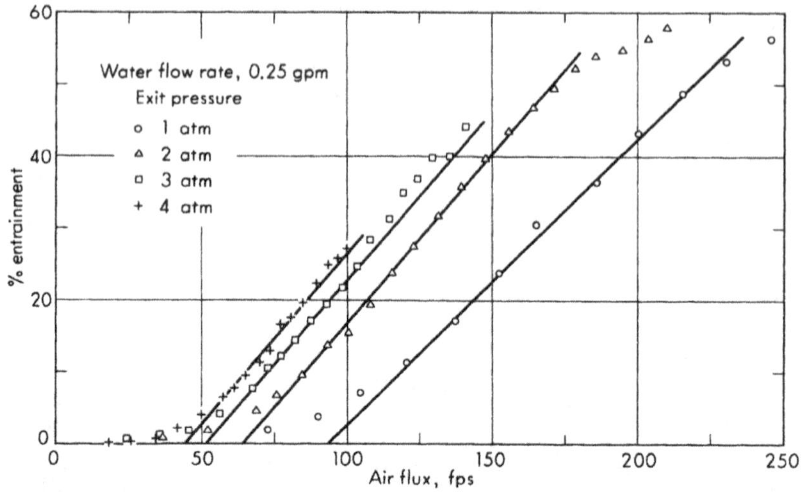

Fig. 12.8 The effect of pressure on entrainment. (*Steen.*[36])

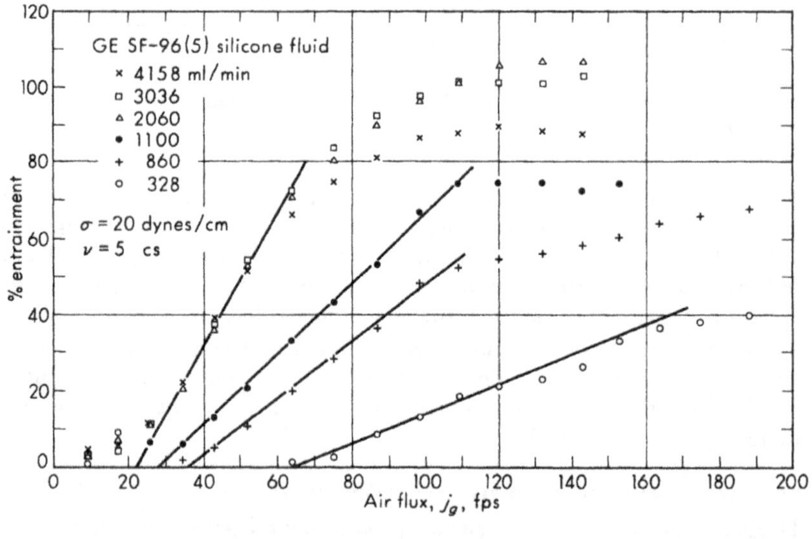

Fig. 12.9 Entrainment as a function of air flux for various liquid flow rates. (*Steen.*[36])

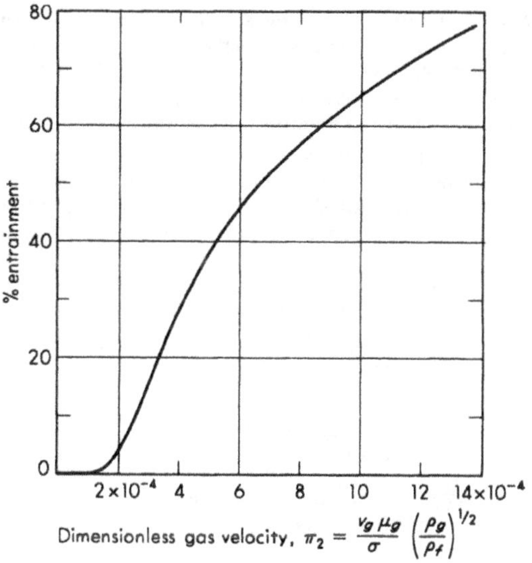

Fig. 12.10 Wallis[43] correlation for equilibrium entrainment.

at Harwell was found to be virtually constant throughout the flow. The corresponding velocity profiles showed a tendency to become more and more parabolic as entrainment increased (Fig. 5.2). The velocity profiles fit Eq. (5.8) remarkably well except that the value of the constant k decreased from 0.4 to as low as 0.1 for the greatest droplet concentration. For a power law form of correlation the velocity defect with entrainment follows approximately a square law with radius as though the flow had become "laminarized," with the fine droplet dispersion resembling the classic kinetic model of a perfect gas.

PROBLEMS

12.1. Water condenses on the horizontal ceiling of a shower stall where the temperature is 90°F. What size droplets will eventually drip down from this surface?

12.2. A garden hose with a spray head discharges 4 gal/min through 30 holes with $\frac{1}{16}$ in. diam. Estimate the droplet size (*a*) in still air and (*b*) in a 60-mph gale.

12.3. Fuel leaks from a small hole in the tank of an aircraft flying at 20,000 ft above sea level at a speed of 550 mph. Estimate the size of droplets which are formed.

12.4. What is the maximum stable size of a falling raindrop? What is its terminal velocity?

12.5. What is the maximum stable size of mercury droplets falling through water?

12.6. A countercurrent flow scrubber to clean flue gases is to be installed in a power station. If the flue-gas velocity is 10 fps at a temperature of 500°F, what size of water droplet will ensure downflow of liquid at the bottom of the scrubber? What is the maximum downward velocity with which these droplets can be sprayed from nozzles if their temperature is constant at 50°F and they are not to break up?

12.7. It is found by experiment that fuel droplets larger than 15-μ diam will fail to "turn the corner" at the point where an automobile carburetor joins the inlet manifold and will be deposited on the wall. This defeats the purpose of the carburetor as a droplet dispersal device and increases the time lag in engine response to a change in throttle setting. Estimate the fraction of the fuel which is supplied to the carburetor which will be deposited on the wall as a function of the air-fuel ratio and the air velocity at the fuel injection point in the carburetor.

12.8. Compare Eqs. (12.25) and (12.27) with the "best fit" normal and log-normal probability distributions of the form

$$p(d) = \frac{1}{\sigma \sqrt{2\pi}} e^{-(d-\bar{d})^2/2\sigma^2}$$

$$p(d) = \frac{1}{\ln \sigma \sqrt{2\pi}} e^{-(\ln d - \ln \bar{d})^2/2 \ln^2 \sigma}$$

$\bar{d}$ is an average diameter and σ measures the "sharpness" of the distribution.

12.9. Drop size spectra can be measured experimentally by taking high-speed photographs. Unfortunately, this technique provides only an instantaneous sample of an area of space and does not give the true size spectrum if the droplets are moving with different velocities. Show how the measured spectrum is distorted (*a*) if drops are being accelerated in a high-speed gas flow and are photographed while they are still accelerating and (*b*) if droplets are falling in still air at their terminal velocities.

Assume that the photograph is taken across the direction of motion.

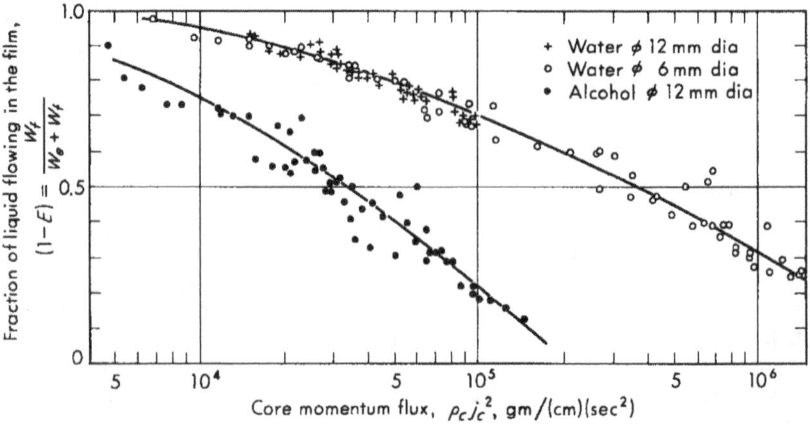

Fig. 12.11 Entrainment correlation of Minh and Huyghe.[48]

Table 12.1 Entrainment and pressure-drop data of Cousins, Denton, and Hewitt.[49] Air and water flowing in a ⅜-in.-diam tube. $W_e + W_f$ = total water flow rate

W_g, lb/hr	$W_e + W_f$, lb/hr	*Distance from injector, in.*						
		6	30	71½	107½	141	180	216
40	100	0.8	3.3	4.0	5.0	4.5	5.1	5.7
	230	5.7	8.6	12.8	16.0	20.9	24.7	33.9
50	100	1.2	5.2	7.2	7.8	8.7	11.1	13.7
	230	10.1	15.0	23.8	29.8	37.2		
60	100	1.6	8.9	14.5	14.4	18.8	20.1	30.0
	230	10.0	23.3	44.7	52.7	70.5		
70	100	2.5	12.8	21.7	24.3	28.1	34.0	
	230	12.7	37.9	67.0	85.6			

(*a*) Entrainment data: the numbers in the table give the values of W_e in pounds per hour.

W_g, lb/hr	$W_e + W_f$, lb/hr	*Distance from injector, in.*						
		9	27	68½	104½	138	177	213
40	100	39.59	38.62	36.62	34.58	33.40	31.63	29.72
	230	39.27	37.62	33.73	30.26	27.43	23.48	19.95
50	100	39.39	38.21	35.76	33.35	31.42	28.25	20.15
	230	38.78	36.87	32.06	27.86	24.47		
60	100	39.20	37.77	34.17	31.63	28.39	25.38	21.52
	230	38.55	36.01	29.78	25.57	21.03		
70	100	39.00	37.14	32.83	29.74	26.55	22.19	
	230	38.21	35.51	27.52	22.82			

(*b*) Pressure-drop data: the numbers in the table give the pressure in pounds per square inch absolute.

12.10. Drop fluidization is a possible reason for the occurrence of a *maximum heat flux* during natural convection boiling from a horizontal surface. Use Eq. (12.40) to deduce the maximum heat flux, assuming that all the heat supplied at the surface forms vapor near the surface and that the liquid is at saturation temperature. [Kutateladze[47] found that the empirical value of the coefficient in Eq. (12.40) in this case is 0.16.]

12.11. Gas is bubbled through a stagnant pool of water. Compare the gas flux needed to cause flooding in bubbly flow with the flux needed to cause drop fluidization.

12.12. What values of droplet diameter and the index n in Eq. (4.14) will give agreement with the right-hand curve in Fig. 12.2?

12.13. Compare the entrainment correlation shown in Fig. 12.10 with the data in Figs. 12.5, 12.8, 12.9, and 11.17.

12.14. A steam-water mixture at 500 psia flows in a 1-in.-diam vertical pipe. Estimate the equilibrium level of entrainment in the annular-mist regime as a function of quality and mass flow rate. Using these values calculate the increase in apparent density of the vapor core due to the presence of droplets.

12.15. A seaman observes that a salt spray is being blown off the tops of waves during a storm. Estimate the minimum wind velocity.

12.16. Figure 12.11 shows the entrainment data of Minh and Huyghe[48] for air-ethyl alcohol and air-water mixtures in annular flow in pipes of 6 and 12 mm diam. Assuming that the temperature was 20°C, for which the property values are $\mu_g = 0.018$ cp, $\rho_f = 1\ \mathrm{g/cm^3}$, $\sigma = 72.8$ dynes/cm for water, and $\rho_f = 0.79\ \mathrm{g/cm^3}$, $\sigma = 22.3$ dynes/cm for alcohol, show that the data are consistent with the correlation scheme shown in Fig. 12.10 but that the actual levels of entrainment are lower. Discuss possible reasons for the disagreement. ρ_c and j_c are the density and volumetric flux for the core flow assuming that it is a homogeneous mixture of droplets and gas.

12.17. One difficulty with entrainment prediction is that equilibrium is never reached because, due to decreasing pressure or phase change, the gas flux increases steadily along the duct. Table 12.1 illustrates this effect using some data of Cousins, Denton, and Hewitt[49] taken with air and water in a ⅜-in.-diam tube. Compare these results with Fig. 12.10 and show how the experimental curves move to the right as non-equilibrium effects become more important.

REFERENCES

1. Rayleigh, Lord: *Proc. London Math. Soc.*, vol. 10, p. 1, 1878. 4, *Proc. Roy. Soc.*, (London), vol. 29, p. 71, 1879. *Phil. Mag.*, vol. 34, p. 177, 1892. For a discussion summarizing the work, see Sir Horace Lamb's "Hydrodynamics," 6th ed., pp. 471–473, Dover Publications, Inc., New York.
2. Lewis, H. C., D. G. Edwards, M. J. Goglia, R. I. Rice, and L. W. Smith: *Ind. Eng. Chem.*, vol. 40, no. 1, p. 67, 1948.
3. Haas, F. C.: *A. I. Ch. E. J.*, vol. 10, pp. M920–924, 1964.
4. Prandtl, L.: "Essentials of Fluid Mechanics," p. 328, Blackie & Son, Ltd., Glasgow, 1953.
5. Hinze, J. O.: *Appl. Sci. Res.*, vol. A1, pp. 263–272, 1948.
6. Hinze, J. O.: *Appl. Sci. Res.*, vol. A1, pp. 275–288, 1948.
7. Masugi Isshiki, N.: Rept. 35, Transportation Technical Research Institute, Tokyo, Japan, July, 1959.
8. Dickerson, R. A., and T. A. Coultas: AIAA paper no. 66-611, June, 1966.
9. Giffen, E., and A. Muraszew: "The Atomisation of Liquid Fuels," Chapman & Hall, Ltd., London, 1953.
10. Nukiyama, S., and Y. Tanasawa: *Trans. Soc. Mech. Engrs.* (Japan), vol. 4, no. 14, p. 86, 1938.
11. Sauter, J.: NACA Rept. TM-518, 1929.
12. Soo, S. L.: *Ind. Eng. Chem. Fundamentals*, vol. 4, pp. 426–433, 1965.
13. Kliegel, J. R.: *Intern. Symp.* Combust, 9th, Academic Press, Inc., New York, pp. 811–826, 1963.
14. Dallavalle, J. M.: "Micromeritics," Pitman Publishing Corporation, New York, 1943.
15. Nukiyama, S., and Y. Tanasawa: *Trans. Soc. Mech. Engrs.* (Japan), vol. 5, no. 18, p. 63, 1939.
16. Ingebo, R. D.: NASA transl. D-290, June, 1960.

17. Tribus, M.: "Thermostatics and Thermodynamics," D. Van Nostrand Company, Inc., Princeton, N.J., 1961.
18. Shapiro, A. H., and A. J. Erickson: *Trans. ASME*, vol. 79, p. 775, 1957.
19. MacVean, S. S.: unpublished work, Dartmouth College, Hanover, N.H., 1967.
20. Levich, V. G.: "Physicochemical Hydrodynamics," pp. 409–429, Prentice-Hall, Inc., Englewood Cliffs, N.J., 1962.
21. Levich, V. G., Ref. 20, p. 431.
22. Pratt, H. R. C., et al.: *Trans. Inst. Chem. Engrs.*, vol. 29, pp. 89–148, 1951; vol. 31, pp. 57–93 and 289–326, 1953; vol. 35, pp. 267–342, 1957.
23. Wallis, G. B.: discussion of paper no. 27, *Intern. Heat Transfer Conf.*, Boulder, Colo., pp. D-70–D-72, ASME, 1962.
24. Wallis, G. B.: paper no. 3, *Two-phase Fluid Flow Symp.*, Institution of Mechanical Engineers, London, 1962.
25. Gaylor, R., N. W. Roberts, and H. R. C. Pratt: *Trans. Inst. Chem. Engrs.*, vol. 31, pp. 57–68, 1953.
26. Blanding, F. H., and J. C. Elgin: *Trans. A. I. Ch. E.*, vol. 38, pp. 305–338, 1942.
27. Thornton, J. D., and H. R. C. Pratt: *Trans. Inst. Chem. Engrs.*, vol. 31, pp. 289–326, 1953.
28. Elgin, J. C., and F. M. Browning: *Trans. A. I. Ch. E.*, vol. 31, p. 639, 1935; vol. 32, p. 105, 1936.
29. Crawford, J. W., and C. R. Wilke: *Chem. Engr. Progr.*, vol. 47, p. 423, 1951.
30. Kutateladze, S. S., and V. N. Moskvicheva: *Zh. Tech. Fiz.*, vol. 29, pp. 1135–1141, 1959.
31. Taylor, G. I.: *Proc. Roy. Soc.* (London), vol. A148, p. 141, 1932.
32. Hanratty, T. J., and J. M. Engen: *A. I. Ch. E. J.*, vol. 3, pp. 229–304, 1957.
33. Wallis, G. B., J. M. Turner, I. Bemberis, and D. Kaufman: AEC Rept. NYO-3114-4, 1964.
34. Hanratty, T. J., and A. Hershman: *A. I. Ch. E. J.*, vol. 7, pp. 488–497, 1961.
35. Chung, H. S., and W. Murgatroyd: *Symp. Two-phase Flow*, Exeter, England, vol. 2, pp. A201–A214, June, 1965.
36. Steen, D. A., and G. B. Wallis: AEC Rept. NYO-3114-2, 1964.
37. Van Rossum, J. J.: *Chem. Eng. Sci.*, vol. 11, pp. 35–52, 1959.
38. Wallis, G. B.: General Electric Company, Rept. 62GL127, Schenectady, N.Y., 1962.
39. Wicks, M., and A. E. Dukler: *A. I. Ch. E. J.*, vol. 6, pp. 463–468, 1960.
40. Gill, L. E., G. F. Hewitt, and J. W. Hitchon: UKAEA Rept. AERE-R3954, 1962.
41. Zhivaikin, L. I.: *Khim. Mashinostr.*, vol. 6, p. 25, 1961.
42. Mozarov, N. A.: *Teploenergetica*, vol. 4, p. 60, 1961.
43. Wallis, G. B.: discussion of Ref. 44, *Intern. J. Heat Mass Transfer*, vol. 11, pp. 783–785, 1968.
44. Paleev, I. I., and B. S. Filippovich: *Intern. J. Heat Mass Transfer*, vol. 9, pp. 1089–1093, 1966.
45. Gill, L. E., G. F. Hewitt, and P. M. C. Lacey: UKAEA Rept. AERE-R3955, 1963.
46. Cravarolo, L., A. Hassid, and E. Pedrocchi: CISE R-109, Milan, Italy, 1964.
47. Kutateladze, S. S.: *Izv. Akad. Nauk SSSR, Otd. Tekhn. Nauk*, no. 4, p. 529, 1951.
48. Minh, T. Q., and J. D. Huyghe: *Symp. Two-phase Flow*, Exeter, England, vol. 1, pp. C201–C212, June, 1965.
49. Cousins, L. B., W. H. Denton, and G. F. Hewitt: *Symp. Two-phase Flow*, Exeter, England, vol. 2, pp. C401–C430, June, 1965.

Appendix A

A.1 SOME USEFUL CONVERSION FACTORS

Length	1 cm = 0.394 in. = 0.0328 ft
Volume	1 cm^3 = 3.51×10^{-5} ft^3
Mass	1 g = 2.205×10^{-3} lb_m
Density	1 g/cm^3 = 62.43 lb_m/ft^3
Force	1 dyne = 2.248×10^{-6} lb_f
Pressure	1 atm = 14.7 psia = 1.013×10^6 $dynes/cm^2$
Viscosity	1 lb_m/(sec)(ft) = 14.88 g/(sec)(cm) (or poise)
Surface tension	1 dyne/cm = 6.85×10^{-5} lb_f/ft
Energy	1 kwhr = 3413 Btu
	1 $watt/cm^2$ = 3171 Btu/(hr)(ft^2)

A.2 SOME USEFUL NUMBERS AND PROPERTIES

Gravitational constant on earth	981 cm/sec^2, 32.2 ft/sec^2
Surface tension ol water in air at 20°C	72.6 dynes/cm
Gas constant for air	53.3 (ft)(lb_f)/(lb_m)(°F)

Press, psia	Temp., °F	Volume, ft³/lb$_m$			Enthalpy, Btu/lb$_m$			Entropy, Btu/(lb$_m$)(°F)		
		Water v_f	Evap. v_{fg}	Steam v_g	Water h_f	Evap. h_{fg}	Steam h_g	Water s_f	Evap. s_{fg}	Steam s_g
3208.2	705.47	0.05078	0.00000	0.05078	906.0	0.0	906.0	1.0612	0.0000	1.0612
3000.0	695.33	0.03428	0.05073	0.08500	801.8	218.4	1020.3	0.9728	0.1891	1.1619
2500.0	668.11	0.02859	0.10209	0.13068	731.7	361.6	1093.3	0.9139	0.3206	1.2345
2000.0	635.80	0.02565	0.16266	0.18831	672.1	466.2	1138.3	0.8625	0.4256	1.2881
1500.0	596.20	0.02346	0.25372	0.27719	611.7	558.4	1170.1	0.8085	0.5288	1.3373
1000.0	544.58	0.02159	0.42436	0.44596	542.6	650.4	1192.9	0.7434	0.6476	1.3910
700.0	503.08	0.02050	0.63505	0.65556	491.6	710.2	1201.8	0.6928	0.7377	1.4304
500.0	467.01	0.01975	0.90787	0.92762	449.5	755.1	1204.7	0.6490	0.8148	1.4639
400.0	444.60	0.01934	1.14162	1.16095	424.2	780.4	1204.6	0.6217	0.8630	1.4847
300.0	417.35	0.01889	1.52384	1.54274	394.0	808.9	1202.9	0.5882	0.9223	1.5105
200.0	381.80	0.01839	2.26890	2.28728	355.5	842.8	1198.3	0.5438	1.0016	1.5454
100.0	327.82	0.017740	4.4133	4.4310	298.5	888.6	1187.2	0.4743	1.1284	1.6027
80.0	312.04	0.017573	5.4536	5.4711	282.1	900.9	1183.1	0.4534	1.1675	1.6208
60.0	292.71	0.017383	7.1562	7.1736	262.2	915.4	1177.6	0.4273	1.2167	1.6440
40.0	267.25	0.017151	10.4794	10.4965	236.1	933.6	1169.8	0.3921	1.2844	1.6765
20.0	227.96	0.016834	20.070	20.087	196.27	960.1	1156.3	0.3358	1.3962	1.7320
14.696	212.00	0.016719	26.782	26.799	180.17	970.3	1150.5	0.3121	1.4447	1.7568
10.0	193.21	0.016592	38.404	38.420	161.26	982.1	1143.3	0.2836	1.5043	1.7879
8.0	182.86	0.016527	47.328	47.345	150.87	988.5	1139.3	0.2676	1.5384	1.8060
6.0	170.05	0.016451	61.967	61.984	138.03	996.2	1134.2	0.2474	1.5820	1.8294
4.0	152.96	0.016358	90.63	90.64	120.92	1006.4	1127.3	0.2199	1.6428	1.8626
2.0	126.07	0.016230	173.74	173.76	94.03	1022.1	1116.2	0.1750	1.7450	1.9200
1.0	101.74	0.016136	333.59	333.60	69.73	1036.1	1105.8	0.1326	1.8455	1.9781
0.50	79.586	0.016071	641.5	641.5	47.623	1048.6	1096.3	0.0925	1.9446	2.0370
0.20	53.160	0.016025	1526.3	1526.3	21.217	1063.5	1084.7	0.0422	2.0738	2.1160
0.10	35.023	0.016020	2945.5	2945.5	3.026	1073.8	1076.8	0.0061	2.1705	2.1766

Table A.2 Properties of air at low pressures†,‡

T, °R	T, °F	c_p, Btu/(lb)(°F)	c_v, Btu/(lb)(°F)	$\gamma = \frac{c_p}{c_v}$	c, fps	$\mu \times 10^7$, lb_m/(sec)(ft)	k, Btu/(hr)(ft)(°F)	Pr = $c_p\mu/k$ K
500	40.3	0.2396	0.1710	1.401	1096.4	118	0.0143	0.71
550	90.3	0.2399	0.1713	1.400	1149.6	126	0.0156	0.70
600	140.3	0.2403	0.1718	1.399	1200.3	135	0.0168	0.70
650	190.3	0.2409	0.1723	1.398	1248.7	143	0.0180	0.69
700	240.3	0.2416	0.1730	1.396	1295.1	151	0.0191	0.68
750	290.3	0.2424	0.1739	1.394	1339.6	158	0.0202	0.68
800	340.3	0.2434	0.1748	1.392	1382.5	166	0.0213	0.68
900	440.3	0.2458	0.1772	1.387	1463.6	179	0.0237	0.67
1000	540.3	0.2486	0.1800	1.381	1539.4	192	0.026	0.66
1100	640.3	0.2516	0.1830	1.374	1610.8	205	0.028	0.66
1200	740.3	0.2547	0.1862	1.368	1678.6	218	0.030	0.66
1300	840.3	0.2579	0.1894	1.362	1743.2	230	0.032	0.66
1400	940.3	0.2611	0.1926	1.356	1805.0	242	0.035	0.65
1500	1040.3	0.2642	0.1956	1.350	1864.5	253	0.037	0.65
1600	1140.3	0.2671	0.1985	1.345	1922.0	264	0.039	0.65
1700	1240.3	0.2698	0.2013	1.340	1977.6	274	0.041	0.65
1800	1340.3	0.2725	0.2039	1.336	2032	284	0.043	0.65
1900	1440.3	0.2750	0.2064	1.332	2084	293	0.045	0.65
2000	1540.3	0.2773	0.2088	1.328	2135	302	0.046	0.65
2100	1640.3	0.2794	0.2109	1.325	2185	311	0.048	0.65
2200	1740.3	0.2813	0.2128	1.322	2234	320	0.050	0.65
2300	1840.3	0.2831	0.2146	1.319	2282	329	0.052	0.65
2400	1940.3	0.2848	0.2162	1.317	2329			

† The value of the gas constant R is 53.3 (ft)(lb_f)/(lb_m)(°F).

‡ From J. H. Keenan and J. Kaye, "Gas Tables," John Wiley & Sons, Inc., New York, 1948 (by permission).

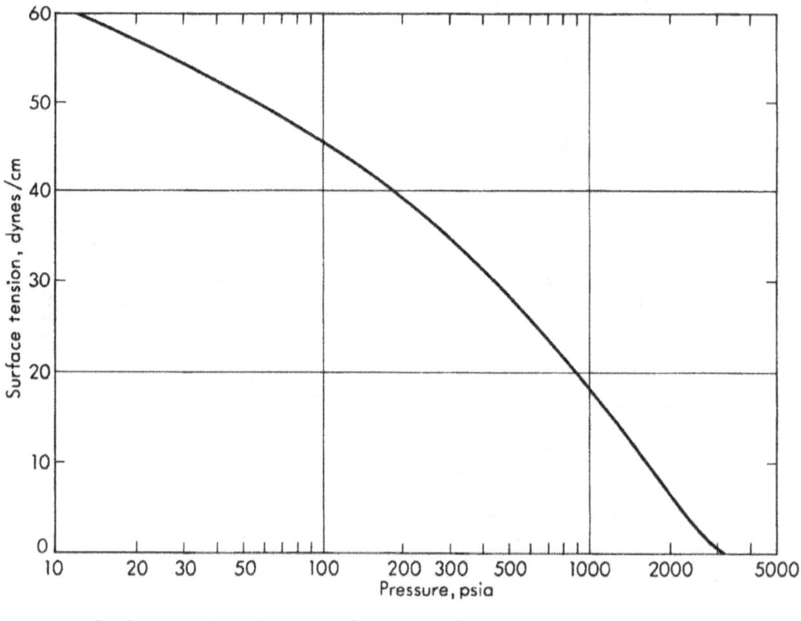

Fig. A.1 Surface tension of saturated water (1 dyne/cm = 6.85×10^{-5} lb/ft).

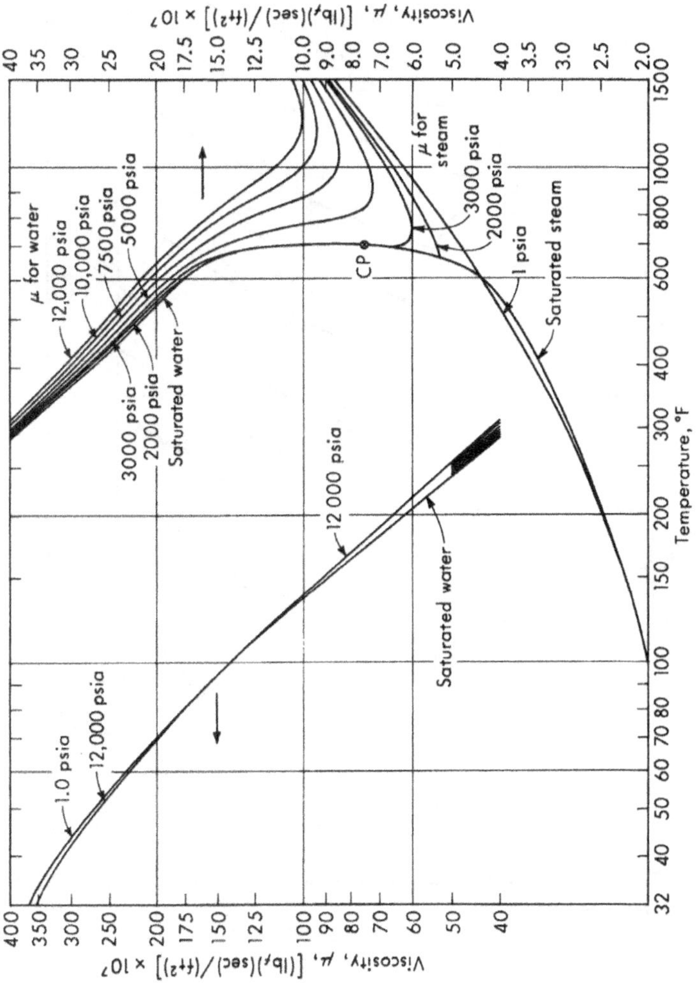

Fig. A.2 Viscosity of steam and water. *(From 1967 ASME Steam Tables, by permission.)*

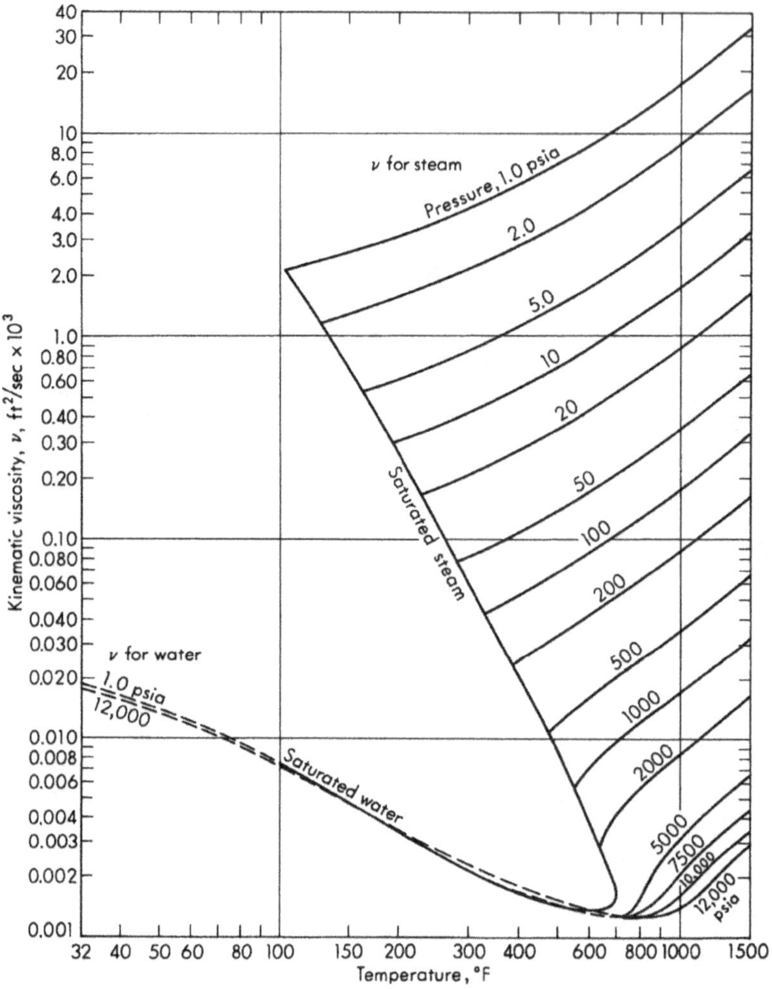

Fig. A.3 Kinematic viscosity of steam and of water. *(From 1967 ASME Steam Tables, by permission.)*

Appendix B
Interpretation of Figures

Lengths in any direction on a two-dimensional surface, such as a piece of paper or a screen, may be measured by a choice of scale such as centimeters. Features on a figure with Cartesian axes x,y are physically represented on this same scale. Areas have units of cm^2, and angles are dimensionless.

On a graph such as Figure 6.2(*b*), one axis is labeled t. Time cannot be a measure of length on the figure. This must be represented by a suitable conversion or scaling factor, S_y, such that

$$t = S_y y \qquad \text{(B.1)}$$

If time is chosen to be in hours, then S_y is 1 hour/cm.

The x-axis may also require scaling, as when events occur over long distances. Then

$$z = S_x x \qquad \text{(B.2)}$$

with S_x perhaps being 5 m/cm.

The scale of a map is often represented by a statement such as 1 inch = 1 mile, which is false if taken literally but is a measure of S_x and S_y. A similar statement such as 1 cm = 1 hour may be useful for interpreting some simple features of a graph but can lead to incongruous results.

The slopes of lines on a figure are also basically given in *x,y* units. To obtain the actual speed of a wave from Fig. 6.2(*b*), one must use (B.1) and (B.2) to obtain

$$V_w = dz/dt = (S_x/S_y)\ dx/dy \text{ or } (S_x/S_y)/(dy/dx) \text{ m/hr} \tag{B.3}$$

The wave velocities V_w and V_s are represented as arc tangent or arc cotangent in figures in Chapter 6 and subsequent chapters. They are dimensionless measures of slopes on the graph and need scaling factors, as in (B.3), in order to derive the physical wave speed. The direction of wave motion is determined by the sign of the slope, which is not directly evident from the angle itself.

Since the scaling factors depend on the particular situation being analyzed, as well as arbitrary choices by the user, it has been chosen to represent all wave velocities in this dimensionless form for display purposes on figures. This leads to a general representation independent of the particular application. It should not affect the utility of the approach, since the user is unlikely to make measurements of slopes on a graph but rather will compute a wave speed such as V_w directly from the relationship between flux j and concentration α.

Appendix C
Added Mass

The force postulated in equation (8.157) is solely of academic value for investigating the consequences that it might have. There are several reasons why it is inappropriate for describing real situations.

The convection acceleration term $v_s\partial(v_s\text{-}v_f)/\ \partial z$ is invalid. It cannot describe the force on a stationary particle in a weak converging potential flow. This may be modeled by the flow around a sphere produced by a dipole far away pointing away from the sphere and producing a reflection in the sphere. For this case, I obtained the resulting force per unit volume as $3/2\rho_f\, v_f\ dv_f/dz$. Subtracting the force due to the imposed overall pressure gradient leaves a coefficient of ½, as expected from the added mass coefficient. This suggests that the appropriate second term in (8.157) should be the relative acceleration $v_s\,\partial v_s/\ \partial z\text{-}\ v_f\partial v_f/\ \partial z$.

The potential flow solution is only valid for sufficiently low viscosity. In this case, the Reynolds Number will be large for any significant flow around the particle, generating a separated wake and invalidating the potential flow. Moreover, a converging or diverging flow will influence boundary layer separation, changing the size of the wake and influencing the drag. The effects of added mass and drag can neither be borrowed from some simpler circumstances nor added to each other, as in equation (8.158). Their combined effect must be determined.

In a realistic accelerating flow with finite particle or bubble void fraction, there is the additional effect of changes in the morphology of the dispersed array. For example, an array that is initially cubical–or resembles those studied analytically by Wallis, Cai, and Luo[1] and experimentally by Cai and Wallis[2]–will be compressed laterally and expanded in the flow direction, changing both the drag and the added mass. Even in a potential flow, this would lead to changes in the fluid momentum attributable to the relative motion and therefore to a force component related to the rate of change of the added mass coefficient.

It would be a substantial task to model all these effects empirically.

1. Wallis, G. B., X. Cai, and C. Luo: *Chem. Eng. Commun.,* vol. 118, p. 141, 1992.
2. Cai, X., and G. B. Wallis: *Phys. Fluids A.,* vol. 5, no. 7, pp. 1614–1629, 1993.

Index